Controlling Chaos

Theoretical and Practical Methods in Non-linear Dynamics

Tomasz Kapitaniak
Professor of Mechanics
Technical University of Lodz
Lodz
Poland

Academic Press

Harcourt Brace & Company, Publishers

London   San Diego   New York   Boston   Sydney   Tokyo   Toronto

ACADEMIC PRESS LIMITED
24–28 Oval Road
LONDON NW1 7DX

U.S. Edition Published by
ACADEMIC PRESS INC.
San Diego, CA 92101

This book is printed on acid free paper

Typeset by Phoenix Photosetting, Chatham, Kent

Printed and bound in Great Britain by
Butler & Tanner Ltd, Frome and London

To my parents

Contents

Preface

More than two decades of intensive studies on nonlinear dynamics have posed the question on the practical applications of chaos. One of the possible answers is to control chaotic behavior in such a way as to make it predictable. Indeed, nowadays the idea of controlling chaos, which we try to explain in this book, is an appealing one.

This book is organized as follows. In Part I (Chapters 1–5) we describe basic methods of controlling chaos, while in Part II we reprint fundamental contributions to this field.

In Chapter 1 we give basic information about controlling procedures. Ideas of feedback and nonfeedback methods as well as of chaos synchronization are explained. Additionally, we describe Chua's circuit (a very simple electronic device) which will be used in several examples in the following chapters.

Chapter 2 describes feedback controlling methods in which unstable periodic orbits embedded in the chaotic attractor are stabilized. We discuss the Ott–Grebogi–Yorke and Pyragas methods and their connections with classical controlling methods. The method of controlling chaos by chaos, in which chaotic behavior can be modified by coupling via a feedback loop with another chaotic system, is also mentioned.

Nonfeedback methods are explained in Chapter 3. This approach is inevitably much less flexible than feedback methods, but in many practical systems it is easier to apply. We give methods in which chaos can be controlled through operating conditions or by system design. Additionally, taming chaos, entrainment and migration control procedures are discussed.

The synchronization chaos procedures of Chapter 4 allow two chaotic systems to have exactly the same response. The procedures of Pecora and Carroll and of continuous control are described. We also explain the idea of secure communication.

In Chapter 5 we discuss the problem of the selection and engineering implementation of the chaos controlling method for a particular practical problem.

Finally, in the references are listed the most important works on chaos controlling and synchronization.

Of the methods given in Chapters 2–5, it is mainly those which the author took a small part in developing that are described in detail. Other methods are only briefly described: full details of these can be found in the papers reprinted in Part II. The author has tried to select papers which, in his opinion, have had very significant impact on the development of the field.

This book is generally for those who have some introductory knowledge of nonlinear dynamics and who are interested in its potential applications. Knowledge of the classical control theory is not necessary to understand chaos controlling methods but could be of benefit to readers.

Finally, I would like to acknowledge the valuable comments of J. Brindley, C. Grebogi, L.O. Chua, M.S. El Naschie, L. Kocarev, V.S. Anishchenko, M. Ogorzalek and J. Wojewoda who have helped me in preparation of this work.

Tomasz Kapitaniak
Rosanów, 1996

Acknowledgements

The following have kindly granted permission to reprint the following papers.

Paper 1   *Phys. Rev. Lett.*, 1990, **64**, 1196. © 1990 The American Physical Society

Paper 2   *Physica* D, 1992, **58**, 165. © 1992 Elsevier Science Publishers B.V.

Paper 3   *Phys. Rev. Lett.*, 1992, **68**, 1. © 1991 The American Physical Society

Paper 4   *Phys. Rev. Lett.*, 1990, **65**, 3211. Work of the U.S. Government. Not subject to U.S. copyright

Paper 5   *J. Phys.* A, 1991, **24**, L1359. © 1991 IOP Publishing Ltd

Paper 6   *Phys. Rev. Lett.*, 1990, **65**, 3215. © 1990 The American Physical Society

Paper 7   *Phys. Lett.* A, 1992, **170**, 421. © 1992 Elsevier Science Publishers B.V.

Paper 8   *Physica* D, 1991, **50**, 341. © 1991 Elsevier Science Publishers B.V. (North-Holland)

Paper 9   *Phys. Rev. Lett.*, 1990, **64**, 821. Work of the U.S. Government. Not subject to U.S. copyright

Paper 10   *Phys. Lett.* A, 1993, **181**, 203. © 1993 Elsevier Science Publishers B.V.

Paper 11   *Phys. Rev. Lett.*, 1993, **71**, 65. © 1993 The American Physical Society

Paper 12   *Phys. Rev. Lett.*, 1995, **74**, 1970. © 1995 The American Physical Society

Paper 13   *Phys. Rev. Lett.*, 1995, **74**, 5028. © 1995 The American Physical Society

PART I:
General Outlook

1 Introduction

Chaos occurs widely in engineering and natural systems; historically it has usually been regarded as a nuisance and is designed out if possible. It has been noted only as irregular or unpredictable behavior, often attributed to random external influences. More recently, there have been examples of the potential usefulness of chaotic behavior, and we describe some of its potential usefulness in this book.

In Chapters 2 and 3 we review a number of methods by which undesirable chaotic behavior may be controlled or eliminated. More speculatively, we indicate ways in which the existence of chaotic behavior may be directly beneficial or exploitable.

We can divide chaos controlling approaches into two broad categories: firstly those in which the actual trajectory in the phase space of the system is monitored and some *feedback* process is employed to maintain the trajectory in the desired mode, and secondly *nonfeedback* methods in which some other property or knowledge of the system is used to modify or exploit chaotic behavior. Feedback methods do not change the controlled systems and stabilize unstable periodic orbits on strange chaotic attractors, while nonfeedback methods slightly change the controlled system, mainly by a small permanent shift of control parameter, changing the system behavior from chaotic attractor to periodic orbit which is close to the initial attractor. The main idea of both methods is illustrated in *Figure 1.1*.

We describe several methods by which chaotic behavior in a dynamical system may be modified, displaced in parameter space or removed. The Ott–Grebogi–Yorke (OGY) method (Ott *et al.*, 1990 – *Paper 1*) is extremely general, relying only on the universal property of chaotic attractors, namely that they have embedded within them infinitely many unstable periodic orbits (or even static equilibria). On the other hand, the method requires following the trajectory and employing a feedback control system which must be highly flexible and responsive; such a system in some experimental configurations may be large and expensive. It has the additional disadvantage that small amounts of noise may cause occasional large departures from the desired operating trajectory.

The nonfeedback approach is inevitably much less flexible, and requires more prior knowledge of equations of motion. On the other hand, to apply such a method, we do not have to follow the trajectory. The control procedures can be applied at any time and we can switch from one periodic orbit to another without returning to the chaotic behavior, although after each switch, transient chaos may be observed. The lifetime of this transient chaos strongly

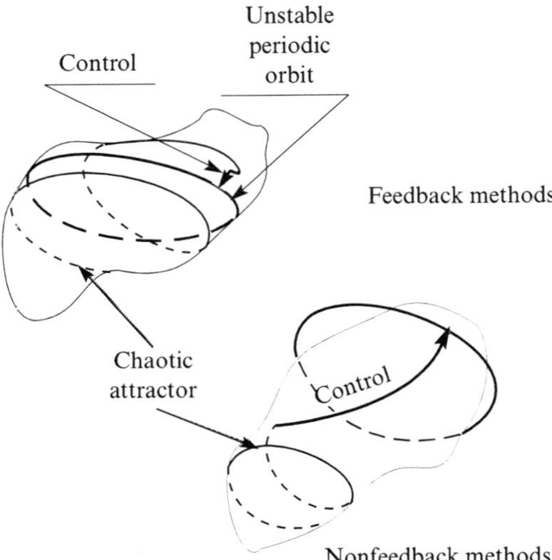

Figure 1.1 General idea of feedback and nonfeedback controlling methods.

depends on initial conditions. Moreover, in a nonfeedback method we do not have to wait until the trajectory is close to an appropriate unstable orbit; in some cases this time can be quite long. The dynamic approach can be very useful in mechanical systems, where feedback controllers are often very large (sometimes larger than the control system). In contrast, a dynamical absorber having a mass of order 1% of that of the control system is able, as we will show in the example of Chapter 3, to convert chaotic behavior to periodic over a substantial region of parameter space. Indeed, the simplicity by which chaotic behavior may be changed in this way, and the possibility of an easy access to different periodic orbits, may actually motivate the search for, and exploitation of, chaotic behavior in practical systems. This prompts us to pose a final question – how can we exploit chaos in real systems? The OGY method, at least in theory, gives access to the wide range of possible behavior encompassed by the unstable periodic (and other) orbits embedded in a chaotic attractor. Moreover, the sensitivity of the chaotic regime to both initial conditions and parameter values means that the desired effects may be produced by fine tuning. Thus, we may actually wish to design chaos into a system, in order to exploit this adaptability. Nonfeedback methods can, in principle, give us advice on the design, whether we wish to design chaos out or in. Additionally, they enable us to choose regions of design parameter space or operating parameter space within which chaos will occur and will be acceptable. An example of practical use might be the minimalization of metal fatigue by switching from a necessary strictly periodic operation of the fully loaded conditions, where repeated stresses are applied at certain places, to a noisy periodicity (rather like a healthy heartbeat) under idling conditions.

The essential property of a chaotic trajectory is that it is not asymptotically stable. Closely correlated initial conditions have trajectories which quickly become uncorrelated. Despite this obvious disadvantage, it has been established that control leading to the synchronization of two chaotic systems is possible. In Chapter 4 we describe basic synchronization procedures and discuss its potential application to secure communications.

Methods described in Chapters 2–4 are illustrated by the example of the controlling chaos in Chua's circuit (Chua *et al.*, 1986; Chua, 1993) shown in *Figure 1.2*. Chua's circuit contains three linear energy-storage elements (an inductor and two capacitors), a linear resistor, and a single nonlinear resistor N_R, namely Chua's diode with a three-segment piecewise-linear v–i characteristic defined by

$$f\left(v_{C_1}\right) = m_0 v_{C_1} + \tfrac{1}{2}\left(m_1 - m_0\right)\left(\left|v_{C_1} + 1\right| - \left|v_{C_1} - 1\right|\right) \tag{1.1}$$

where the slopes in the inner and outer regions are m_0 and m_1 respectively (*Figure 1.3*).

In this case the state equations for dynamics of our scheme of *Figure 1.2* are as follows:

$$C_1 \frac{dv_{C_1}}{dt} = G\left(v_{C_2} - v_{C_1}\right) - f\left(v_{C_1}\right)$$

$$C_2 \frac{dv_{C_2}}{dt} = G\left(v_{C_1} - v_{C_2}\right) + i_L \tag{1.2}$$

$$L \frac{di_L}{dt} = v_{C_2}$$

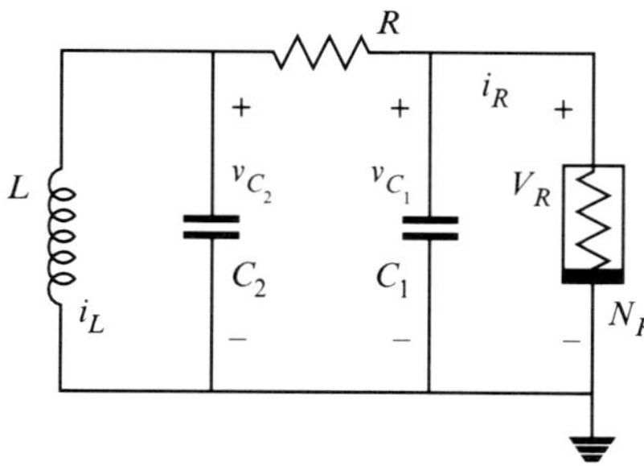

Figure 1.2 Chua's circuit.

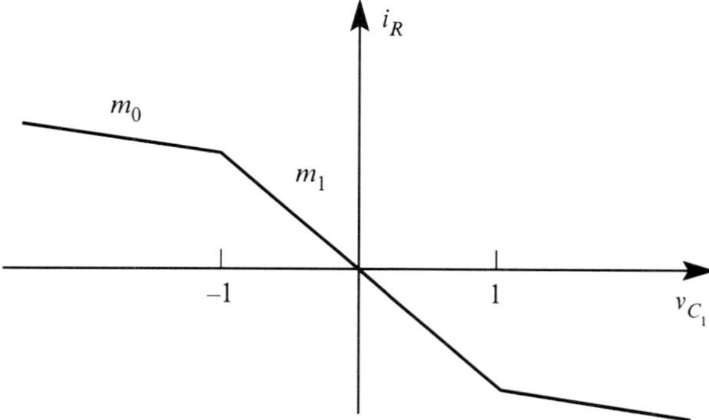

Figure 1.3 i_R–v_{C_1} characteristic of non-linear resistor.

where $G = 1/R$.

Introducing new variables, $x = v_{C_1}$, $y = v_{C_2}$, $z = i_L/G$, $\alpha = C_2/C_1$ and $\beta = C_2/LG^2$, we can rewrite *Equation (1.2)* in dimensionless form:

$$\dot{x} = \alpha\left[y - x - f(x)\right]$$

$$\dot{y} = x - y + z \qquad\qquad (1.3)$$

$$\dot{z} = -\beta y$$

It is well-known that for $R = 1.64\ \text{k}\Omega$, $C_1 = 10\ \text{nF}$, $C_2 = 99.34\ \text{nF}$, $m_1 = -0.76$ mS, $m_0 = 0.41$ mS and $L = 18.46$ mH, Chua's circuits operate on the chaotic double-scroll Chua's attractor shown in *Figure 1.4*.

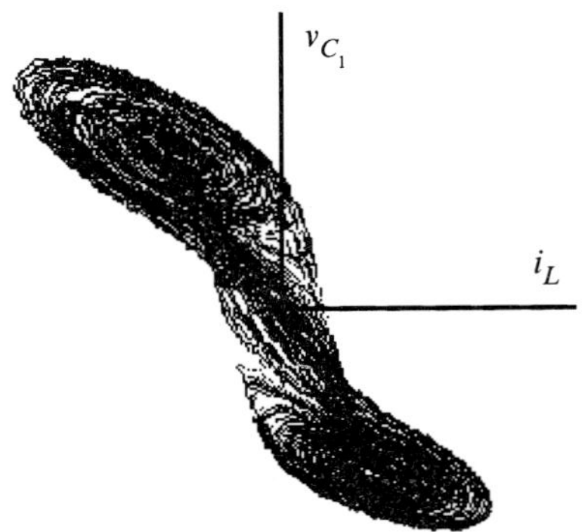

Figure 1.4 Double scroll attractor.

The chaotic dynamics of Chua's circuit have been widely investigated (e.g. Madan, 1993). One of the main advantages of this system is the very good accuracy between numerical simulations of *Equation (1.3)* and experiments on real electronic devices. Experiments with this circuit are very easy to perform, even for nonspecialists.

The problem of controlling chaos in engineering systems is discussed in Chapter 5. Some basic modifications of controlling procedures which allow their direct practical implementation are described.

In Part II we have reprinted a selection of important contributions to the problem of controlling and synchronization of chaotic systems.

2 Controlling chaos through feedback

2.1 Ott–Grebogi–Yorke method

Ott, Grebogi and Yorke (Ott *et al.*, 1990 – *Paper 1*; Romeiras *et al.*, 1992 – *Paper 2*) have, in an important series of papers, proposed and developed a method by which chaos can always be suppressed by shadowing one of the infinitely many unstable periodic orbits (or perhaps steady states) embedded in the chaotic attractor.

Basic assumptions of this method are as follows.

(a) The dynamics of the system can be described by an *n*-dimensional map of the form.

$$\zeta_{n+1} = f(\zeta_n, p) \tag{2.1}$$

This map, in the case of continuous-time systems, can be constructed, e.g. by introducing a transversal surface of section for system trajectories (Poincaré map).

(b) p is some accessible system parameter which can be changed in some small neighborhood of its nominal value p^*.

(c) For this value p^* there is a periodic orbit within the attractor around which we would like to stabilize the system.

(d) The position of this orbit changes smoothly with changes in p, and there are small changes in the local system behavior for small variations of p.

Let ζ_F be a chosen fixed point of the map f of the system existing for the parameter value p^*. In the close vicinity of this fixed point with good accuracy we can assume that the dynamics are linear and can be expressed approximately by

$$\zeta_{n+1} - \zeta_F = M(\zeta_n - \zeta_F) \tag{2.2}$$

The elements of the matrix M can be calculated using the measured chaotic time series and analyzing its behavior in the neighborhood of the fixed point. Further, the eigenvalues λ_s, λ_u and eigenvectors e_s, e_u of this matrix can be found. These eigenvectors determine the stable and unstable directions in the small neighborhood of the fixed point.

Denoting by f_s, f_u the contravariant eigenvectors ($f_s e_s = f_u e_u = 1$, $f_s e_u = f_u e_s = 0$) we can find the linear approximation valid for small $|p_n - p^*|$:

$$\zeta_{n+1} = p_n g + (\lambda_u e_u f_u + \lambda_s e_s f_s)(\zeta_n - p_n g) \tag{2.3}$$

where

$$g = \left.\frac{\partial \zeta_F(p)}{\partial p}\right|_{p+p^*}$$

Because ζ_{n+1} should fall on the stable manifold of ζ_F, choose p_n such that $f_u\zeta_{n+1} = 0$:

$$p_n = \frac{\lambda_n\zeta_n f_u}{(\lambda_u - 1)g f_u} \tag{2.4}$$

The OGY algorithm is schematically explained in *Figure 2.1*, and its main properties are as follows.

(a) No model of dynamics is required. One can use either full information from the process or a delay coordinate embedding technique using single variable experimental time series. An extremely interesting development in this direction has been described by Dressler and Nitsche (1992 – *Paper 3*).
(b) Any accessible variable (controllable) system parameter can be used as the control parameter.
(c) In the absence of noise and error, the amplitude of applied control signal must be large enough (exceed a threshold) to achieve control.
(d) Inevitable noise can destabilize the controlled orbit, resulting in occasional chaotic bursts.
(e) Before settling into the desired periodic mode, the trajectory exhibits chaotic transients, the length of which depends on the actual starting point.

In Ogorzalek (1993b) the OGY method has been applied to control chaos in Chua's circuit (*Figure 1.2, Equation (1.2)*). Using a specific software package (Dabrowski *et al.*, 1992), unstable periodic orbits embedded in the

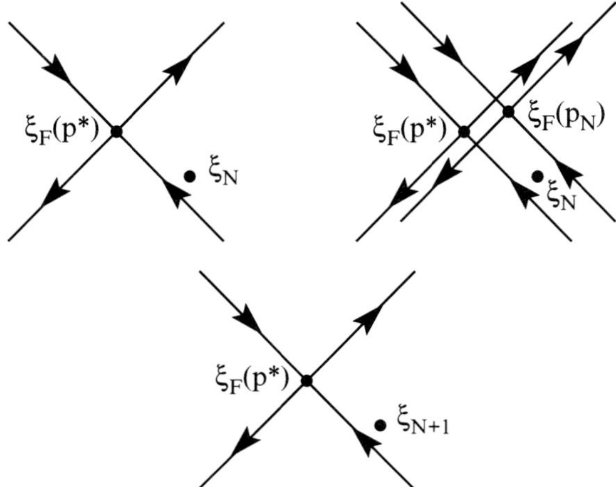

Figure 2.1 Idea of Ott–Grebogi–Yorke method.

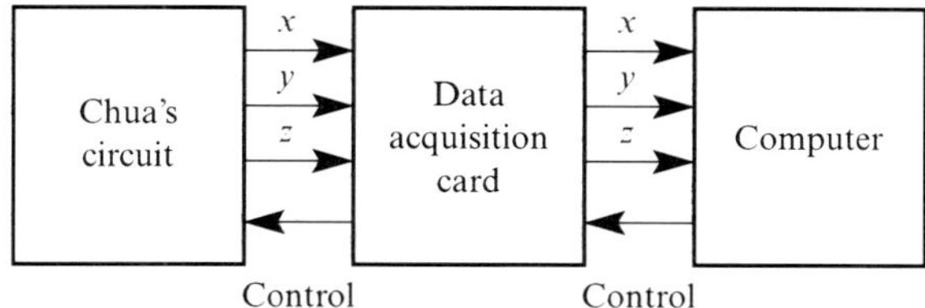

Figure 2.2 Practical implementation of OGY method.

double scroll attractor which could serve as goals of control were found. The controlling method was implemented in the way shown in *Figure 2.2*.

The computer was used for data acquisition, identification of the chaotic system in terms of unstable periodic orbits and calculation of the control signal. *Plate 1* shows the results of stabilization of period-one and period-two unstable periodic orbits. Before control is achieved, the trajectories exhibit chaotic transients (shown in red). Actual controlled trajectories are shown in yellow.

When applying the OGY method to control chaos in a real electronic circuit, the main problem encountered was the noise introduced due to inevitable noise of the circuit elements, A/D and D/A conversion of signals (quantification), etc. The method was found to be very sensitive to the noise level – very small signals sometimes are hidden within the noise, and control is impossible (Ogorzalek, 1993).

The OGY approach has stimulated a good deal of research activity, both theoretical and experimental. The efficiency of the technique has been demonstrated by Ditto *et al.* (1990 – *Paper 4*), in a periodically forced system, converting its chaotic behavior into period-one and period-two orbits, and the application of the method to stabilize higher periodic orbits in a chaotic diode resonator has been demonstrated by Hunt (1991). Another interesting application of the method is the generation of a desired aperiodic orbit (Mehta and Henderson, 1991); Tél (1991 – *Paper 5*) has been able to demonstrate controlled transient chaos.

Though the OGY theory has been proposed in the context of low dynamical systems, and most of the experimental or observation investigations have been concerned with clearly low order mechanical or electrical contexts, the interesting experiments by Singer *et al.* (1991) demonstrate its potential for fluid (and perhaps fluid–solid) mechanical phenomena. The experiments succeeded in achieving regular laminar flow in previously unstable thermal convection loops by use of a thermostat-type feedback. Theoretical generalization of OGY method for higher-dimensional systems can be found in Auerbach *et al.* (1992).

Generally, the experimental application of the OGY method requires a permanent computer analysis of the state of the system. The changes of the parameters, however, are discrete in time since the method deals with the Poincaré map. This leads to some serious limitations. The method can stabilize only those periodic orbits which maximal Lyapunov exponent is small compared to the reciprocal of the time interval between parameter changes. Since the corrections of the parameter are rare and small, the fluctuation noise leads to occasional bursts of the system into the region far from the desired periodic orbit, especially in the presence of noise.

One of the disadvantages of this method is that the control procedure can be applied only if the controlled trajectory is in the neighborhood of the appropriate unstable orbit or unstable fixed point. To increase the effectiveness of the OGY method the initial targeting procedure which allows one to direct a trajectory to the desired state (neighborhood of appropriate unstable attractor) has been introduced (Shinbrot *et al.*, 1990 – *Paper 6*; Shinbrot *et al.*, 1992). The idea of targeting and its difference from controlling is described in *Figure 2.3*. Targeting in finite *a priori* known time is discussed in Kapitaniak (1993). Higher-dimensional targeting is described in Kostelich *et al.* (1993).

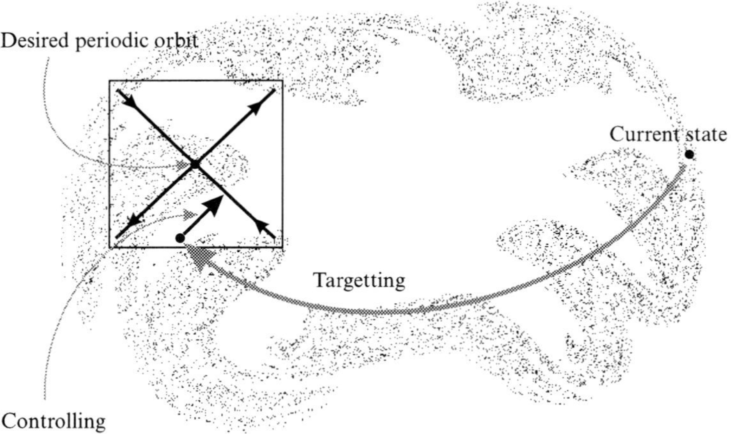

Figure 2.3 General idea of targeting procedure.

2.2   Pyragas's and classical control methods

A different approach to feedback control which allows the above-mentioned problems to be avoided, the method of a time-continuous control, was proposed by Pyragas (1992 – *Paper 7*). This method is based on the construction of a special form of a time-continuous perturbation, which does not change the form of the desired unstable periodic orbit, but under certain conditions can stabilize it. Two feedback controlling loops, shown in *Figure 2.4*, have been proposed.

A combination of feedback and periodic external force is used in the first method (*Figure 2.4(a)*). The second method (*Figure 2.4(b)*) does not require any external source of energy and it is based on self-controlling delayed

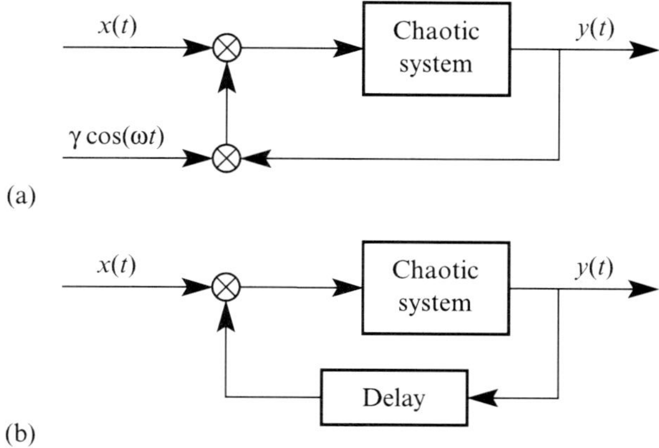

(a)

(b)

Figure 2.4 Feedback controlling loops; (a) control by periodic external perturbation, (b) control by time delay.

feedback. If the period of external force $2\pi/\omega$ or a time delay τ is equal to the period of one of the unstable periodic orbits embedded in the chaotic attractor it is possible to find a constant K which allows stabilization of the unstable periodic orbit. This approach, being noise resistant, can easily be used in experimental systems (Pyragas and Tamasevicius, 1993).

The first of Pyragas's methods (*Figure 2.4(a)*) can be considered as the special case of the direct application of classical controlling methods (e.g. Klamka, 1991) to the problems of controlling chaos.

The dynamical system

$$\dot{x} = f(x) \tag{2.5}$$

where $x \in R^n$, is controllable if there exists a control function $u(t)$, such that

$$\dot{x} = f(x) + u(t) \tag{2.6}$$

allows to move trajectory $x(t)$ from point x_0 at time t_0 to the desired point x in finite time T.

The controllability concept can be applied to the chaos controlling problems. For example Chen and Dong (1992) obtained the following result on the controllability of Chua's circuit, *Equation (1.3)*.

Theorem: Let $(\bar{x},\bar{y},\bar{z})$ be the unstable periodic orbit of the system *Equation (1.3)*. Then, the chaotic trajectory (x,y,z) of the circuit can be driven to reach the periodic orbit by a simple canonical linear feedback control of the form

$$\begin{bmatrix} v_1 \\ v_2 \\ v_3 \end{bmatrix} = -K \begin{bmatrix} \xi - \bar{\xi} \\ \psi - \bar{\psi} \\ \zeta - \bar{\zeta} \end{bmatrix} = \begin{bmatrix} K_{11} & 0 & 0 \\ 0 & K_{22} & 0 \\ 0 & 0 & K_{33} \end{bmatrix} \begin{bmatrix} \xi - \bar{\xi} \\ \psi - \bar{\psi} \\ \zeta - \bar{\zeta} \end{bmatrix} \tag{2.7}$$

with

$$K_{11} \geq -\alpha m_1, \qquad K_{22} \geq 0, \qquad K_{33} \geq 0$$

where the control can be applied to the trajectory at any time.

The closed-loop feedback control configuration of the system is shown in *Figure 2.5*.

In the controlling procedure, $K_{11} = 0$, $K_{22} = 2$ and $K_{33} = 0$ have been used, so the equations of the controlled circuit become

$$\dot{x} = \alpha(y - x - f(x))$$
$$\dot{y} = x - y + z - K_{22}(y - \bar{y}) \tag{2.8}$$
$$\dot{z} = -\beta y$$

A similar approach is considered in Chen and Dong (1993a).

The main advantages of this method are as follows.

(a) Any solution of the original system can be a goal of the control (fixed point, unstable periodic orbit, etc.).
(b) The controller has a very simple structure.
(c) Access to system parameters is not required.
(d) It is immune to small parameter variations.

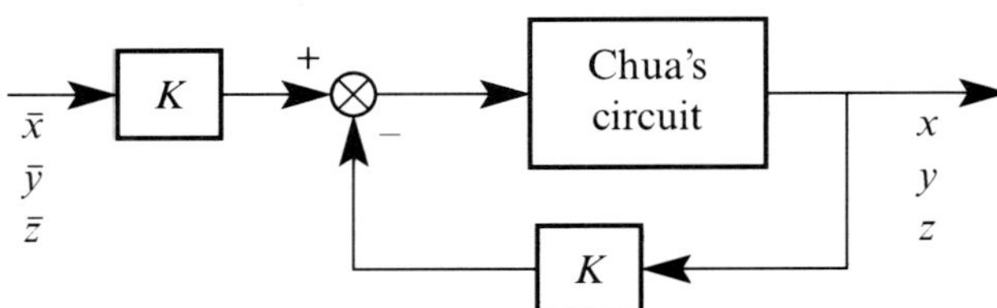

Figure 2.5 Closed-loop feedback control configuration.

2.3 Controlling chaos by chaos

In this section, we show that the chaotic behavior of one system can be controlled by coupling it with another one which can also be chaotic. We focus on an example of coupled chaotic systems in which the effect of the coupling is to enhance predictability in the coupled system, relative to its separate components. We follow the broad approach of Pyragas's continuous chaos control scheme described in the previous section.

Thus we consider two chaotic systems, which we call A and B respectively,

$$\dot{x} = f(x)$$
$$\dot{y} = g(y) \tag{2.9}$$

where $x, y \in \mathbf{R}^n$, and we use the controlling strategy which is schematically illustrated in *Figure 2.5*; the two systems are coupled through the operators λ, μ, which have a very simple linear form. We assume that some or all state variables of both systems A and B can be measured, so that we can measure signal $x(t)$ from the system A and signal $y(t)$ from B, and that the systems are coupled in such a way that the differences $D_{1,2}(t)$ between the signals $x(t)$ and $y(t)$ are

$$F_1(t) = \lambda[x(t) - y(t)] = \lambda D_1(t)$$
$$F_2(t) = \mu[y(t) - x(t)] = \mu D_2(t)$$

(2.10)

used as control signals introduced respectively into each of the chaotic systems A and B as negative feedback. We take $\lambda, \mu > 0$ to be experimentally adjustable weights of the perturbation.

Using the coupling schematically shown in *Figure 2.6*, we have shown that one chaotic system coupled with the other one can significantly change the behavior of one of them (unidirectional coupling, i.e. λ or $\mu = 0$) or of both systems (mutual coupling, i.e. $\lambda, \mu \neq 0$). This property allows us to describe the above procedure as the 'controlling chaos by chaos' method. In Kocarev and Kapitaniak (1995) and Brindley *et al.* (1995), rigorous conditions are given, under which chaotic attractors of systems A and B are equivalent, or the evolution of one of them is forced to take place on the attractor of the other one. Here we describe some applications of controlling chaos by chaos in geophysical systems.

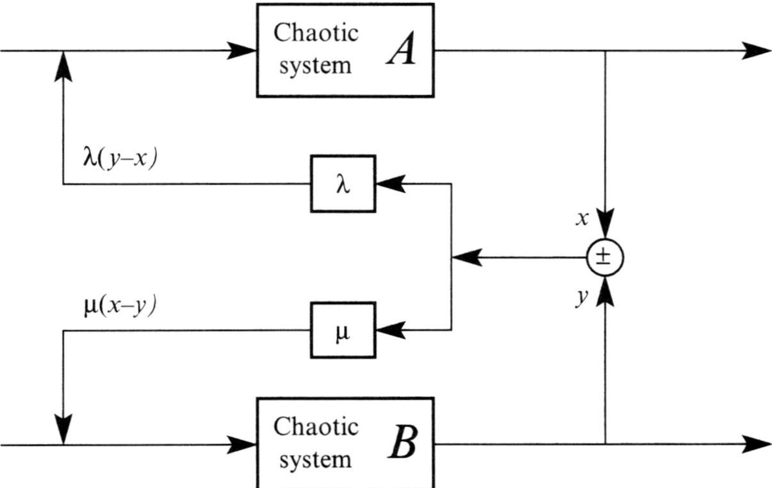

Figure 2.6 Controlling chaos-by-chaos scheme.

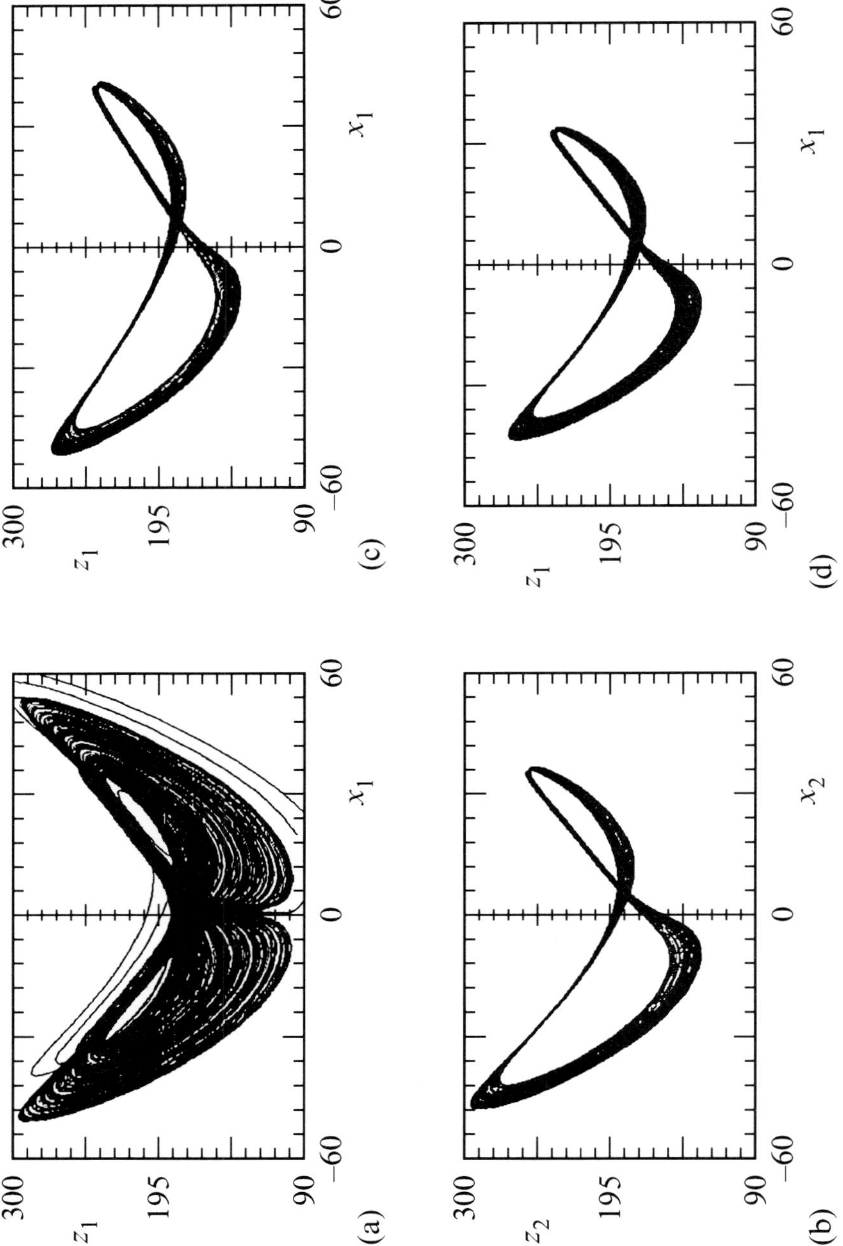

Figure 2.7 Attractors of *Equations (2.11)* (a)(b) uncoupled systems, (c)(d) coupled systems.

In our numerical examples we first consider two Lorenz models (Lorenz, 1963) mutually coupled in the following way:

$$\dot{X}_1 = -\sigma X_1 + \sigma Y_1 + \lambda(X_2 - X_1)$$
$$\dot{Y}_1 = -X_1 Z_1 + r_1 X_1 - Y_1 + \lambda(Y_2 - Y_1)$$
$$\dot{Z}_1 = X_1 Y_1 - b Z_1 + \lambda(Z_2 - Z_1)$$
$$\dot{X}_2 = -\sigma X_2 + \sigma Y_2 + \mu(X_1 - X_2)$$
$$\dot{Y}_2 = -X_2 Z_2 + r_2 X_2 + \mu(Y_1 - Y_2)$$
$$\dot{Z}_2 = X_2 Y_2 - b Z_2 + \mu(Z_1 - Z_2)$$

$$(2.11)$$

where σ, $r_{1,2}$ and b are constants. The Lorenz model has often been proposed as a paradigm for the 'chaotic' extra-tropical atmospheric circulation. The variables X, Y and Z then represent, in some broad sense, Rossby wave components of the extra-tropical general circulation. The coupling between two Lorenz models introduced in *Equations (2.11)* might then be interpreted as mutual interdependence of extra-tropical circulations in two regions characterized by different r parameter values, say an intensive storm track and a relatively stable anticyclonic region. The concept of teleconnections of this kind, achieved through the mechanism of quasi-linear Rossby trains, has both theoretical and observational support.

In *Figures 2.7(a)* and *(b)* we show the chaotic attractors of single Lorenz models (λ, $\mu = 0$) for $\sigma = 10.0$, $r_1 = 197.4$, $b = 8/3$ (*Figure 2.7(a)*) and $r_2 = 211.0$ (*Figure 2.7(b)*). These attractors are characterized by the following spectra of Lyapunov exponents $\lambda_1 = 1.87$, $\lambda_2 = 0$, $\lambda_3 = -15.54$ (*Figure 2.7(a)*) and $\lambda_1 = 0.78$, $\lambda_2 = 0$, $\lambda_3 = -14.44$. In *Figure 2.7(c)* we show the behavior when the above-mentioned Lorenz systems are mutually coupled with $\lambda = 100$ and $\mu = 1$, while in *Figure 2.7(d)* Lorenz systems are unidirectionally coupled with $\lambda = 100$ and $\mu = 0$. Although both attractors are still chaotic ($\lambda_1 = 0.79$, $\lambda_2 = 0$, $\lambda_3 = -14.34$), trajectory behavior on them is more predictable as both Lyapunov dimensions, $d_L = 2.053$ and $d_L = 2.044$ are smaller than this dimension of the original attractor ($d_L = 2.121$).

This dimension increase is produced by a significant decrease of positive Lyapunov exponent ($\lambda_1 = 0.79$ in comparison with $\lambda_1 = 1.87$ of the original attractor). In *Figure 2.8* we show the plots of the measure of predictability

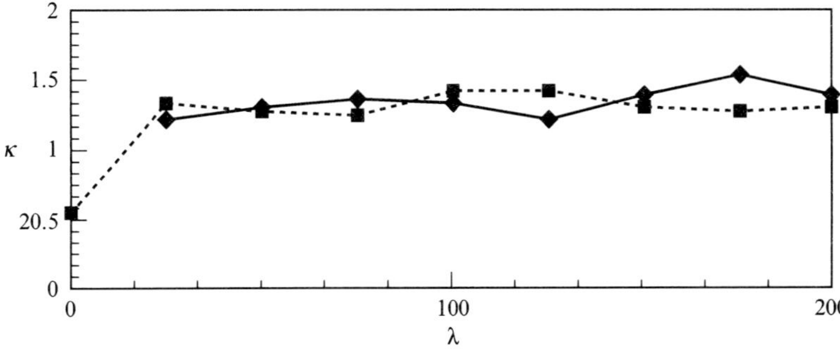

Figure 2.8 Measure of predictability α versus coupling stiffness λ.

$\kappa = 1/\lambda_{max}$, where λ_{max} is the largest Lyapunov exponent, versus coupling coefficient λ for both mutual and unidirectional coupling. The analysis of *Figure 2.8* shows that for sufficiently large λ the predictability is significantly increased both in the case of mutual coupling ($\blacklozenge$) and unidirectional coupling ($\blacksquare$).

This method does not stabilize unstable periodic orbits like other methods described in this chapter. Here the goal of controlling is to modify chaotic behavior although there are examples where coupling (*2.10*) results in the periodic or steady state locking (Stefanski and Kapitaniak, 1996).

3   Controlling chaos without feedback

3.1   Control through operating conditions

Virtually all engineering and most natural systems are subjected during operation to external forcing. This forcing will contain (and hopefully be dominated by) planned and intentional components; it will also almost invariably contain unintentional 'noise'. Judicious design and control of this forcing is often able to annihilate, or shift to a harmless region of parameter space, an unwanted chaotic behavior (in some circumstances, as we remark later, exactly the reverse process may be desirable, so that we may wish to produce chaotic behavior).

We can illustrate this idea in the context of Duffing's oscillator:

$$\ddot{x} + a\dot{x} + bx + c^3 = B_0 + B_1\cos(\Omega t) \tag{3.1}$$

where a, b, c, B_0, B_1 and Ω are constants.

It is well-known that *Equation (3.1)* shows chaotic behavior for certain values of the parameters (Ueda 1979, 1991; Sato *et al.*, 1983; Kapitaniak, 1991a,b). In many cases it can be shown that chaotic behavior is obtained via period doubling bifurcation (Ueda, 1991; Sato *et al.*, 1983; Kapitaniak, 1991a,b). Recently, there have been some attempts to create an analytical criterion which allows us to estimate the chaotic domain in the parameter space (Szemplinska-Stupnicka, 1989; Kapitaniak, 1988; 1991a,b). Boundaries of the chaotic zone have been obtained using classical approximate theory of nonlinear oscillations, by examining approximate periodic solutions and studying particular types of higher order instabilities which precede the destruction of a periodic attractor in the variational Hill-type equation (Hayashi, 1964). Now we adopt a similar procedure (particularly harmonic balance method) to control chaotic behavior.

First consider the first approximate solution in the form

$$x(t) = C_0 + C_1\cos(\Omega t + \zeta) \tag{3.2}$$

where C_0, C_1 and ζ are constants. Substituting *Equation (3.2)* into *Equation (3.1)* it is possible to determine these constants (Kapitaniak, 1991; Hayashi, 1964). To study the stability of the solution *(3.2)* a small variational term $\delta x(t)$ is added to *Equation (3.2)* as

$$x(t) = C_0 + C_1\cos(\Omega t + \zeta) + \delta x(t) \tag{3.3}$$

After some algebraic manipulations, the linearized equation with periodic coefficients for $\delta x(t)$ is obtained,

$$\delta\ddot{x} + a\delta\dot{x} + \delta x \left(\lambda_0 + \lambda_1 \cos \Xi + \lambda_2 \cos 2\Xi\right) = 0 \tag{3.4}$$

where $\lambda_0 = 3C_0^2 + (3/2)C_1^2$, $\lambda_1 = 6C_0C_1$, $\lambda_2 = (3/2)C_1^2$, $\Xi = \Omega t + \zeta$. In the derivation of *Equation (3.4)*, for simplicity it was assumed without loss of generality that $b = 0$. As we have a parametric term of frequency $\Omega - \lambda_1 \cos \Xi$, the lowest order unstable region is that which occurs close to $\Omega/2 \approx \sqrt{\lambda_0}$, and at its boundary we have the solution

$$\delta x = b_{1/2} \cos\left(\frac{\Omega}{2}t + \zeta\right) \tag{3.5}$$

To determine the boundaries of the unstable region we insert *Equation (3.5)* into *Equation (3.4)*, and the conditions of nonzero solution for $b_{(1/2)}$ lead us to the following criterion to be satisfied at the boundary:

$$\left(\lambda_0 - \frac{\Omega^2}{4}\right)^2 + a^2 \frac{\Omega^2}{4} - \frac{\lambda_1^2}{4} = 0 \tag{3.6}$$

From *Equation (3.6)* one obtains the interval $(\Omega_1^{(2)}, \Omega_2^{(2)})$ within which period-two solutions exist. Further analysis shows that at Ω_2 we have a stable period-doubling bifurcation for decreasing Ω and at Ω_1 an unstable period-doubling bifurcation for increasing Ω. In this interval we can consider the period-two solution of the form

$$x(t) = A_0 + A_{1/2} \cos\left(\frac{\Omega}{2}t + \eta\right) + A_1 \cos(\Omega t) \tag{3.7}$$

where A_0, $A_{1/2}$, A_1 and η are constants to be determined. Again, to study the stability of the period-two solution we have to consider a small variational term $\delta x(t)$ added to *Equation (3.7)*. The linearized equation for $\delta x(t)$ has the following form:

$$\delta\ddot{x} + a\delta\dot{x} + \delta x \left[\lambda_0^{(2)} + \lambda_{1/2c} \cos\left(\frac{\Omega}{2}t\right) + \lambda_{3/2} \cos\left(\frac{\Omega}{2}t + \eta\right)\right.$$

$$\left. + \lambda_{1c}^{(2)} \cos(\Omega t) + \lambda_{1s}^{(2)} \sin(\Omega t) + \lambda_2^{(2)} \cos(2\Omega t)\right] = 0 \tag{3.8}$$

where

$$\lambda_0^{(2)} = 3\left(A_0^2 + \tfrac{1}{2}A_{1/2}^2 + \tfrac{1}{2}A_1^2\right), \qquad \lambda_{1/2c} = 3A_{1/2}\left(2A_0 + A_1\right) \cos \eta$$

$$\lambda_{1/2s} \sin\left(\frac{\Omega}{2}t\right) = 3A_{1/2}\left(A_1 + 2A_0\right) \sin \eta, \qquad \lambda_{3/2} = 3A_1A_{1/2}$$

$$\lambda_{1c}^{(2)} = 6A_0A_1 + \tfrac{3}{2}A^2_{1/2c} \cos 2\eta, \qquad \lambda_{1s}^{(2)} = -\tfrac{3}{2}A^2_{1/2s} \sin 2\eta, \qquad \lambda_2^{(2)} = \tfrac{3}{2}A_1^2$$

The form of *Equation (3.8)* enables us to find the range of existence of a period-four solution, represented by

$$\delta x = b_{1/4} \cos\left(\frac{\Omega}{4}t + \eta\right) + b_{3/4} \cos\left(3\,\frac{\Omega}{4}t + \eta\right) \tag{3.9}$$

After inserting *Equation (3.9)* into *Equation (3.8)* the condition of nonzero solution for $b_{1/2}$ and $b_{3/4}$ gives us the following set of nonlinear algebraic equations for Ω, $\cos\eta$, and $\sin\eta$ to be satisfied for existence:

$$\left(\lambda_{1/2s} + \lambda_{1s}^{(2)}\right) - 0.5\left(\lambda_{1/2c} + \lambda_{1c}^{(2)}\right)\left(-a\frac{\Omega}{2} + \lambda_{1/2s} - \lambda_{3/2}\sin\eta\right)$$

$$\left(\tfrac{9}{8}\Omega^2 + 0.5\lambda_0 + \lambda_{3/2}\cos\eta\right) - 0.5\left(\lambda_{1/2c} + \lambda_{1c}^{(2)}\right)\left(\lambda_{1s}^{(2)} + \lambda_{1/2c}\right) = 0 \tag{3.10}$$

$$\left(-\tfrac{3}{2}a\Omega - \lambda_{3/2}\sin\eta\right) - 0.5\left(\lambda_{1/2c} + \lambda_{1c}^{(2)}\right)\left(\lambda_{1s}^{(2)} + \lambda_{1/2c}\right) = 0$$

Solving *Equations (3.10)* by a numerical procedure it is possible to obtain $\Omega_1^{(4)}$ and $\Omega_2^{(4)}$, the frequencies of stable and unstable period-four bifurcations.

Now we assume that the Feigenbaum model (Feigenbaum, 1978) of period doubling is valid for our system, i.e.

$$\lim_{n\to\infty} \frac{\Omega_{1,2}^{2^n} - \Omega_{1,2}^{2^{n-1}}}{\Omega_{1,2}^{2^{n+1}} - \Omega_{1,2}^{2^n}} \to \delta \tag{3.11}$$

where $\delta = 4.669\ldots$ is the universal Feigenbaum constant and $n = 1, 2, \ldots$.

Although it has not been proved that all period-doubling bifurcations fulfill the Feigenbaum model, there are examples where this model can be taken as a good approximation to the real phenomena (Isomaki *et al.*, 1987; Steeb and Louw, 1986).

To obtain approximate values of the limits of period-doubling bifurcations (accumulation points) we replaced the limit in *Equation (3.11)* by an equality. After it let us indicate

$$\Delta\Omega_{1,2}^{2^n} = \Omega_{1,2}^{2^n} - \Omega_{1,2}^{2^{n-1}} \tag{3.12}$$

Now it is easy to show that $\Delta\Omega_{1,2}^{2^n}$, $\Omega_{1,2}^{2^{n-1}}$, $\ldots$ form infinite geometrical series with a ratio $1/\delta$. With both stable and unstable period-two and period-four boundaries using the above described approximation one obtains

$$\Omega_1^{\infty} = \Omega_1' + \frac{\Delta\Omega_1}{1 - \dfrac{1}{\delta}}$$

$$\Omega_2^{\infty} = \Omega_2' + \frac{\Delta\Omega_2}{1 - \dfrac{1}{\delta}} \tag{3.13}$$

where $\Delta\Omega_1 = \Omega_1^{(4)} - \Omega_1^{(2)}$ and $\Delta\Omega_2 = \Omega_2^{(2)} - \Omega_2^{(4)}$.

The domain where chaotic behavior can occur is proposed to be between the limits of unstable and stable period-doubling cascades, in the interval

$(\Omega_1^{(\infty)}, \Omega_2^{(\infty)})$ and of course to expect chaos one must have

$$\Omega_1^{(\infty)} < \Omega_2^{(\infty)}$$

More details about this method can be found in Kapitaniak (1992). This approach, besides estimating the chaotic region in parameter space, evaluates analytically the approximate unstable orbits or at least regions in parameter space where they exist. The above analysis can be used to control *Equation (3.1)* by changing parameter Ω in the range $\Omega \in [\Omega - \Omega^*, \Omega - \Omega^*]$. It should be noted here that the frequency Ω is this parameter which can be very easily changed in real experimental systems modeled by *Equation (3.1)*.

From *Figure 3.1* one can find that by changing Ω by $\Omega^* < 0.12$ one can obtain different types of periodic behavior; from period-one to theoretically period -2^n ($n = 1, 2, \ldots$). Periodic orbits with higher order ($n > 4$) are difficult to obtain as the Ω intervals of the existence of these solutions are very small. The example of controlling a few of periodic orbits is shown in *Figure 3.2*.

We plot coordinate x of the Poincaré map as a function of discrete time

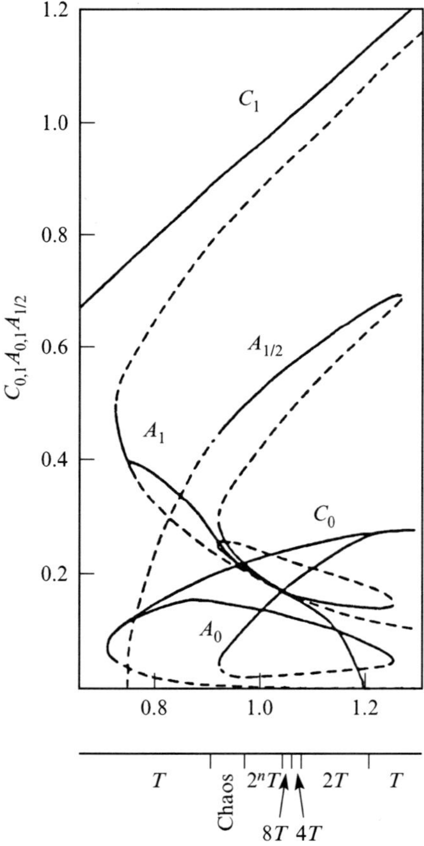

Figure 3.1 Parameters of period-one and period-two approximate solutions and the Ω intervals of period-four and period-eight solutions; $a = 0.05$, $b = 0$, $c = 1$, $B_0 = 0.03$, $B_1 = 0.16$, solid line indicates stable solutions while broken line indicates unstable solutions.

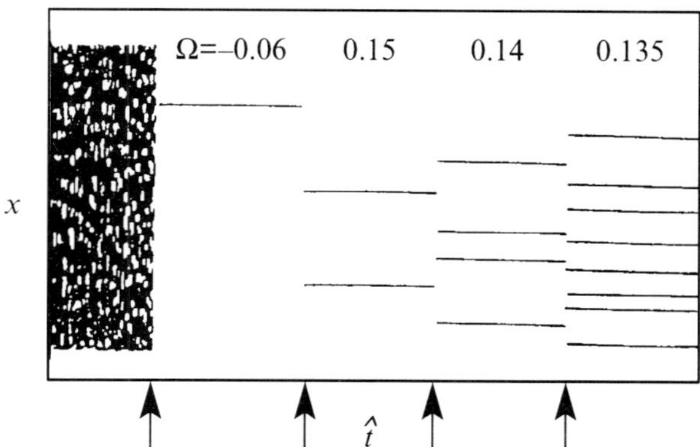

Figure 3.2 Successive control of period-one, -two, -four, -eight orbits, the arrows indicate time switching.

$\hat{t} = 2\pi n/(\Omega + \Omega^*)$, $n = 1, 2, \ldots$. The frequency perturbations we programmed to control four different periodic orbits successively. The times at which we switched the control from stabilizing one periodic orbit to stabilize another are shown by arrows in *Figure 3.2*.

The procedure described in this section is based on the direct change of one of the system parameters to shift system behavior from chaotic to periodic, close to the chaotic attractor. It cannot be called a controlling method in the sense of methods described in Chapter 2, but it illustrates that having a system designed as chaotic, we obtain easy access to different types of periodic behavior.

3.2   Control by system design

In this section we explore the idea of modifying or removing chaotic behavior by appropriate system design. It is clear that, to a certain extent, chaos may be 'designed out' of a system by appropriate modification of parameters, perhaps corresponding to modification of mass or inertia of moving parts. Equally clearly, there exist strict limits beyond which such modifications cannot go without seriously affecting the efficiency of the system itself.

In this section we describe a method for controlling chaos in which the chaos effect is achieved by coupling the chaotic main system to a simpler autonomous system (controller), usually linear, as shown in *Figure 3.3*. Our method is developed for chaotic systems in which for some (e.g. technological) reason it is difficult, if not impossible, to change any parameter of the main system. In particular consider the coupling of the chaotic system

$$\dot{\mathbf{x}} = f(\mathbf{x}, \mu) \tag{3.14}$$

where $\mathbf{x} \in \mathbf{R}^n$, $n \geq 3$ and $\mu \in \mathbf{R}$ is a system parameter, to another (simpler)

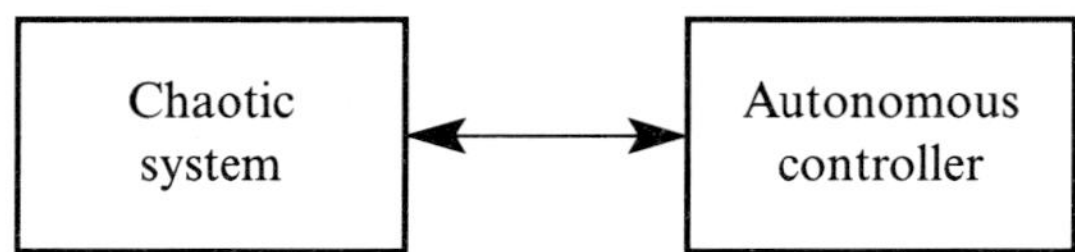

Figure 3.3 Coupling scheme.

asymptotically stable system (controller) described by

$$\dot{\mathbf{y}} = \mathbf{g}(\mathbf{y},e) \tag{3.15}$$

where $\mathbf{y} \in \mathbf{R}^m$, e is a vector denoting the controller's parameters, where at least one of the parameters e_i can be easily changed. For practical reasons, the dimension m of the controller system (3.15) should be chosen as low as possible. Since our method was mainly designed for controlling chaos in mechanical systems, we choose $m = 2$, i.e. one-degree-of-freedom controller (the simplest mechanical system). The equations for the augmented system are

$$\dot{\mathbf{x}} = f(\mathbf{x},\mu) + \zeta_y y$$
$$\dot{\mathbf{y}} = \mathbf{g}(\mathbf{y},e) + \zeta_x x \tag{3.16}$$

where ζ_x and ζ_y are the coupling matrices. When $\zeta_x = 0$ and $\zeta_y = 0$, the x and y subsystems in *Equation (3.16)* are uncoupled and for small $\|\zeta_x\|$ and $\|\zeta_y\|$ (i.e. all entries are small) only small additional signals are injected into the main system. Since the y subsystem is asymptotically stable, the role of the controller is to change the behavior of the system from chaotic to some desired periodic, possibly constant, operating regime. If $\|\zeta_x\|$ and $\|\zeta_y\|$ are sufficiently small, $\mathbf{x}(t)$ of the coupled *Equation (3.16)* will evolve in a small neighborhood of the original attractor of *Equation (3.14)* thereby preserving the qualitative dynamics of the main system. As the evolution of $\mathbf{x}(t)$ given by *Equation (3.15)* takes place on the attractor, there exists a vector $M \in \mathbf{R}^n$ such that $\|x_i(t)\| \leq M_i$ for all time t. Our control will be effective if $\mathbf{x}(t)$ given by *Equation (3.16)* fulfills the relation $\|x_i(t)\| \leq M_i + \varepsilon$, where ε is a small parameter.

The idea of this method is similar to that of the so-called dynamical vibration absorber. A dynamical vibration absorber is a one-degree-of-freedom system, usually mass on a spring (sometimes viscous damping is also added), which is connected to the main system as shown in *Figure 3.4*.

The additional degree of freedom introduces shifts resonance zones, and in some cases can eliminate oscillations of the main mass. Although such a dynamical absorber can change the overall dynamics substantially, it need usually only be physically small in comparison with the main system, and does not require an increase of excitation force. It can be easily added to the existing system without major changes of design or construction. This contrasts with devices based on feedback control, which can be large and costly.

(a)

(b)

Plate 1 Stabilization of period-one (a) and period-two (b) orbits.

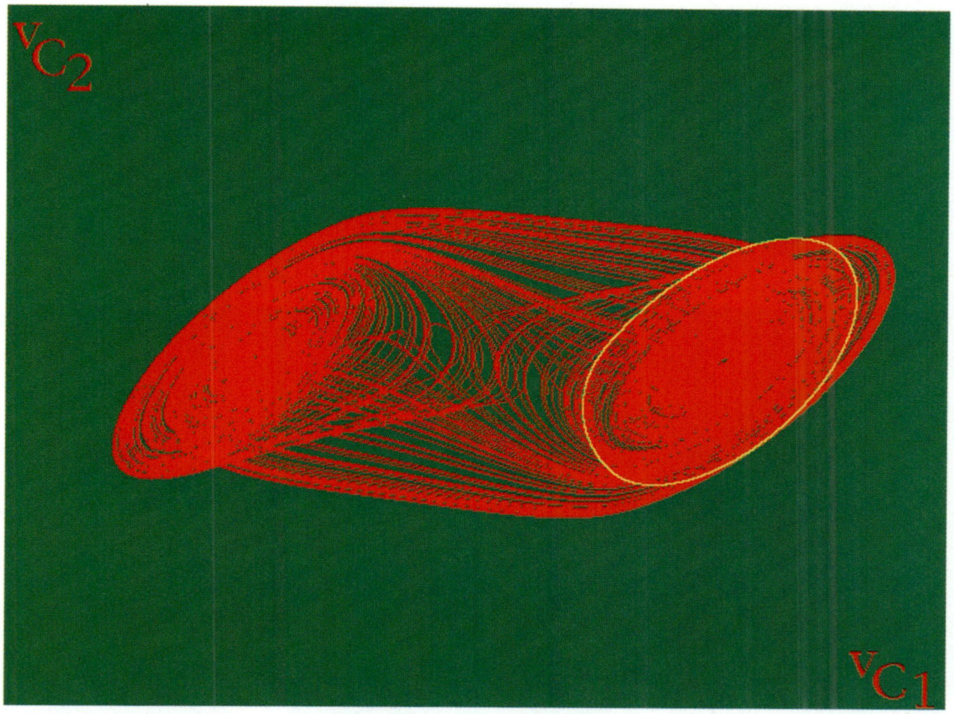

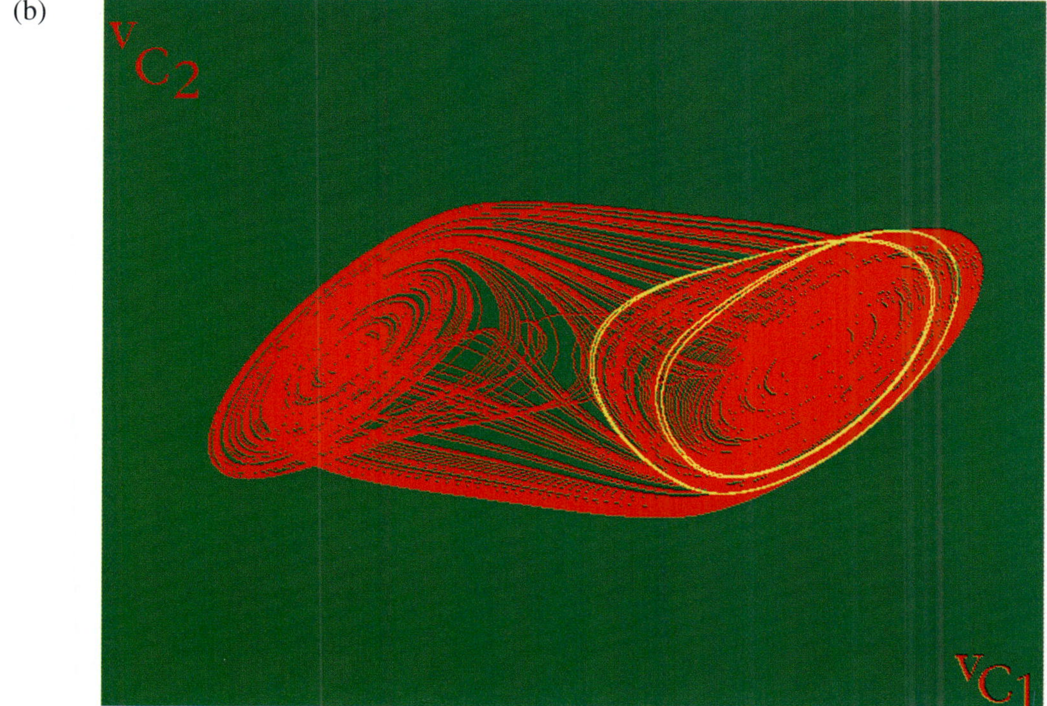

Plate 2 Controlling periodic orbit located on spiral chaotic attractor.

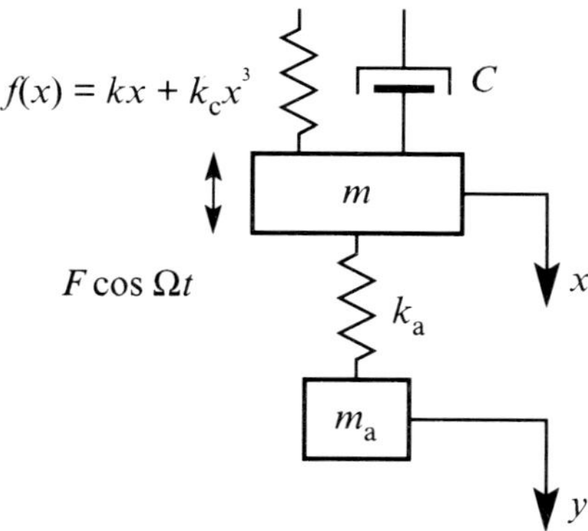

Figure 3.4 Dynamical damper as chaos controller.

To explain the role of dynamical absorbers in controlling chaotic behavior let us consider the Duffing oscillator, coupled with an additional linear system:

$$\ddot{x} + a\dot{x} + bx + cx^3 + d(x - y) = B_0 + B_1 \cos \Omega t \qquad (3.17a)$$

$$\ddot{y} + e(y - x) = 0 \qquad (3.17b)$$

where a, b, c, d, e, B and Ω are constants. Here d and e are the characteristic parameters for the absorber, and we take e as the control parameter. The parameters of *Equation (3.17)* are related to those of *Figure 3.4* in the following way:

$$a = \frac{c}{m\Omega^2}, \; b = \frac{k}{m\Omega^2}, \; c = \frac{k_c}{m\Omega^2}, \; d = \frac{k_a}{m\Omega^2}, \; e = \frac{k_a}{m\Omega^2}, \; B_0 = \frac{F_0}{m\Omega^2}, \; B_1 = \frac{F_1}{m\Omega^2}$$

It should be noted here that parameters d and e are related to each other through the absorber stiffness k_a. For simplicity, in the rest of this section we assume that d is constant and consider e as a control parameter, i.e. we take constant stiffness k_a and allow the absorber mass m_a to vary.

It is well-known that the uncoupled *Equation (3.17a)* (i.e. without the dynamical absorber) shows chaotic behavior for certain parameter regions (Ueda, 1991; Kapitaniak 1991a,b). As has been mentioned in the previous section, in many cases the route to chaos proceeds via a sequence of period-doubling bifurcations, and in such cases our method provides an easy way of switching between chaotic and periodic behavior.

To analyze the system with the absorber (present d, $e \neq 0$), we first assume that all parameters of *Equation (3.17)*, excluding the forcing frequency Ω, are constant, and estimate the Ω-domain where chaos exists. The application of the harmonic balance method enables us to determine the stability domain of appropriate $2\pi/\Omega = T$ periodic solutions, i.e.

$$x = C_0 + C_1 \cos(\Omega t + \psi)$$
$$y = D_0 + D_1 \cos(\Omega t + \gamma) \qquad (3.18)$$

and $2T$ periodic solutions

$$x = A_0 + A_{1/2} \cos\left(\frac{\Omega}{2}t + \rho\right) + A_1 \cos \Omega t$$

$$(3.19)$$

$$y = E_0 + E_{1/2} \cos\left(\frac{\Omega}{2}t + \beta\right) + E_1 \cos \Omega t$$

where C_0, C_1, D_0, D_1, A_0, $A_{1/2}$, A_1, E_0, $E_{1/2}$, E_1, ψ, γ, ρ and β are constants which are determined by substituting *Equations (3.19)* or *(3.18)* into *Equation (3.17)*. Approximate boundaries of stability as functions of forcing frequency Ω for each solution can be estimated by adding small perturbations dx and dy to x and y, and considering an appropriate Hill's equation. The whole procedure is similar to the one described in the previous section, so we omit the details here. Knowing the period-doubling bifurcation values Ω_1' and Ω_2' at which we have bifurcation from $T \to 2T$ periodic solutions, and Ω_1'' and Ω_2'' at which we have bifurcation from $2T \to 4T$ periodic solutions, we can obtain approximate values for the accumulation points Ω_1^∞ and Ω_2^∞ from *Equation (3.13)* and the interval $[\Omega_1^\infty, \Omega_2^\infty]$ can be considered as an approximation of the Ω frequency domain for which chaos exists.

The above procedure can be easily performed using any symbolic algebra system (we used Mathematica) and by following it for different values of e we are able to obtain a map of behavior of *Equations (3.17)* as a function of two parameters: the frequency Ω and the dynamical absorber control parameter e, as shown for example in *Figure 3.5* (solid lines).

Other parameters of *Equation (3.18)* have been fixed at the values $a = 0.077$, $b = 0$, $c = 1.0$, $B_0 = 0.045$ and $B_1 = 0.16$. This plot is in good agreement with numerically obtained behavior domains as shown in *Figure 3.5* (broken lines). Numerical results were obtained using a fourth-order Runge–Kutta method with a time step $\pi/200\Omega$, and to determine chaotic behavior the Lyapunov exponents were calculated using the algorithm of Wolf *et al.* (1985).

From *Figure 3.5* it is clear that, for fixed Ω, we can obtain different types of periodic behavior by making slight changes in e. As an example, consider a system with $\Omega = 0.98$. For $e < 0.09$, the system is chaotic, but by changing e from 0.01 to 0.16 it is possible to obtain easily T, $2T$, $4T$ and $8T$ periodic orbits. Theoretically, the orbits of higher periods are also possible, but their narrow range of existence makes them difficult to be found either experimentally or numerically. What is of vital significance is that the values of parameter $e \in [0.01, 0.16]$ can be obtained with absorber mass m_a approximately 100 times smaller than the main mass (*Figure 3.4*). A similar controlling effect can be obtained by varying the absorber stiffness, i.e. by simultaneous changes of parameters d and e.

As the next example consider controlling chaos in Chua's circuit (*Equation (1.3)*) by coupling it to a second-order linear circuit, as shown in *Figure 3.6*. The dimensionless state equations of the augmented circuit are

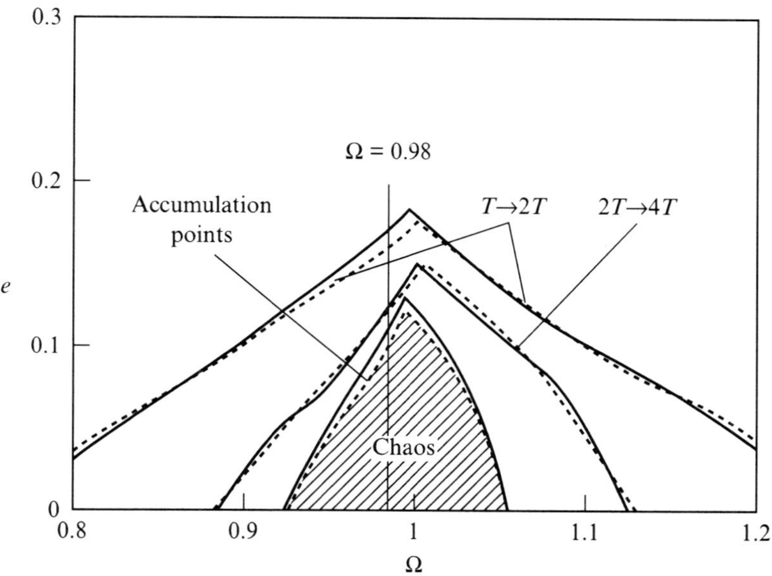

Figure 3.5 Behavior of *Equation (3.17)*: $a = 0.077$, $b = 0$, $c = 1$, $B_0 = 0.045$ and $B_1 = 0.16$; analytical approximation: solid line, numerical simulation: broken line.

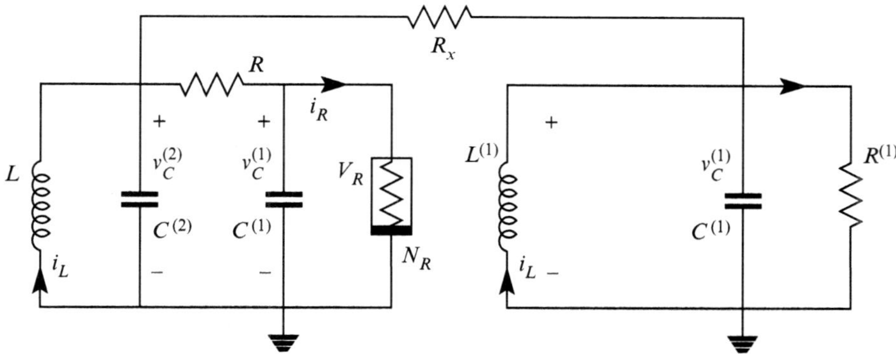

Figure 3.6 Control of Chua's circuit by second order autonomous circuit.

given by

$$\dot{x} = \alpha\left[y - x - f(x)\right]$$

$$\dot{y} = x - y + z + \varepsilon(y^{(1)} - y)$$

$$\dot{z} = -\beta y \tag{3.20}$$

$$\dot{y}^{(1)} = \alpha^{(1)}\left[-\gamma^{(1)}y^{(1)} + z^{(1)} + \varepsilon(y - y^{(1)})\right]$$

$$\dot{z}^{(1)} = -\beta^{(1)}y^{(1)}$$

where $y^{(1)} = v^{(1)}$, $\alpha^{(1)} = C_2/C^{(1)}$, $\beta^{(1)} = C_2/L^{(1)}G^2$, $\gamma^{(1)} = R/R^{(1)}$ and $\varepsilon = R/R_x$, ε denotes the coupling stiffness. Note that if $\varepsilon = 0$, *Equation (3.20)* describe two uncoupled systems. When ε is sufficiently small, the dynamics of *Equation (3.20)* are closely related to the dynamics of the original system.

The results of experimental control to obtain different types of periodic orbits are shown in *Figure 3.7(a)–(f)*. Observe that a small change in the coupling stiffness ε allows us to obtain a periodic orbit ($\varepsilon = 0.302$) as shown in *Figure 3.7(a)*, a period-two orbit ($\varepsilon = 0.148$) as shown in *Figure 3.7(b)*, a period-four orbit ($\varepsilon = 0.134$) as in *Figure 3.7(c)*, a period-five orbit ($\varepsilon = 0.105$) as in *Figure 3.7(d)* and a period-three orbit ($\varepsilon = 0.097$) as shown in *Figure 3.7(e)*. In *Figure 3.7(f)* ($\varepsilon = 0.322$) this methods allows stabilization of the fixed point.

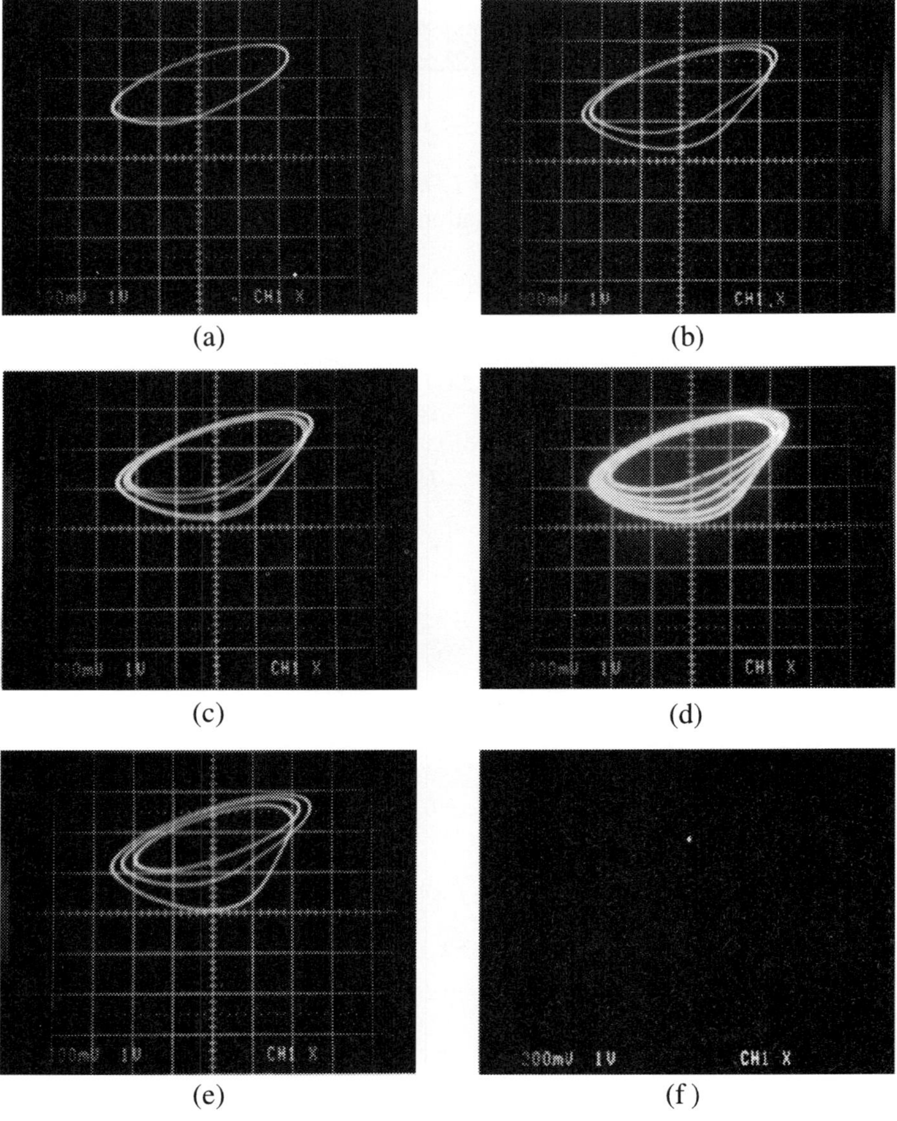

(a)   (b)   (c)   (d)   (e)   (f)

Figure 3.7 Control of different periodic orbits and fixed point.

As this method is designed mainly for experimental applications, we shall now briefly suggest some guidelines for applying it.

(1) The coupled system has to be as simple as possible.
(2) The coupling stiffness ε should be chosen as small as possible.
(3) If it is possible one should couple the controller in such a way that the locations of the fixed points of the original system are not changed. For example, consider the system

$$\dot{x}_1 = f_1(x_1, x_2, x_3)$$
$$\dot{x}_2 = f_2(x_1, x_2, x_3) \tag{3.21}$$
$$\dot{x}_3 = f_3(x_1, x_2, x_3)$$

Suppose that the fixed points of the system (3.21) are such that the second coordinate is zero: $x_2 = 0$. Then the following coupled system

$$\dot{x}_1 = f_1(x_1, x_2, x_3)$$
$$\dot{x}_2 = f_2(x_1, x_2, x_3) + \varepsilon(y_1 - x_2)$$
$$\dot{x}_3 = f_3(x_1, x_2, x_3) \tag{3.22}$$
$$\dot{y}_1 = a_{11}y_1 + a_{12}y_2 + \varepsilon(x_2 - y_1)$$
$$\dot{y}_2 = a_{21}y_1 + a_{22}y_2$$

will have the same fixed points as those of *Equation (3.21)* as can be seen from *Equation (3.22)*. In fact, the coupling in the example of controlling Chua's circuit is chosen in this way. The advantage of such a coupling is obvious: with relatively small ε we can stabilize the exact fixed point of the original system.

More details about this method can be found in Kapitaniak *et al.* (1993).

3.3   Taming chaos

In papers by Steeb *et al.* (1986) and Kapitaniak *et al.* (1987) it was first demonstrated that chaos in a dynamical system can be reduced (the largest Lyapunov exponent is decreased) or replaced by regular behavior by applying a weak external periodic signal. Periodic perturbation can be introduced to the system as external force (Steeb *et al.*, 1986; Braiman and Goldhirsch, 1991) or as a perturbation of one of the internal system parameters (Kapitaniak *et al.*, 1987; Pettini, 1989). Given an external perturbation, it is possible to show that chaotic system is capable of finding an appropriate orbit.

In Kapitaniak (1988) and Herzel (1988) it was shown that the chaotic system can be set into regular motion (periodic motion perturbed by noise – 'noisy periodicity') by addition of suitable random noise. A similar approach is described in Braiman and Goldhirsch (1991) and Rajasekar and

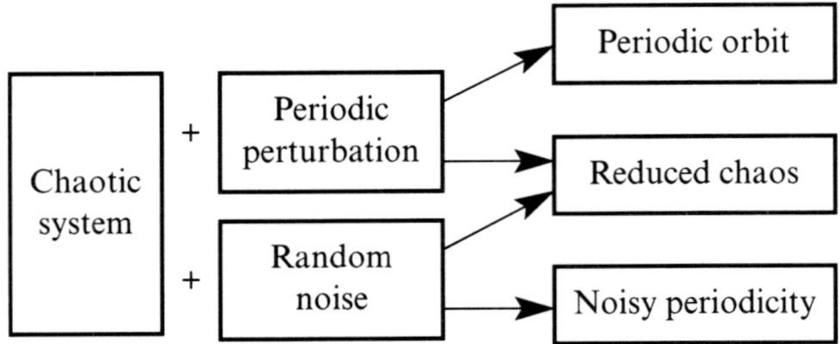

Figure 3.8 Idea of taming chaos.

Lakshamanan (1992).

Later both these approaches were called taming chaos. Possible outputs of taming chaos procedures are sketched in *Figure 3.8*.

3.4 Entrainment and migration control

Another representative approach for the control of chaos is connected with entrainment and migration control methods (Jackson, 1990, *Paper 8*; Hübler, 1987, 1989; Hübler and Luscher, 1989; Jackson and Hübler, 1990; Jackson and Kodogeorgiou, 1991). This approach has been successfully applied to many complex dynamical systems.

The concept of the entrainment–goal control method is based on the existence of certain convergent regions in the phase space of multi-attractor system. In each of these convergent regions, all the nearby orbits converge locally towards each other. An important observation is that although many of these attractors have positive Lyapunov exponents, they nevertheless have such regions in their basins of attraction where nearby orbits generally converge, meaning that the Lyapunov exponents are merely an average measure of their dynamics. This property is fully described in Kapitaniak (1995).

The purpose of migration–goal control is to transfer the system dynamics from one convergent region to another. There are many reasons for this kind of state transfer. For example, among the many coexisting attractors of a complex system, some may have very different types of dynamics.

Both entrainment and migration control is fully described in *Paper 8*.

4 Synchronization of chaos

4.1 Pecora and Carroll's approach

As chaotic trajectory is not asymptotically stable, closely correlated initial conditions have trajectories which quickly become uncorrelated. Despite this obvious disadvantage, it has been established that synchronization of two chaotic systems is possible (Pecora and Carroll, 1990 – *Paper 9*).

This synchronization procedure can be described as follows: suppose that an n-dimensional dynamical system

$$\dot{u} = h(u), \qquad u = u(x,y) \tag{4.1}$$

can be divided into two subsystems

$$\dot{x} = f(x,y)$$
$$\dot{y} = g(x,y) \tag{4.2}$$

where

$$x = [u_1, \ldots, u_m]^\mathrm{T}, f = [h_1(u), \ldots, h_m(u)]^\mathrm{T}, y = [u_{m+1}, \ldots, u_n]^\mathrm{T}, g = [h_{m+1}(u), \ldots, h_n(u)]^\mathrm{T}$$

Let us create a new subsystem z identical to the y subsystem, substitute the set of variables x for the corresponding x' in the function g, and augment *Equation (4.2)* with this new system, giving

$$\dot{x} = f(x,y)$$
$$\dot{y} = g(x,y) \tag{4.3}$$
$$\dot{z} = g(x,z)$$

The first two equations of *(4.3)* are called a driving subsystem and the third one a response subsystem. Lyapunov exponents of the response subsystem for a particular input $x(t)$ are called conditional Lyapunov exponents. Let $y(t)$ be a chaotic trajectory with initial condition $y(0)$ and $z'(t)$ be a trajectory starting in the nearby point $z'(0)$. It was shown that the necessary and sufficient condition for

$$|z(t) - y(t)| \to 0$$

that is, the two subsystems being synchronized, is that all of the conditional Lyapunov exponents must be negative. De Sousa Viera *et al.* (1992) showed that the boundary of possible synchronization and nonsynchronization is strictly connected with the transition from chaotic to hyperchaotic behavior that is characterized by at least two positive Lyapunov exponents (Rossler, 1979; Kapitaniak and Steeb, 1991).

We can describe this procedure using an example of Chua's circuit introduced in Chapter 1. Its dimensionless *Equation (1.3)* can be decomposed in three different ways:

(1) *x*-drive configuration where the state equations become

$$\begin{array}{l} \dot{x} = f(x,y) \\ \dot{y} = g(x,y) \end{array} \qquad \left. \begin{array}{l} \dot{x} = \alpha(y - x - f(x)) \\ \dot{y} = x - y + z \\ \dot{z} = \beta y \end{array} \right\} \quad \text{driving subsystem}$$

$$\dot{z} = g(x,z) \qquad \left. \begin{array}{l} \dot{y}' + x - y' + z' \\ \\ \dot{z}' = -\beta y' \end{array} \right\} \quad \text{response subsystem}$$

(2) *y*-drive configuration where the state equations become

$$\begin{array}{l} \dot{x} = f(x,y) \\ \dot{y} = g(x,y) \end{array} \qquad \left. \begin{array}{l} \dot{x} = \alpha(y - x - f(x)) \\ \dot{y} = x - y + z \\ \dot{z} = \beta y \end{array} \right\}$$

$$\dot{z} = g(x,z) \qquad \left. \begin{array}{l} \dot{y}' + x - y' + z' \\ \dot{z}' = -\beta y' \end{array} \right\}$$

(3) *z*-drive configuration where the state equations become

$$\begin{array}{l} \dot{x} = f(x,y) \\ \dot{y} = g(x,y) \end{array} \qquad \left. \begin{array}{l} \dot{x} = \alpha(y - x - f(x)) \\ \dot{y} = x - y + z \\ \dot{z} = \beta y \end{array} \right\}$$

$$\dot{z} = g(x,z) \qquad \left. \begin{array}{l} \dot{y}' + x - y' + z' \\ \dot{z}' = -\beta y' \end{array} \right\}$$

It can be shown that for $\alpha = 10.0$, $\beta = 14.87$, $m_0 = -1.27$ and $m_1 = -0.68$, subsystems $y = g(x,y)$ and $z = g(x,z)$ can synchronize only in x and y configurations, as conditional Lyapunov exponents for each configuration are respectively (Chua *et al.*, 1993)

$$\left[\lambda^1_{(c)} = -0.05, \lambda^2_{(c)} = -0.05\right]$$

$$\left[\lambda^1_{(c)} = -2.5 \pm 0.05, \lambda^2_{(c)} = 0\right]$$

$$\left[\lambda^1_{(c)} = -5.42 \pm 0.02, \lambda^2_{(c)} = 1.23 \pm 0.03\right]$$

4.2   Synchronization by continuous control

To synchronize two chaotic systems A and B, the dynamics of which are described respectively by

$$\dot{x} = f(x,a) \tag{4.4a}$$

$$\dot{y} = f(y,a) \tag{4.4b}$$

where $x, y \in \mathbf{R}^n$, $a \in \mathbf{R}$ describes the control parameters, we use the strategy which is schematically illustrated in *Figure 4.1*. We assume that some state variables of both systems A and B can be measured. Let us say that we can measure signal $x_i(t)$ from the system A and signal $y_i(t)$ from B where $i \in \{1, 2, \ldots, n\}$. Chaotic systems A and B are coupled unidirectionally in such a way that the difference $D(t)$ between the signals $x_i(t)$ and $y_i(t)$ is used as a control signal,

$$F(t) = K[x_i(t) - y_i(t)] = KD(t) \tag{4.5}$$

which is applied to one of the chaotic systems (A in *Figure 4.1*) as negative feedback.

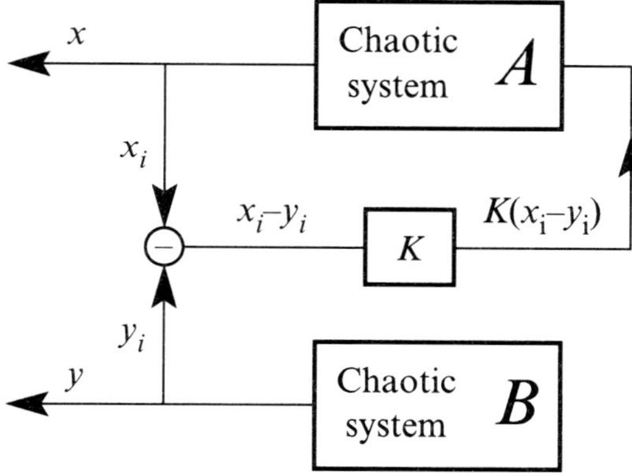

Figure 4.1 Synchronization by continous control.

The parameter $K > 0$ is an experimentally adjustable weight of the perturbation and we shall discuss its selection later. Experimental realization of such feedback presents no difficulties for many practical systems. An important feature of the perturbation signal (*Equation (4.5)*) is that it does not change the solution of *Equation (4.4b)*. When synchronization is achieved $F(t)$ becomes zero, so the chaotic systems A and B become practically uncoupled.

We illustrate our synchronization procedure using two identical unidirectionally-coupled Chua's circuits, as shown in *Figure 4.2*.

The state equations for the circuit of *Figure 4.2* are as follows:

$$C_1 \frac{dv_{C_1}^{(1)}}{dt} = G\left(v_{C_2}^{(1)} - v_{C_1}^{(2)}\right) - f\left(v_{C_1}^{(1)}\right) \tag{4.6a}$$

$$C_2 \frac{dv_{C_2}^{(1)}}{dt} = G\left(v_{C_1}^{(1)} - v_{C_2}\right) + i_L^{(1)} - K\left(v_{C_2}^{(2)} - v_{C_2}^{(1)}\right) \tag{4.6b}$$

$$Ldi_L^{(1)} = v_{C_2}^{(1)} \tag{4.6c}$$

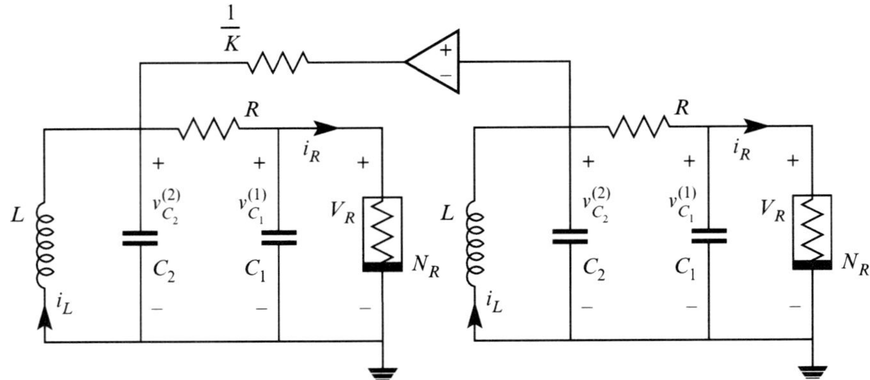

Figure 4.2 Synchronization scheme for two Chua's circuits.

$$C_1\frac{\mathrm{d}v_{C_1}^{(2)}}{\mathrm{d}t} = G\!\left(v_{C_2}^{(2)} - v_{C_1}^{(2)}\right) - f\!\left(v_{C_1}^{(2)}\right) \tag{4.6d}$$

$$C_2\frac{\mathrm{d}v_{C_2}^{(2)}}{\mathrm{d}t} = G\!\left(v_{C_1}^{(2)} - v_{C_2}^{(1)}\right) + i_L^{(1)} \tag{4.6e}$$

$$L\mathrm{d}i_L^{(2)} = v_{C_2}^{(2)} \tag{4.6f}$$

where $G = 1/R$. As shown in Chapter 1 for $C_1 = 10$ nF, $B_p = 1$ V, $C_2 = 99.34$ nF, $m_1 = -0.76$ mS, $m_0 = -0.41$ mS, $L = 18.46$ mH, $R = 1.64$ kΩ and $K = 0$, both Chua's circuits operate on the chaotic double-scroll attractor shown in *Figure 1.4*. If $x(0)$ is slightly different from $y(0)$ $\left(x = \left[v_{C_1}^{(1)}, v_{C_2}^{(1)}, i_L^{(1)}\right]^{\mathrm{T}}, y = \left[v_{C_1}^{(2)}, v_{C_2}^{(2)}, i_L^{(2)}\right]^{\mathrm{T}}\right)$ the two trajectories $x(t)$ and $y(t)$ diverge exponentially from each other. To achieve synchronization we add the perturbation signal

$$K\!\left(v_{C_2}^{(2)} - v_{C_1}^{(1)}\right) \tag{4.7}$$

to *Equation (4.4b)*.

In *Figures 4.3(a)* and *(b)* we show two v_{C_1} versus v_{C_2} plots of system *(4.6)*. Observe from *Figure 4.3(a)* that for $K = 1.2$ this plot is a straight line which indicates that synchronization of both Chua's circuits is achieved. For smaller values of K, for example $K = 1.1$, the structure of the $v_{C_1}^{(1)}$ versus $v_{C_1}^{(2)}$ plot is more complicated, and the two circuits in *Figure 4.3(b)* are not synchronized. Our experimental results are in good agreement with the numerical simulations of *Equation (4.6)*, as can be seen in *Figure 4.4(a)* and *(b)* where we present the corresponding $v_{C_1}^{(1)}$ versus $v_{C_1}^{(2)}$ numerically obtained plots. To summarize our experimental results we found that synchronization can be achieved for $K < 1.17$ (Kapitaniak *et al.*, 1994).

The example presented above has shown that our method is a very convenient way to synchronize multidimensional systems by feeding back a single variable. Generally, it has to be one of the state variables described by a drive subsystem (in the previously described classification of Pecora and Carroll) of the chaotic systems *A* and *B*, since feeding back variables from the driven subsystem does not result in a successful continuous chaos control

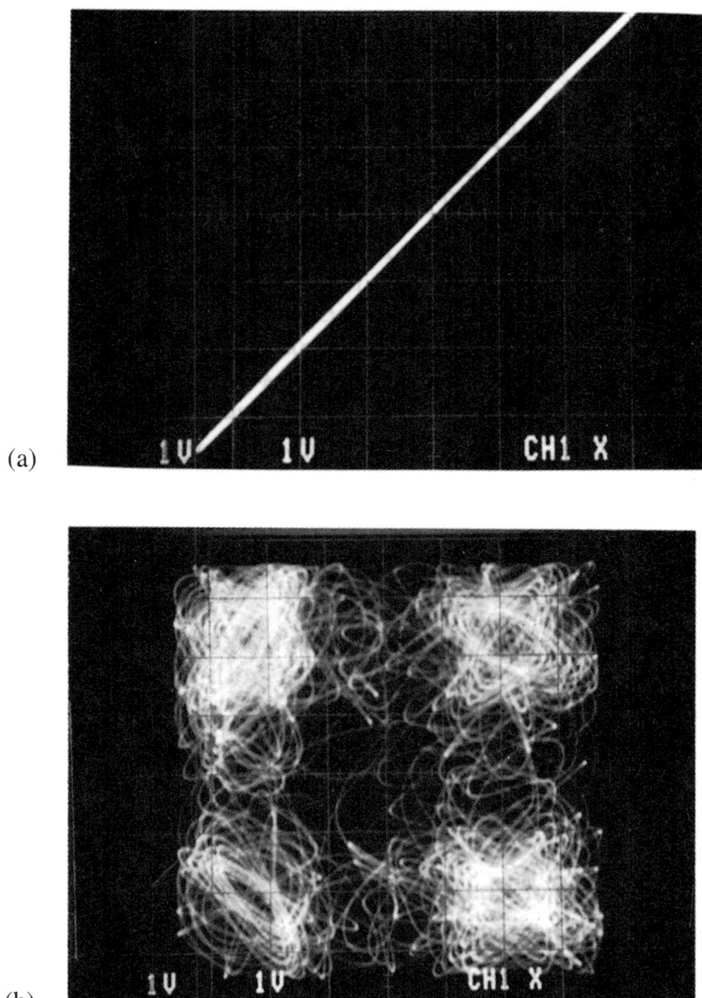

Figure 4.3 Experimental $v_{C_1}^{(1)}$ versus $v_{C_1}^{(2)}$ plots of unidirectionally-coupled Chua's circuits; (a) synchronized state, (b) unsynchronized state.

method (Qu *et al.*, 1993). However, due to the limitations of continuous chaos control methods (Qu *et al.*, 1993) feeding back only one state variable does not always result in a successful synchronization. One can easily show that synchronization can be achieved only if the number of positive Lyapunov exponents of the 'composite' coupled system is equal to the number of positive Lyapunov exponents of the component system. In our example only one positive Lyapunov exponent in the spectrum of *Equations (4.6)* is allowed in order to achieve synchronization. Knowing the equations of chaotic systems A and B we can easily check the above condition by direct computation of the Lyapunov exponents. If the Lyapunov exponents

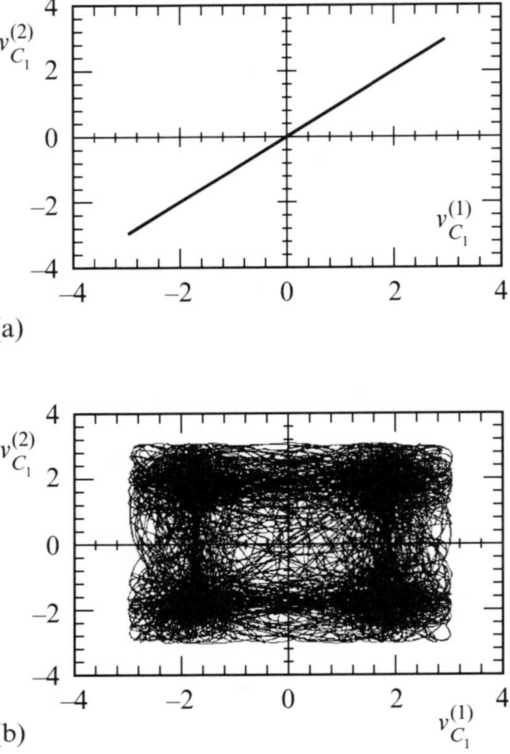

(a)

(b)

Figure 4.4 Numerical $v_{C_1}^{(1)}$ versus $v_{C_1}^{(2)}$ plots; (a) synchronized state, (b) unsynchronized state.

condition is fulfilled, then the coupled system *(4.6)* will evolve on the same manifold on which both chaotic systems evolve and this is why synchronization can be obtained. When it is not fulfilled, the coupled system will evolve on a higher-dimensional manifold on which hyperchaotic attractor exists, and according to De Sousa *et al.* (1992), synchronization cannot be obtained. The region of synchronization in the K-parameter space can be enlarged if we simultaneously feedback more state variables of chaotic systems A and B. Such an example is shown in *Figures 4.5(a)* and *(b)*.

In this figure we show experimental and numerical plots of system trajectories for $K = 1.1$ when, besides the perturbation *(4.7)* added to *Equation (4.6b)*, an additional perturbation

$$K\left(v_{C_1}^{(2)} - v_{C_1}^{(1)}\right) \tag{4.8}$$

is added to *Equation (4.6a)*. It can be seen that by feeding back the two state variables, v_{C_1} and v_{C_2}, synchronization is achieved.

In contrast, observe that for the situation in *Figures 4.3(b)* and *4.4(b)* (with $K = 1.1$ and only v_{C_2} is feedback), the two Chua's circuits are not

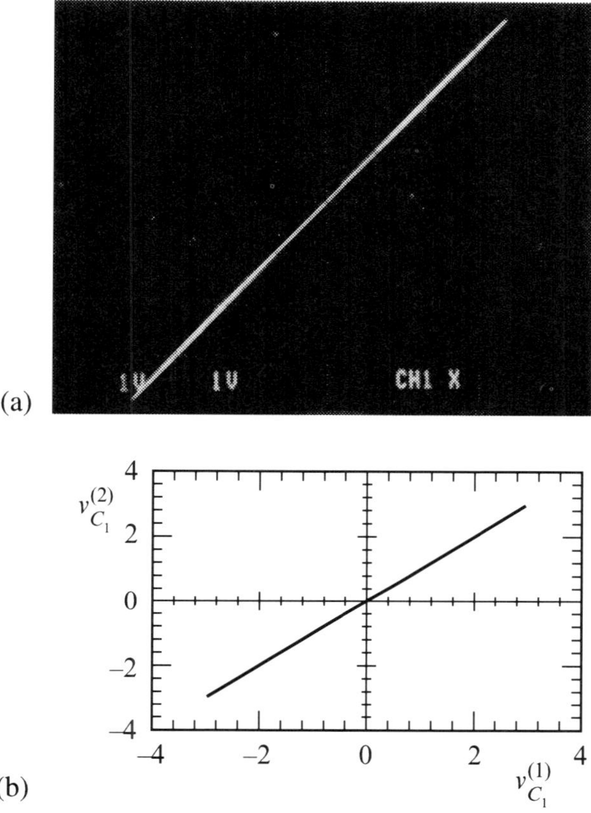

Figure 4.5 $v_{C_1}^{(1)}$ versus $v_{C_1}^{(2)}$ plots of unidirectionally-coupled Chua's circuit, both v_{C_1} and v_{C_2} are fed back; (a) experimental, (b) numerical.

synchronized. By feeding back v_{C_1} and v_{C_2}, it is possible to synchronize both Chua's circuits for $K < 0.9$ (Kapitaniak *et al.*, 1994).

Similar synchronization approach is described in Pyragas (1993 – *Paper 10*).

The described method does not require monitoring of the chaotic trajectories, or applying some targeting procedure that was necessary in other methods (Lai and Grebogi, 1993). Due to the continuity of our controlling scheme, our synchronization method is noise resistant and can be easily applied to the experimental systems.

4.3   Monotonic synchronization

Define $e(t) = x(t) - y(t)$ as an error function. Chaotic system *(4.4b)* synchronizes with chaotic evolution of the system *(4.4a)* if

$$\lim_{t \to \infty} \| e(t) \| = 0 \qquad (4.9)$$

that is the dynamics $e(t)$ are asymptotically stable for $e = 0$.

If we additionally assume that $\left|e_i(t)\right| = \left|x_i(t) - y_i(t)\right|$ is a monotone nonincreasing function of time, we have monotone synchronization of ith state variables. This type of synchronization is investigated in this section. If monotonicity of $\left|e_i(t)\right|$ holds for all i ($i = 1, \ldots, n$), we say that the chaotic system can synchronize monotonically with chaotic system (4.4b) (Kapitaniak *et al.*, 1996).

As an example of monotone synchronization let us consider a pair of unidirectionally coupled identical Chua's circuits, for which the combined equations of motion are

$$\dot{x} = \alpha\left(y - x - f(x)\right) \tag{4.10a}$$

$$\dot{y} = x - y + z \tag{4.10b}$$

$$\dot{z} = -\beta y \tag{4.10c}$$

$$\dot{u} = \alpha\left(v - u - f(u)\right) + K_1\left(u - x\right) \tag{4.10d}$$

$$\dot{v} = u - v + w + K_2\left(y - v\right) \tag{4.10e}$$

$$\dot{w} = -\beta v + K_3\left(z - w\right) \tag{4.10f}$$

where $f(x)$ is given by *Equation (1.1)*, variables u, v and w in the second circuit are equivalent to variables x, y and z of the first, and α and β are constants. The second Chua's circuit (*Equations (4.10d–f)*) is coupled with the first (*Equations (4.10a–c)*) in such a way that the differences between the signals x, y, z and u, v, w,

$$g_1 = K_1\left(x - u\right)$$
$$g_2 = K_2\left(y - v\right) \tag{4.11}$$
$$g_3 = K_3\left(z - w\right)$$

are introduced into the second circuit as negative feedback. $K_{1-3} > 0$ are the stiffnesses of the perturbations which we consider as control parameters.

In our investigation we have considered again the following parameter values: $\alpha = 10.0$, $\beta = 14.87$, $m_0 = -1.27$ and $m_1 = -0.68$, i.e. in the case of $K_{1-3} = 0$ (no coupling) the dynamics of both Chua's circuits evolve along the double-scroll Chua's attractor (Chua *et al.*, 1986; Chua, 1993). We choose significantly different initial conditions for both circuits, $x(0) = 1.0$, $u(0) = 0.011$ and $y(0) = z(0) = v(0) = w(0) = 0$.

We have considered three special cases.

(a) $K_1 \neq 0$, $K_2 = K_3 = 0$: i.e. only the first state variables x and u are fed back. For values $K_1 > 6.6$ the chaotic trajectories of both Chua's circuits of system (*4.10*) can synchronize, but this synchronization is not monotone. For larger values of K_1 ($K_1 > 18.0$) we observe monotone synchronization.

In *Figures 4.6(a)* and *(b)* we show the x–u projections of system trajectories on the x–u plane. In *Figure 4.6(a)* we present the evolution of the trajectory in the case of nonmonotone synchronization for $K_1 = 6.6$, while in *Figure 4.6(b)* we can observe monotone synchronization for $K_1 = 20.0$.

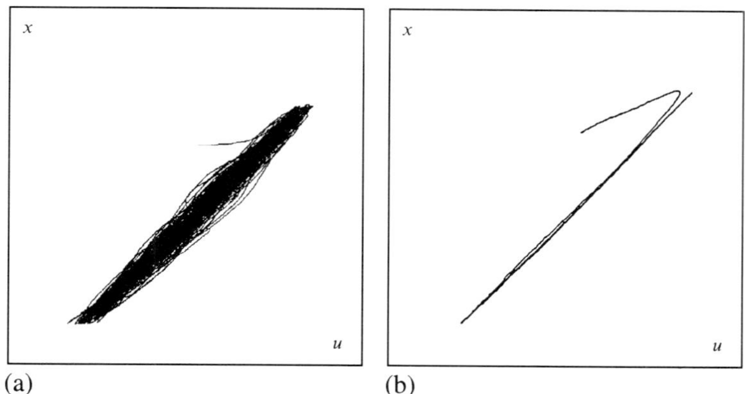

(a)   (b)

Figure 4.6 Nonmonotone (a) and monotone (b) synchronization.

In *Figure 4.7* we show the dependence of the synchronization time T_s on the coupling constant K_1. The synchronization time T_s has been defined as the time taken to reach the synchronization with the assumed precision 10^{-4}, i.e. $\left| x(t) - u(t) \right| < 10^{-4}$. For K in the interval $(17.0, 20.0)$ we estimated T_s for 1000 randomly chosen initial conditions and averaged them. At the transition to monotonic synchronization at $K_1 = 17.8$ we observe a significant downward jump in the synchronization time.

(b) $K_2 \neq 0$, $K_1 = K_3 = 0$: i.e. only the second state variables y and v are fed back. As was shown in Kapitaniak and Chua (1994), for values of $K_2 > 1.17$ the chaotic trajectories of both Chua's circuits of system (4.10)

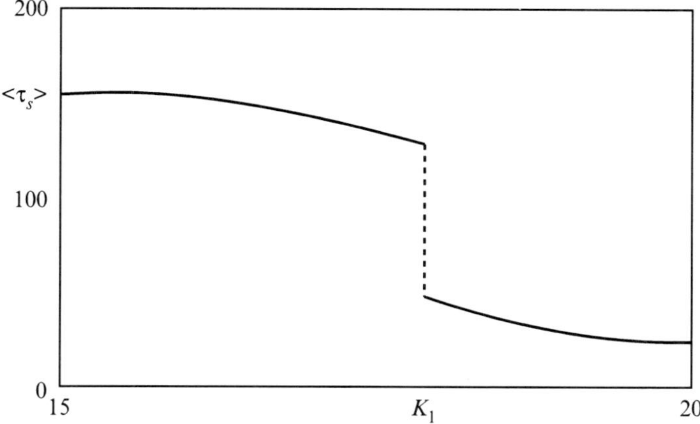

Figure 4.7 Averaged time to achieve synchronization $\langle t_s \rangle$ versus coupling stiffness K_1.

can synchronize, but this synchronization is not monotone. For much larger values of K_2 ($K_2 > 250.0$) we observe monotone synchronization. In *Figure 4.8* we show the evolution of the trajectory in the case of monotone synchronization for $K_1 = 260.0$.

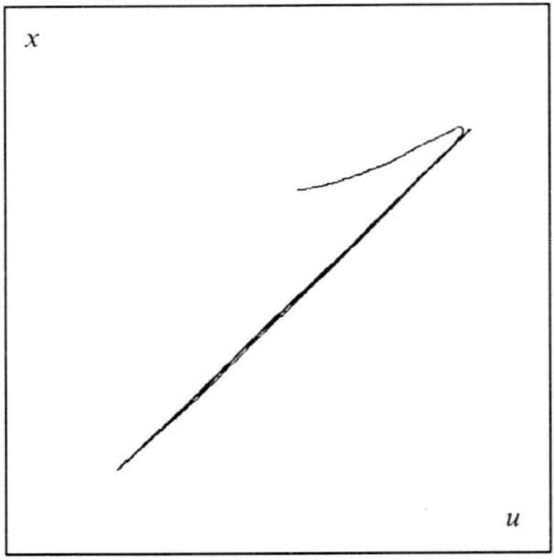

Figure 4.8 Example of monotone synchronization in case (b).

(c) $K_{1-3} \neq 0$, $K_1 = K_2 = K_3$: i.e. all state variables are fed back. As was proved in Fujisaka and Yamada (1983), in this case synchronization is possible for typical initial conditions if $K_{1-3} > \lambda/2$, where λ is the largest Lyapunov exponent of system (*4.4a*). With all control signals (*4.11*) nonzero, it is possible to observe monotone synchronization for smaller values of K_{1-3} than in the previous cases. For example, in *Figure 4.9* we present monotone synchronization for $K_1 = 2.0$, $K_2 = 2.0$, $K_3 = 2.0$.

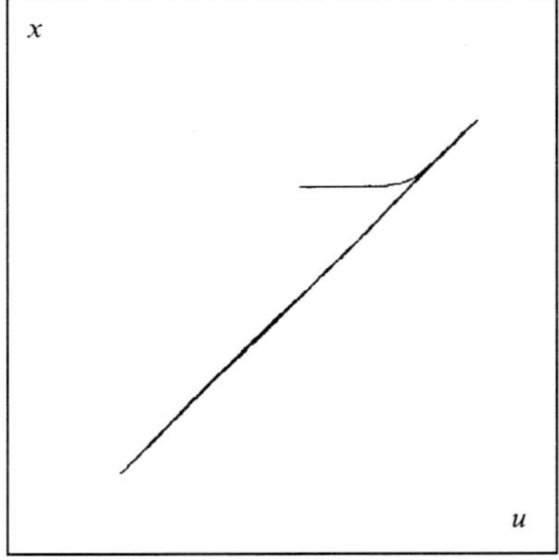

Figure 4.9 Example of monotone synchronization in case (c).

All the above examples have shown that monotone chaos synchronization is possible and that it is relatively easily obtained using the synchronization procedure based on continuous chaos control. Synchronization time (time to achieve synchronization) is much smaller in the case of monotone synchronization.

Monotonic synchronization considered in this section is a stronger case of monotonic stability of the attractor. Let $[x_0, y_0]^\mathrm{T}$ be the initial state of the trajectory of the system (4.4) and let us define the distance of the trajectory $[x(t), y(t)]^\mathrm{T}$ from the attractor A as

$$d(t) = \sqrt{(x_i - y_i)^2 + \cdots + (x_n - y_n)^2}$$

If $d(t)$ is a monotonically decreasing function of time t then attractor A is monotonically stable.

It should be mentioned here that monotonic stability depends on a metric $d(t)$, and in this sense monotonicity is a quantitative property of the attractor and may depend on the observables (Kapitaniak and Thylwe, 1996).

4.4   Practical synchronization

As it is practically impossible to have two identical chaotic systems, in this section we consider two slightly different chaotic systems $(4.4a)$ and

$$\dot{y} = f(a + \delta a, y) \tag{4.12}$$

where δa is a vector describing differences in parameter values of both chaotic systems. Chaotic system (4.12) cannot synchronize with chaotic system $(4.4a)$ in such a way that relation (4.9) is fulfilled, so we have to replace it by weaker relation:

$$\lim_{t \to \infty} \| e(t) \| \leq \varepsilon \tag{4.13}$$

where ε is a vector of small parameters ($\varepsilon_i \ll 1$, $i = 1, 2, \ldots, n$). If for any ε there exists a vector K such that relation (4.13) is fulfilled, then chaotic systems $(4.4a)$ and (4.12) are practically (or noisily) synchronized.

The difference between synchronization defined by relation (4.9) and practical synchronization defined by relation (4.13) is easily visible in the x_i–y_i plot shown in *Figure 4.10*. In the case of ordinary synchronization this plot is a straight line (for example *Figure 4.3(a)*) while in the case of practical synchronization the evolution of the systems is limited to the domain defined by the relation $x_i + \varepsilon_i \leq y_i \leq x_i + \varepsilon_i$ (*Figure 4.10*). As shown in the previous section, if the condition is fulfilled (*Equation (4.9)*) the coupled systems $(4.4a)$ and $(4.4b)$ evolve on the same manifold on which both chaotic systems evolve (phase space is reduced to subspace $x = y$) and that is why synchronization can be obtained. When it is not fulfilled the coupled systems evolve on higher-dimensional manifold on which a hyperchaotic attractor exists. In the case of practical synchronization the attractor of coupled systems $(4.4a)$ and (4.12) is hyperchaotic, evolution of the systems takes place in the six-dimensional phase space but it is restricted to the neighborhood of three-dimensional subspace $x = y$.

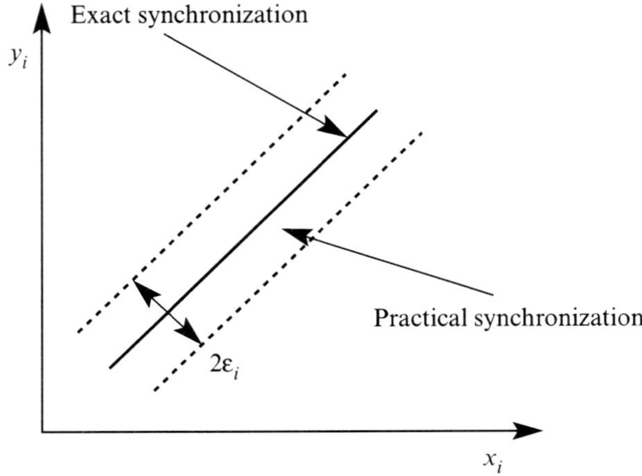

Figure 4.10 Exact and practical synchronization.

As an example of practical synchronization we consider a coupled pair of Lorenz systems, given by

$$\dot{X}_1 = -\sigma X_1 + \sigma Y_1 \tag{4.14a}$$

$$\dot{Y}_1 = -X_1 Z_1 + r_1 X_1 - Y_1 \tag{4.14b}$$

$$\dot{Z}_1 = X_1 Y_1 - bZ_1 \tag{4.14c}$$

$$\dot{X}_2 = -\sigma X_2 + \sigma Y_2 + d(X_1 - X_2) \tag{4.14d}$$

$$\dot{Y}_2 = -X_2 Z_2 + r_2 X_2 + d(Y_1 - Y_2) \tag{4.14e}$$

$$\dot{Z}_2 = X_2 Y_2 - bZ_2 + d(Z_1 - Z_2) \tag{4.14f}$$

where σ, $r_{1,2}$ and b are constants. All state variables of both systems are coupled linearly with equal coupling strength d; parameters σ and b are held fixed at $\sigma = 10.0$, $b = 8/3$, and r_1, r_2 are used as control parameters.

For certain ranges of r_i, each individual system can be on one of two attractors, which are mirror images of each other (*Figures 4.11(a)* and *(b)*); for other ranges of r_i only a single symmetric (butterfly) attractor exists (*Figure 4.11(c)*). Choosing $r_1 = r_2 = 211.0$, we have such a situation, and without coupling (i.e. $d = 0$) we can choose initial conditions so that system 1 (*Equations (4.14a–c)*) is on attractor A_1, say, whilst system 2 (*Equations (4.14d–f)*) is on attractor A_1.

When we introduce coupling, even at a very weak level, we find that synchronization occurs on attractor A_1 (*Figures 4.12(a and b)*). In *Figure 4.12(a)* the transient evolution towards the synchronized state of *Figure 4.12(b)* is shown.

Thus synchronization is normal ($x = y$, $x = [X_1,Y_1,Z_1]^T$, $y = [X_2,Y_2,Z_2]^T$) in coupled Lorenz systems having *identical* values of r. When the values of r_1, r_2 are different, synchronization, by definition, cannot occur. However, as we show in *Figure 4.13(a)*, for a range of r_1, r_2 a practical synchronization takes place. Indeed, there is a noisy modulation about synchrony (*Figure 4.13(b)*) which persists in the case illustrated over the range $211.0 < r_2 < 215.0$. For

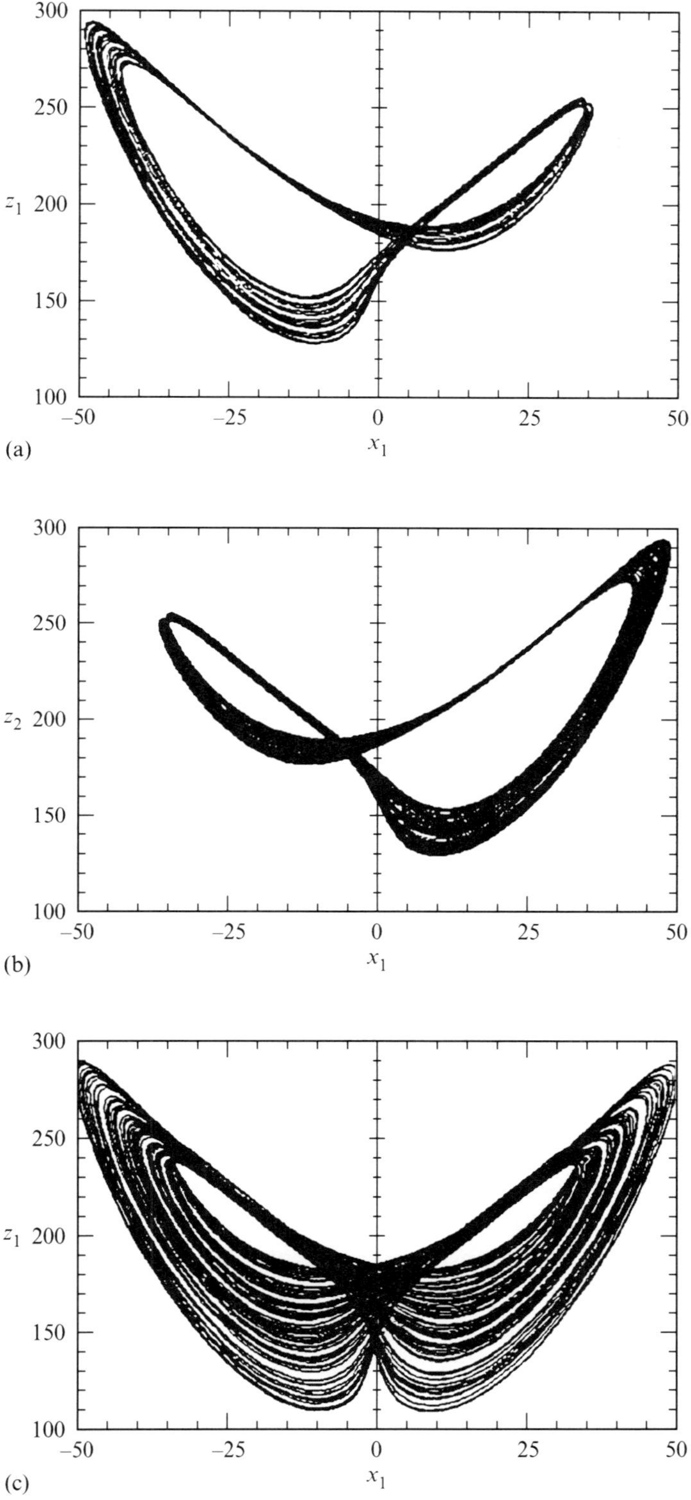

Figure 4.11 Attractors of uncoupled *Equation (4.14)*; (a)(b) two co-existing attractors for $r = 211$, (c) symmetrical attractor for $r = 219$.

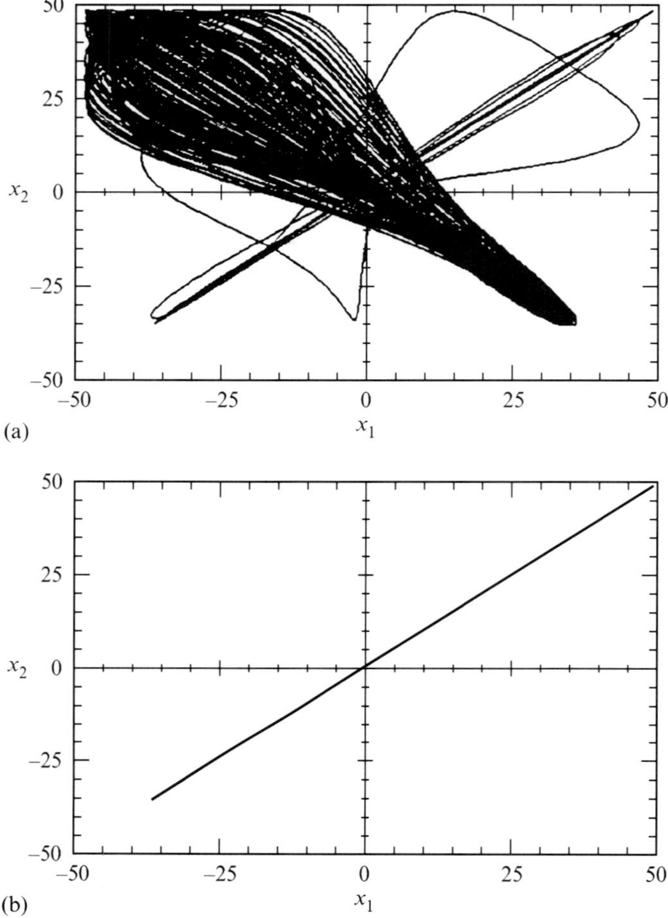

Figure 4.12 Evolution towards synchronized state (a) and final synchronized regime (b).

larger values of r_2 the systems evolve in the neighborhood of a synchronized state for a significantly long periods of time, occasionally bursting out of this neighborhood, as can be seen in *Figure 4.13(b)*. This final collapse of synchronization is associated with the replacement of A_1 and A_2 by a single symmetric attractor B (*Figure 4.12(c)*).

Equation (4.14) for a considered range of parameters represents a quasi-hyperbolic system for which different chaotic attractors coexist. Generally, in such a case synchronization is not straighforward when both systems operate on different attractors. We discuss this problem in the following section.

4.5   Synchronization in quasi-hyperbolic systems

The result of *Sections 4.2* and *4.3* hold if B is the only possible asymptotically stable attractor of systems (*4.4a*) and (*4.4b*) for a_1. However, when *Equations*

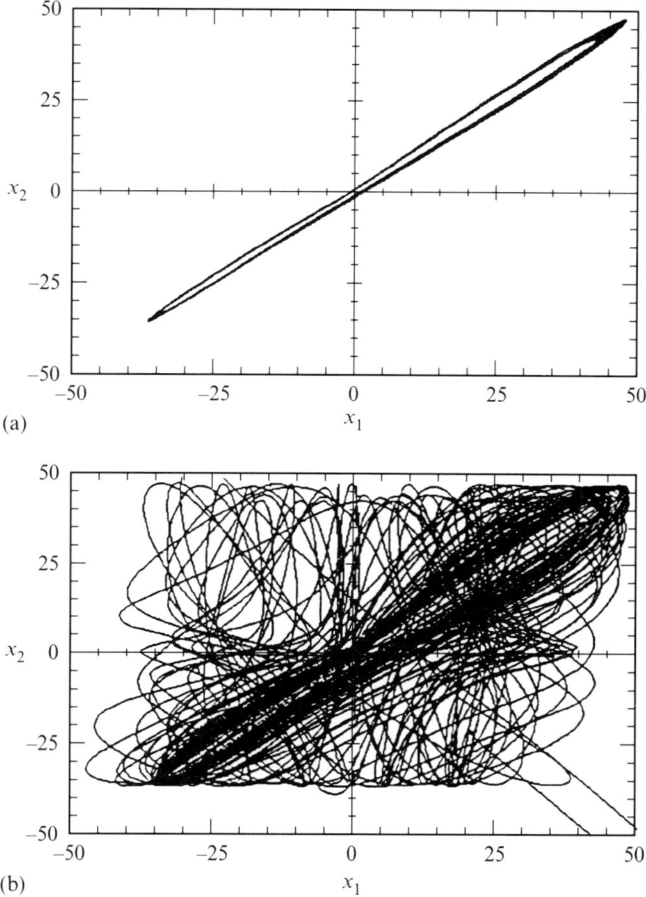

Figure 4.13 Practical synchronization (a) and break of synchronization (b).

(4.4a) and *(4.4b)* describe a quasi-hyperbolic system with at least two coexisting attractors, the synchronization procedure is not straightforward. If the trajectory of one system is on the attractor A_1 and the trajectory of the other is on the coexisting attractor A_2, to achieve synchronization one of the trajectories, say, that on the attractor A_1 has to be perturbed in such a way that it goes to the basin of attraction $b(A_2)$ of the other attractor A_2. Let $e(A_1)$ be the region of the phase space in which the perturbed trajectory $x(t)$ evolves. Necessary condition for synchronization can be given by

$$e(A_1) \cap b(A_2) \neq \varnothing \tag{4.15}$$

(see *Figure 4.14*). In some cases to fulfill relation *(4.15)* strong perturbation, which could be difficult to realize in practice, is necessary.

In this section we discuss the problems of synchronization and desynchronization of two quasi-hyperbolic systems using the procedure of *Section 4.2* with small negative feedback. We introduce the method which under additional conditions allows synchronization even when relation *(4.15)* is not initially fulfilled.

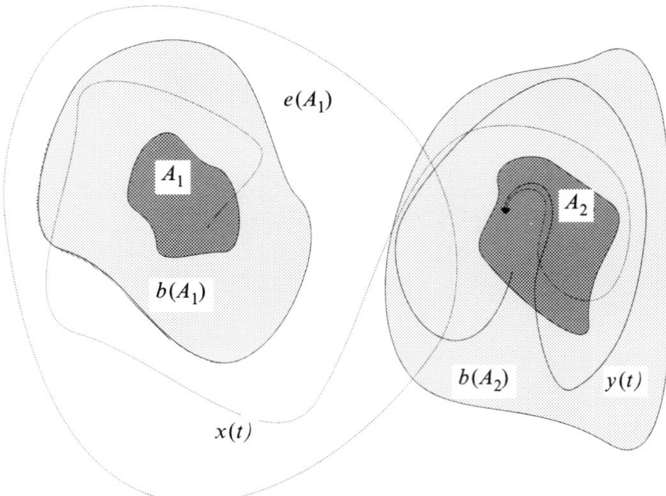

Figure 4.14 Necessary condition for synchronization of two chaotic systems.

If for a different value of control parameter a, say $a = a_2$, *Equations (4.4a)* and *(4.4b)* have different chaotic or periodic attractors in different regions of the phase space, small coupling (*Equation (4.5)*) will not result in the synchronized state $x(t) = y(t)$ (relation *(4.9)* will not be fulfilled) and in the $x_i - y_i$, $i = 1, 2, \ldots, n$ plots we observe unclosed curves instead of a straight line.

In this section we describe a simple method which allows us to obtain synchronization of periodic and chaotic trajectories evolving on different coexisting attractors A_1 and A_2 which are close to the single chaotic attractor B.

Let us assume that the 'one attractor' a_1 and 'coexisting attractors' a_2 values of a control parameter a are close together. Then chaotic behavior of *Equations (4.4a)* and *(4.4b)* can be synchronized through the following coupling:

$$\dot{x} = f\left[x, a(t)\right]$$
$$\dot{y} = f\left[y, a(t)\right] + K(x - y) \tag{4.16}$$

where

$$a(t) = \begin{cases} a_1 & t \in [0, \tau_s] \\ a_2 & t > \tau_s \end{cases} \tag{4.17}$$

and τ_s is the synchronization time of chaotic systems *(4.4a)* and *(4.4b)* (time in which the chaotic systems are synchronized). In the synchronization scheme *(4.16)* the value of a is at first fixed to the 'one attractor' value a_1.

When the synchronization state $x(t) = y(t)$ is achieved, parameter a is switched to the 'coexisting attractors' value a_2. Equality $x(\tau_s) = y(\tau_s)$ ensures the same initial conditions for the transient evolution towards one of the coexisting attractors so for $t > \tau_s$ we always have $x(t) = y(t)$ and the synchronization of trajectories is guaranteed.

As an example, consider a pair of unidirectionally coupled identical Chua's circuits, for which the combined equations of motion are

$$\dot{x} = \alpha\left[y - x - f(x)\right] \tag{4.18a}$$

$$\dot{y} = x - y + z \tag{4.18b}$$

$$\dot{z} = -\beta y \tag{4.18c}$$

$$\dot{u} = \alpha\left[v - u - f(u)\right] \tag{4.18d}$$

$$\dot{v} = u - v + w + k_2(y - v) \tag{4.18e}$$

$$\dot{w} = -\beta v \tag{4.18f}$$

where $f(x)$ is given by *Equation (1.1)* and α, β, a and b are constants. The second Chua's circuit (*Equations (4.18d–f)*) is coupled to the first one (*Equations (4.18a–c)*) in such a way that the differences between the signals y and v are ($k = [0,k_2,0]^T$) introduced into the second circuit as negative feedback.

In our investigation we considered the following parameter values: $\beta = 14.87$, $m_0 = -1.27$ and $m_1 = -0.68$. In the case of $k_2 = 0$ (no coupling) and $\alpha_1 = 10.0$ the dynamics of both Chua's circuits evolve along the double-scroll Chua's attractor while for $\alpha_2 = 8.0$ two coexisting Rossler type attractors are possible.

If we start with the value of a fixed to the 'coexisting attractor' value a_2 and two systems (*4.4a*) and (*4.4b*) evolve on different attractors $x(t) \neq y(t)$, where $x(t) = [x(t),y(t),z(t)]^T$ and $y(t) = [u(t),v(t),w(t)]^T$ as can be seen for periodic and chaotic case in *Figures 4.15(a)* and *(b)*.

When we start with the 'one attractor' value a_1 and after achieving synchronization of chaotic trajectories a is switched to a_2 then trajectories $x(t)$ and $y(t)$ evolve on one attractor and are synchronized as can be observed in *Figures 4.16(a)* and *(b)*. *Figure 4.16(a)* presents the transient evolution from the initial state of *Figure 4.15(b)* to the final synchronized state shown in *Figure 4.16(b)*. In this simulation $a_1 = \alpha_1 = 10.0$ and $a_2 = \alpha_2 = 8.0$ have been taken. Our numerical observations show that there is hysteresis in the dynamical behavior of coupled system (*4.18*), as different behavior is obtained when control parameter is increased from when this parameter is decreased.

This poses the question whether or not it is possible to desynchronize the systems again, and by what means. It is clear that any further parameter changes, in which the $a = \alpha$ values of both attractors change simultaneously, will fail to desynchronize two systems. If, however, we switch the value of one of them, $a_1 = \alpha_1$ say, to the single attractor value,

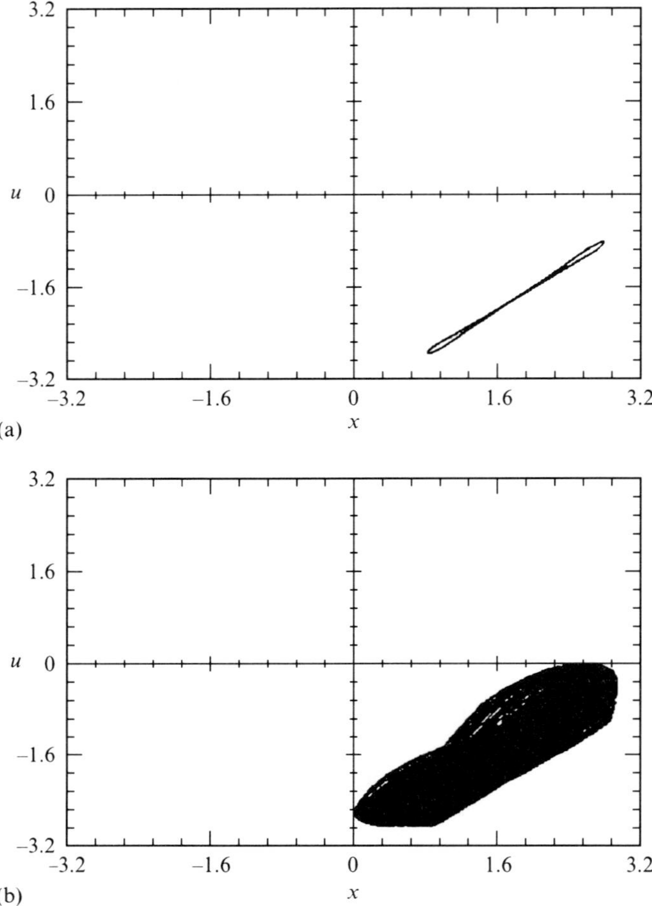

Figure 4.15 Evolution of coupled *Equation (4.18)*, (a) closed Lissajou figure $\alpha = 7$; (b) unclosed curve $\alpha = 8$.

$a_1 = \alpha_1$, leaving the other at $a_2 = \alpha_2$, the two systems must have totally different attractors, B and A_2, say, and hence are desynchronized. When we switch $a = \alpha_1$ back to $a = \alpha_2$, the second system switches either to attractor A_1 or A_2, according to its position on its trajectory at the time of switch (which must be in one of the basins of attraction $b(A_1)$ or $b(A_2)$ of A_1 or A_2). If it switches to A_1, synchronization occurs; if it switches to A_2, the systems remain desynchronized.

We can summarize all these possibilities in the diagram shown in *Figure 4.17*. Thus, probabilities of returning to the original value, $a_2 = \alpha_2$ in either the synchronized or desynchronized state will be proportional to the length of time the trajectory spends in the basins of A_2, A_1 respectively. For long trajectories this is the ratio $b(B) \cap b(A_2):b(B) \cap b(A_1)$, and it follows that, if $b(B) \cap b(A_1)$ is 'small', the system is difficult to desynchronize, i.e. it will tend to persist in synchronized states.

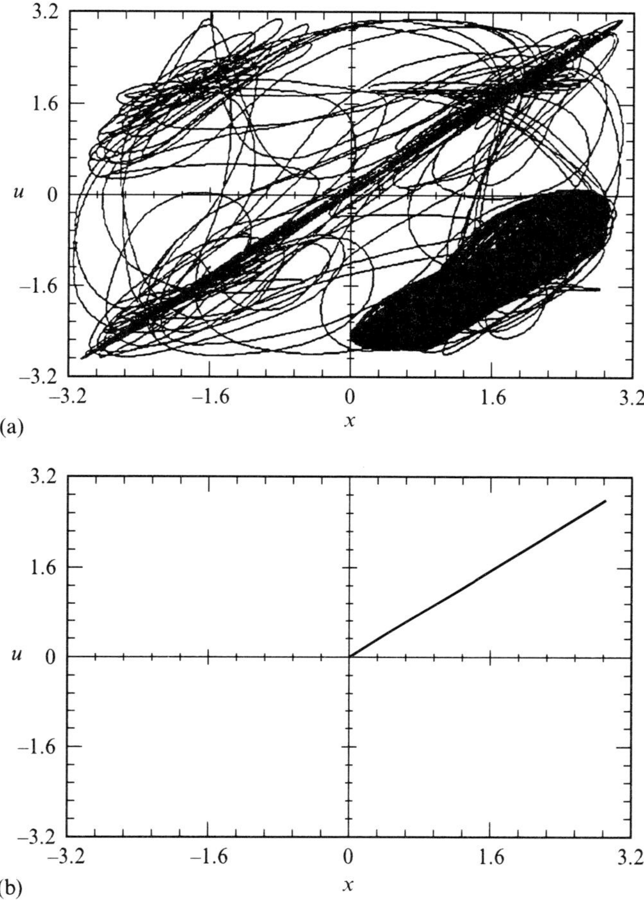

Figure 4.16 Evolution from desynchronized state of *Figure 4.15(b)* towards synchronization using *Equation (4.17)*; (a) transient evolution (b) synchronized state.

4.6   Secure communication

Chaos synchronization allows applications of chaotic systems to mask the information signal $I(t)$ by adding it to a larger chaotic signal $n(t)$ and transmitting the superposition of both signals. Information can be recovered after the comparison of the received signal $I(t) + n(t)$ with the original chaotic noise $n(t)$. In this procedure chaotic signals in the transmitter and receiver systems must be synchronized. As this way of sending information is difficult to unmask, it is called secure communication. The main idea of secure communication is sketched in *Figure 4.18*.

In most of the implementations of secure communications the continuous information signal $s(t)$ is transmitted (Cuomo and Oppenheim, 1993 – *Paper 11*; Kocarev *et al.*, 1992; Halle *et al.*, 1993). The other possible approach to secure communications is to transmit a digital (binary sequence) information

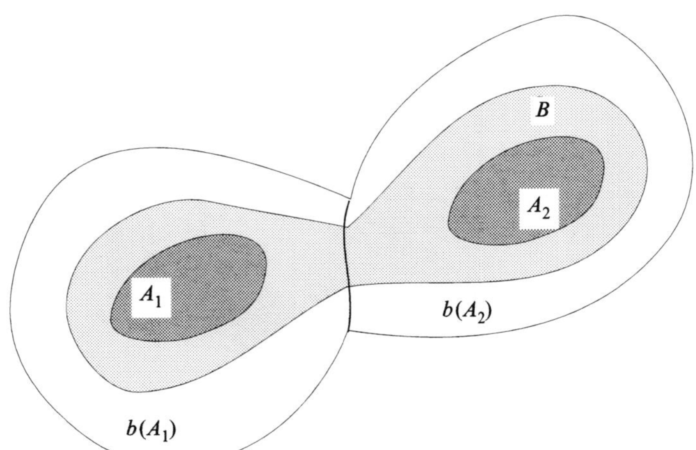

Figure 4.17 Possibilities of desynchronization of *Equation (4.18)*.

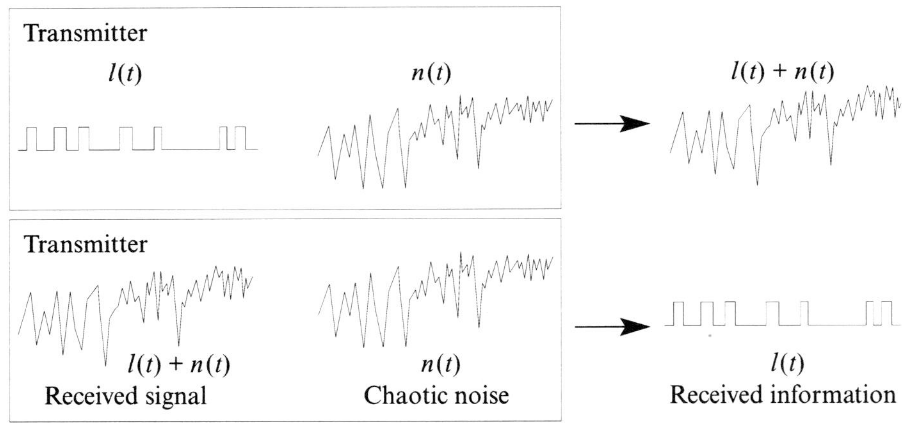

Figure 4.18 Idea of secure communication.

signal. The first experimental application of this way of communication was reported by Parlitz *et al.* (1992).

In this section we show the simpler method of transition of digital signals. We replace the Pecora and Carroll (1990 – *Paper 9*, 1991) method of synchronization, which requires decomposition of a chaotic system into two subsystems, by the monotone synchronization procedure described in *Section 4.3*. We also use a slightly different definition of 1 and 0 in the recovering procedure.

The sketch of our secure communications scheme is shown in *Figure 4.19*.

As is well known, the synchronization regime is characterized by a straight line in the $x_i(t)$–$y_i(t)$ plot. Consider the information signal which is transmitted in the form shown in *Figure 4.20*.

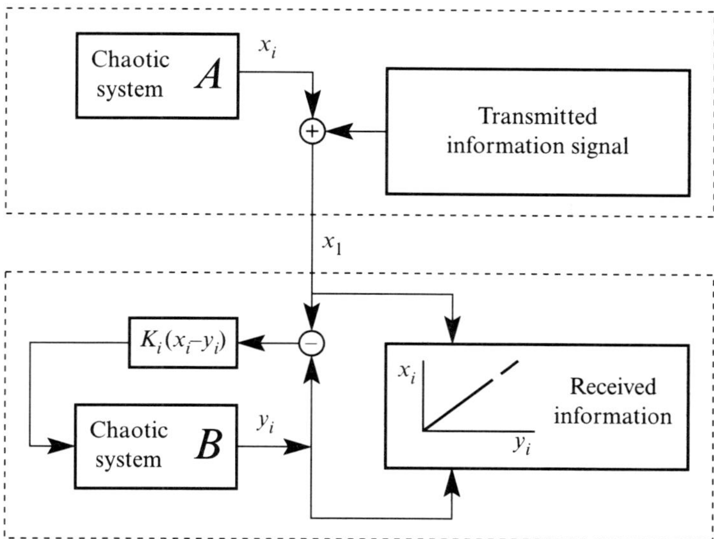

Figure 4.19 Scheme of the method of transmitting binary sequence.

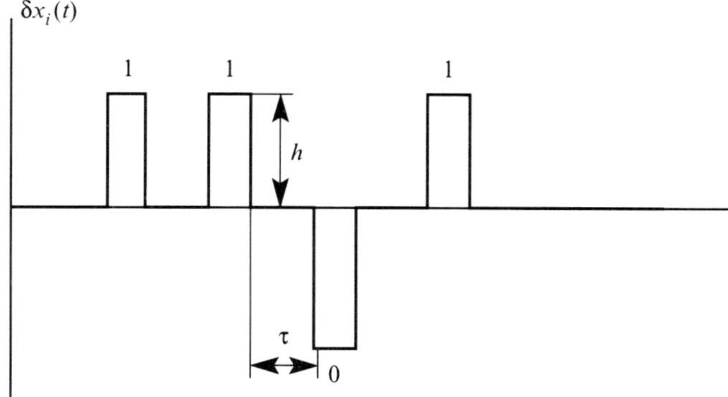

Figure 4.20 Discontinuous information signal $\delta p(t)$.

When the function $\delta x_i(t)$ first jumps to the positive value $h \ll \max(x_i(t))$, (1 in a binary sequence), i.e. the first bit of information is transmitted, the trajectory leaves the synchronization line in the $x_i(t)$–$y_i(t)$ plot and goes into a half plane $x_i(t) > y_i(t)$ (*Figure 4.21*). With this event we can associate the symbol '1' in a recovered sequence. When the value of $\delta p_i(t)$ becomes zero, the system trajectory approaches the neighborhood of the synchronization line and after synchronization time $\tau_s < \tau$ both systems are synchronized again, and we can transmit another bit of information. When a '0' is transmitted, $\delta x_i(t) = -h < 0$, and in the recovering procedure we associate 0 with a departure of trajectory from the synchronization line into the domain $x_i(t) > y_i(t)$ (*Figure 4.21*). Each transmitted binary symbol can be sent after a break of duration τ in which the synchronization of our systems is ensured. τ

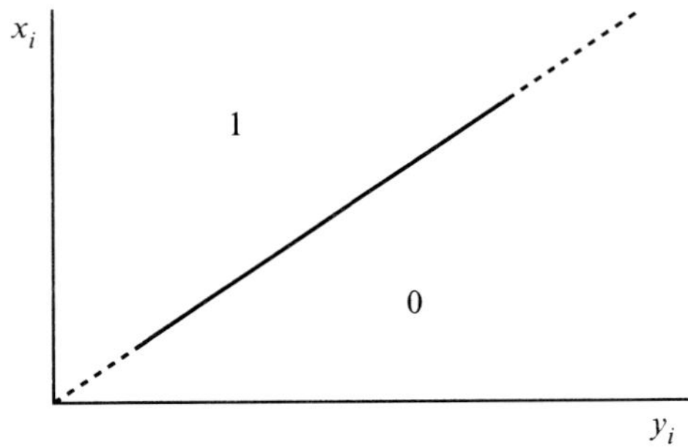

Figure 4.21 $y(t)$–$v(t)$ plot used for recovering information sequence.

has to be evaluated experimentally. As was shown in Kapitaniak (1994), the synchronization time τ_s depends on coupling stiffness K and for large K can be relatively small. In the case of monotone synchronization it is significantly smaller than for nonmonotone synchronization.

In our numerical investigations we used again two identical Chua's circuits, described by *Equation (4.10)* as chaotic systems. Through this system we transmitted the binary sequence of 10^{10} bits with $h = \max(\delta y) = 0.5$, $\tau = 0.02$, $K_2 = 260.0$ and the transmitted signal was recovered in a y–v plot without errors. The procedure of recovering the information is sketched in *Figure 4.22*.

The transmitted signal $\delta x_i(t)$ of the form shown in *Figure 4.20* added to the transmitted signal $x_i(t)$ makes this signal discontinuous. Although with $h \ll \max(x_i(t))$ these discontinuities are hardly visible in the transmitted signal, one can try to localize them using nonlinear time series methods, so perhaps it would be safer to use a continuous information signal of the form shown in *Figure 4.23*. Alternatively one can use higher-dimensional chaotic

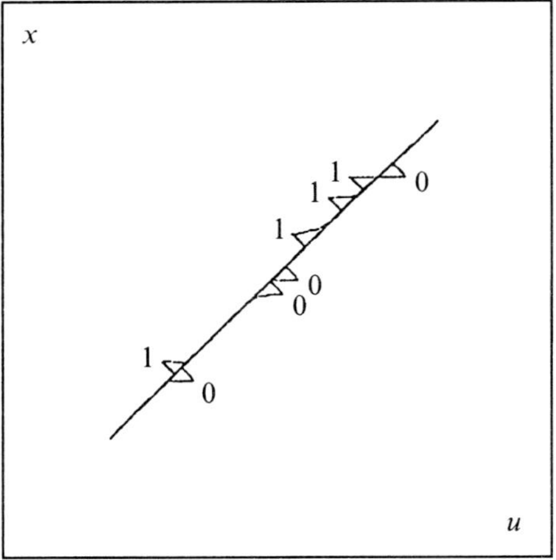

Figure 4.22 Experimental recovering of information sequence.

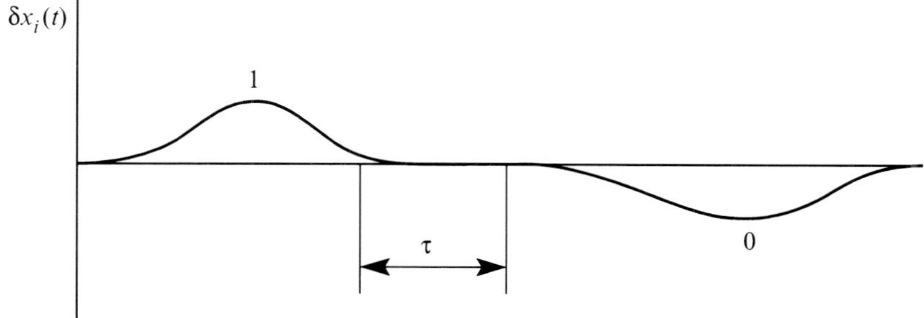

Figure 4.23 Continuous information signal $\delta p(t)$.

systems with more than one positive Lyapunov exponents, as there are no effective methods to analyze higher-dimensional time series.

There are several reasons to claim that the signal transmission system constructed using the monotonic synchronization concept is secure. Firstly, one could claim that the information signal is buried within the chaotic carrier and thus is not distinguishable by an observer. Even for the sender it would be hard, if not impossible, to decode/extract the hidden signal as the transmitted signal is chaotic, broad-band with continuous spectra, having no resemblance to the message signal and at first sight it looks noise-like. So it is quite obvious that from such a 'nonexpert' point of view such a transmission has a substantial measure of security. As a second important security factor, one has to consider the possibilities of decoding the signal – as has been shown already, to reproduce the information signal we must know the parameters of the coder/decoder circuit with sufficient precision (if not exactly). Thus the parameters of the coding/decoding device serve as a secret key. It is obvious that in reality there exists an infinite number of parameter sets which result in chaotic outputs generated by the system. It is truly impossible to explore all these parameter sets to find the right one that enables decoding – such a 'needle in the haystack' would take too long.

Recently Pérez and Cerdeira (1995 – *Paper 12*) have published some results showing that for most primitive methods of signal transmission on a chaotic carrier (direct parameter modulation or masking) they were able to extract the information signal using partial reconstruction techniques using just the transmitted signal without any knowledge about the coder/decoder. Possibly, their approach could also be used to break the code in the scheme considered in this chapter – as the principle we demonstrated was that of simple masking (addition of the info-signal to the chaotic carrier). Furthermore, in such a simple case an intruder might even use such a basic tool as adaptive filtering techniques which could allow him to break the code.

We have to stress here that this example was chosen for the sake of simplicity – our prime interest here was to demonstrate that the monotonic synchronization principle also works in practical applications. We do not claim that any new scheme for increased security has been proposed. We believe, however, that having more sophisticated coding schemes (modulation) and/or more complicated chaotic coders/decoders (e.g. hyperchaotic or cascaded chaotic circuits), which produce signals to be transmitted with virtually no correlation with the information signal, monotonic synchronization can offer several advantages over classical synchronization such as short settling time (which has been demonstrated by examples), robustness to parameter mismatch and channel noise and imperfections. These last-mentioned issues were not addressed in this short chapter and require further study (an interested reader could find some considerations on errors in the recovered signal in Lozi and Aziz-Alaoui (1995) and Carroll (1994)).

General approach to chaotic synchronization based on a decomposition of a given system into active and passive parts has been recently presented by Kocarev and Parlitz (1995 – *Paper 13*). This method allows for an improved method of encoding information signals.

5 Engineering implementations

5.1 Method selection

Although the methods described in the previous chapters have been either developed by physicists or originally described in physical journals, generally most of them can be applied to control of engineering systems.

In particular, the nonfeedback methods can practically always be used. Their applications are straightforward and do not require special complicated controllers to be used. The main disadvantage of these methods is that the goal of controlling, i.e. the appropriate periodic orbit) has to be determined by trial-and-error method.

The motivations for using feedback systems to control chaos are the following: feedback controllers are easy to implement, especially in electrical systems, they can perform the job automatically, and stabilize the overall control system efficiently, and they usually have significant physical meanings. On the other hand, conventional feedback controllers are designed for nonchaotic systems. A chaotic system sensitivity to initial conditions may lead to the impression that in chaotic systems their sensitivity to small errors makes them very difficult, if not impossible, to control using conventional feedback methods over all their phase space. Such an impression may lead to the argument that once the control is initiated there is no need for further monitoring of the system's dynamics, nor feeding back this information in order to sustain the control. Indeed, it turns out that conventional feedback control of chaotic systems is generally difficult, but not impossible, as was shown in *Section 2.3*. Recently, Chen and Dong developed some new ideas and formalized some successful techniques for controlling chaotic systems using a conventional engineering approach (Chen and Dong, 1992, 1993a; Dong and Chen, 1992a,b; Chen, 1993). The approach in which neural network controllers are used in the identification and control of chaotic systems (Chen and Dong, 1995) is especially interesting. In many cases a specially implemented feedback method can guarantee stabilization of the appropriate periodic orbit. To summarize, the selection of the controlling method has to be based on:

(i) the goal of controlling (e.g. if the suppression of chaos is the main goal, nonfeedback methods can be applied in an easier way);

(ii) the level of noise in the system (e.g. if the level of noise is large, Pyragas's methods (*Section 2.2*) can be more effective than the OGY approach (*Section 2.1*);

(iii) the particular characteristics of the system. (Generally, in the electrical systems one can try to use both feedback and nonfeedback methods. In mechanical systems where the suppression of chaos is the main goal of controlling, nonfeedback methods have to be advised.)

5.2   Occasional proportional feedback method

Occasional proportional feedback (OPF) is one-dimensional version of OGY method. The OPF method belongs to the category of classical engineering control approaches and it has been proved to be the most efficient. Following Ogorzalek (1994) we can explain the action of the OPF method in the following way. Let us consider a return map shown in *Figure 5.1*. For nominal parameter values the position of the graph of the map is as shown by the rightmost curve – all periodic points are unstable. In particular, the point P is in unstable equilibrium. Looking at the system operation starting from point v_n, at the next iteration (the next passage of the trajectory through the Poincaré plane), one would obtain v_{n+1}. We would like to direct the trajectories towards the fixed point P. This can be achieved by changing a chosen system parameter in such a way that the graph of the return map moves to a new position as marked on the diagram, thus forcing the next iteration to fall at v_{n+1}^*. After this is done the perturbation can be removed and activated again if necessary.

The OPF controlling procedure is sketched in *Figure 5.2*. It may be applied to any engineering chaotic system (including higher-dimensional ones) where the output can be measured electronically and the control signal can be applied via a single electrical variable. The signal processing is analog and

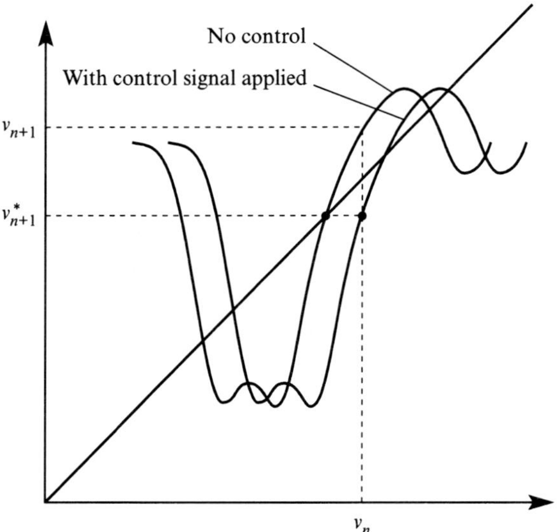

Figure 5.1 Idea of OPF method.

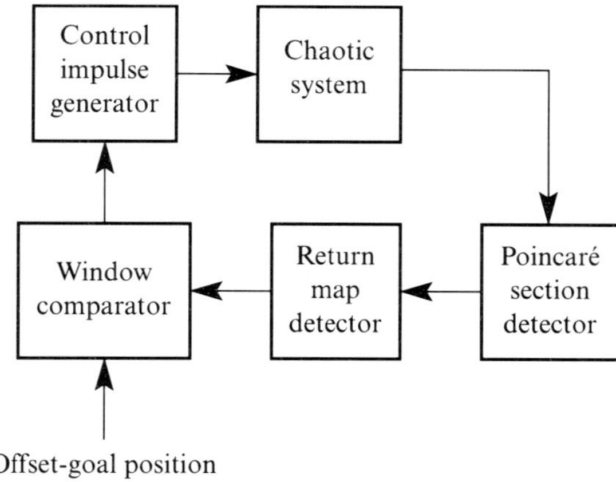

Figure 5.2 Controlling procedure using OPF.

therefore is fast and efficient. Processing here means detecting the position of a one-dimensional projection of a Poincaré map, which can be accomplished by window comparators and track-and-hold circuitry. The accessible goal trajectories have to be determined by a trial-and-error method, so the applicability of this control strategy is limited to systems in which suppression of chaos without more strict requirements is the main goal.

This method has been successfully implemented in a continuous-time analog electronic circuit and used in a variety of applications ranging from stabilization of chaos in laboratory circuits (Hunt, 1991; Johnson *et al.*, 1993; Murphy and Kennedy, 1995) to stabilization of chaotic behaviors in lasers (Corcoran, 1991; Peterson, 1991; Roy *et al.*, 1992).

Another engineering modification of OGY method, the bang-bang control, is described in Galias and Ogorzalek (1995).

5.3   Sampled input waveform method

A very simple, robust and effective method of chaos control in terms of stabilization of an unstable periodic orbit has been proposed by Dedieu and Ogorzalek (1994). A part of the output signal (sampled version), corresponding to a chosen unstable periodic orbit uncovered from the measured signal, is introduced in the chaotic system forcing it to follow this desired trajectory.

The block diagram of this control scheme is shown in *Figure 5.3*. For controlling chaos by this method in Chua's circuit we can use a sampled version of a signal $v'_{C_1}(t)$, where $v'_{C_1}(t) = C^T x'(t)$ and $x'(t)$ is a desired periodic orbit. The injection of a continuous signal $v'_{C_1}(t)$ will force the system to exhibit a solution $x(t)$ which converges to $x'(t)$. This is obvious since forcing $v_{C_1}(t)$ will instantaneously force the current through the piecewise-linear resistance to a desired value $i''_R(t)$. The remaining subcircuit (R, L, C_2) which

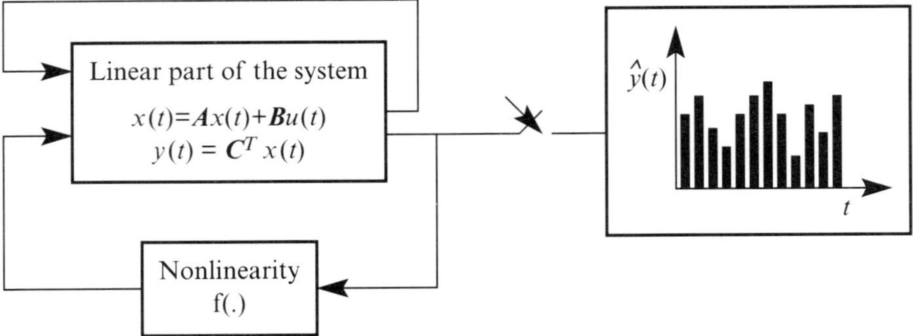

Figure 5.3 Control scheme of sampled input waveform method, with permission from M Ogonatek.

is a *RLC* stable circuit will then exhibit a voltage $v_{C_2}(t)$ and a current $i_L(t)$, which will tend asymptotically towards $v'_{C_2}(t)$ and $i''_L(t)$.

The sampled input control method is very attractive as the goal of control can be specified using the analysis of output time series of the system so the access to system parameters is not required. The control technique is resistant to parameter variations and noise. Unlike in previous methods instead of a controller, we need a generator to synthesize the goal signal. *Plate 2* shows the chaotic spiral attractor ($\alpha = 8.0$, $\beta = 14.87$, $m_0 = -1.27$, $m_1 = -0.68$) and the controlled periodic sample trajectory. More details on this method can be found in Dedieu and Ogorzalek (1994).

The sampled input waveform method is a feedback method based on the idea of a continuous control described in *Section 2.3*. Another approach in the similar direction is a distortion control method in which the periodic trajectory (the goal of the controlling) is approximately estimated using harmonic balance method (Genesio *et al.*, 1993).

5.4   Controlling transient behavior in mechanical systems

Chaos controlling methods, like most of the results in the theory of dynamical systems, have an asymptotic character, i.e. under given conditions a dynamical system with specified parameters behaves in a particular way when $t \rightarrow \infty$. As many real systems are characterized by some values of parameters only for finite time, these results cannot always be used in practice, so there is the necessity for development of efficient controlling methods for the control of transient nonlinear effects.

In this section, we investigate the problem of practical control of Hopf bifurcation. We consider a particular but representative example of such a situation. If $a = a_w$ which is a desired working condition of a dynamical system

$$\dot{x} = f(a, x)$$
(5.1)

where $x \in \mathbf{R}^n$, $a \in \mathbf{R}^m$ ($n, m = 1, 2, \ldots$) is a vector of system parameters, characterized by a steady state behavior ($x = 0$ is a stable fixed point for $a = a_w$). To reach practically the value $a = a_w$ from $a = a_s$ (starting point), a has to go through the interval $[a_1, a_2]$. For $a \in [a_1, a_2]$ steady state $x = 0$ of the dynamical system (5.1) is unstable as system undergoes Hopf bifurcation at $a = a_1$ and reversed Hopf bifurcation at $a = a_2$. We try to answer the question: can we safely reach desired working conditions $a = a_w$ or do we have to restrict ourselves to the parameters $a < a_1$?

The described situation is characteristic for rotor systems, where rotor rotational velocity ω is a system parameter which has to increase continuously from starting velocity $\omega_s = 0$ to $\omega = \omega_w$ (desired working conditions). Unfortunately, in the interval $[\omega_1, \omega_2]$, $\omega_{1,2} < \omega_w$ steady state $x = 0$ is unstable. The passage through this interval, as sketched in *Figure 5.4*, is associated with the increase of the amplitude of rotor vibrations as at ω_1 the self-excited vibrations originate and are added to small oscillations due to the unbalanced ones.

In practice, the limit cycle of self-excited vibrations is not reached (time in which velocity ω increases from ω_1 to ω_2 is not sufficient to reach the attractor). At the value ω_2 the amplitude starts to decrease as self-excited vibrations decay and after a short transient they stabilize at the initial value. In most of the practical systems this interval cannot be passed straightforward as the self-excited oscillations can damage the system as it is described in *Figure 5.5*. The increase of the vibrations amplitude is limited due to the design restrictions and at ω_d our system can be damaged.

The transient behavior in the dangerous zone $[a_1, a_2]$ has to be controlled in

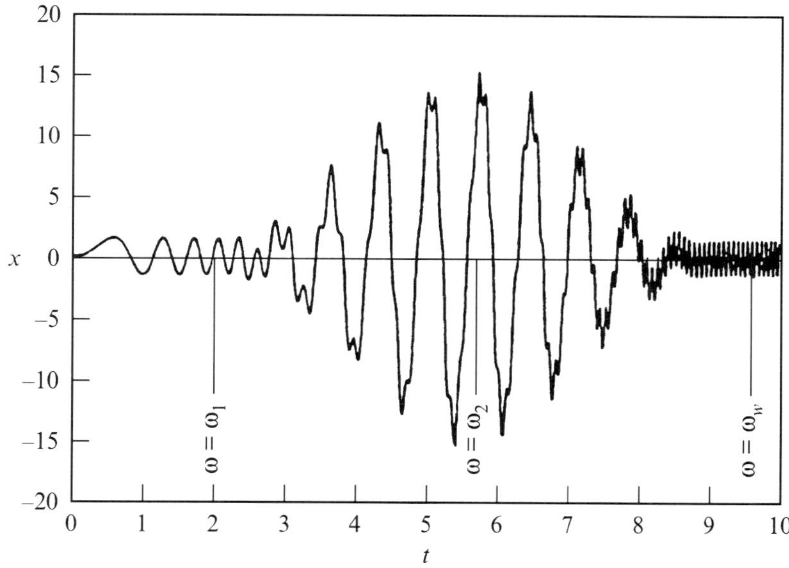

Figure 5.4 Passage through unstable zone.

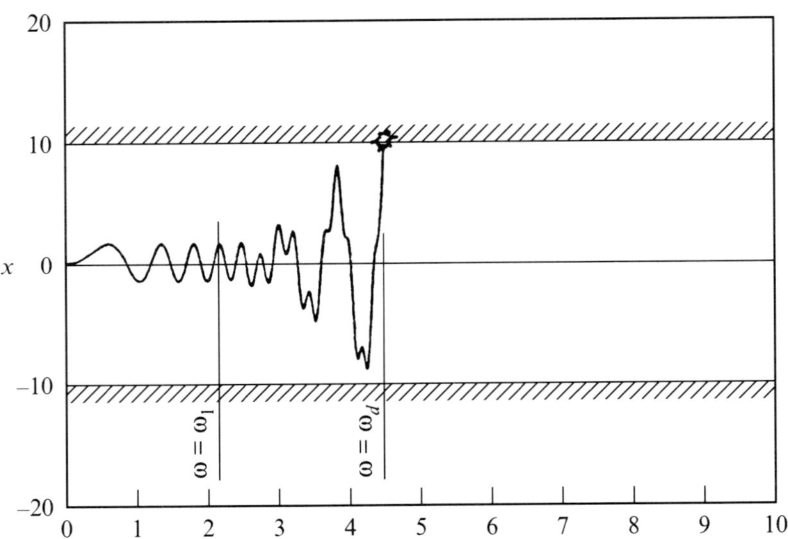

Figure 5.5 Increasement of the amplitude causing damage.

such a way that it could be safely passed. We propose a simple method which is based on the small temporal change of one of the constant components of system parameters vector a, which allows us to stabilize the fixed point $x = 0$ or significantly reduce the amplitude of the self-excited oscillations for $a \in [a_1, a_2]$. When the dangerous zone is passed vector a returns to its initial value. On the controlled system the passage through the dangerous zone is shown in *Figure 5.6*.

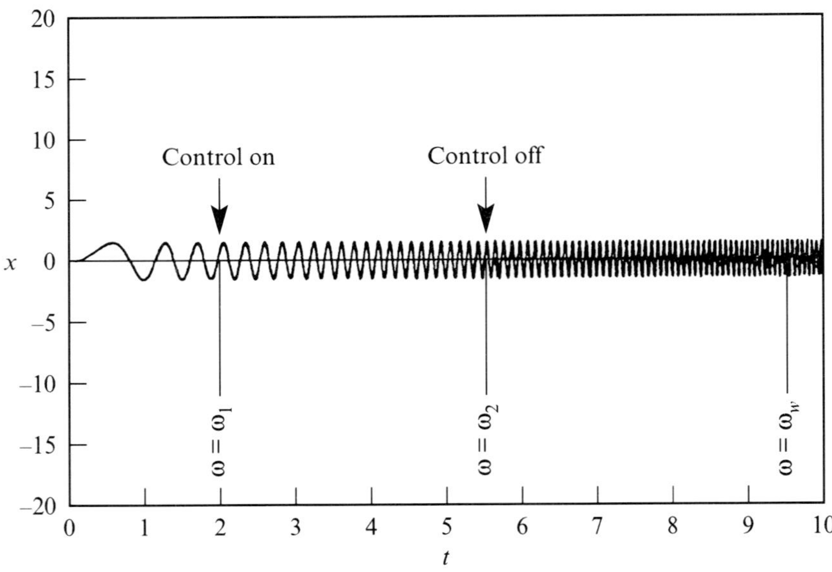

Figure 5.6 Passage through unstable zone with control.

We illustrate this method by the example of controlling rotor system supported in two gas bearings. We consider the system (*Figure 5.7*) which consists of a symmetrical rigid rotor supported on two gas bearings.

The joint base of both the bearing bushes is connected to the casing by means of the linear springs $K_{p1} = 22$, $K_{p2} = 38.5$ (dimensionless) and the viscous dampers C_p. As the springs are connected in series, their global stiffness coefficient is as follows:

$$K_p = \frac{K_{p1}K_{p2}}{K_{p1} + K_{p2}} \tag{5.2}$$

The force $2F_z$ is an external static load of the rotor (e.g. by means of its weight), so F_z is the static response of the bearing to the force $2F_z$. Full dynamics of the considered system have been considered in Czolczynski and Marynowski (1995). Here, we describe only one particular case as the illustration of our controlling method.

Figure 5.8 shows an example of the bearing journal vibrations when the rotational velocity ω increases from 0 to ω_w. As may be seen, for small values $\omega < \omega_1$, when the steady state is stable, the amplitude of the vibrations is small – in this region only the unbalanced vibrations appear. At $\omega = \omega_1$ the system undergoes Hopf bifurcation and the self-excited vibrations grow rapidly. If their amplitude is not limited (by the bush), the increase of ω leads to the decrease of vibrations (at $\omega = \omega_2$ – reversed Hopf bifurcation), and finally the self-excited vibrations vanish. In the gas bearing the increase of the journal vibrations leads to journal–bush contact and, in consequence to the damage of the system. This situation is illustrated by *Figure 5.6*, where the journal hits the bush at $\omega = \omega_d$.

Figure 5.8 shows the stability map (regions of the self-excited vibrations) of the system for two selected values $K_p = 22$ and $K_p = 14$, and for various values of the damping coefficient C_p. As may be seen for $K_p = 22$, when the rotational velocity is sufficiently high ($\omega > \omega_1$) the self-excited vibrations (of conical modes – thin solid lines) appear for any value of C_p, which lead to the damage of the bearings.

We may avoid this danger, changing the stiffness coefficient K_p in the time

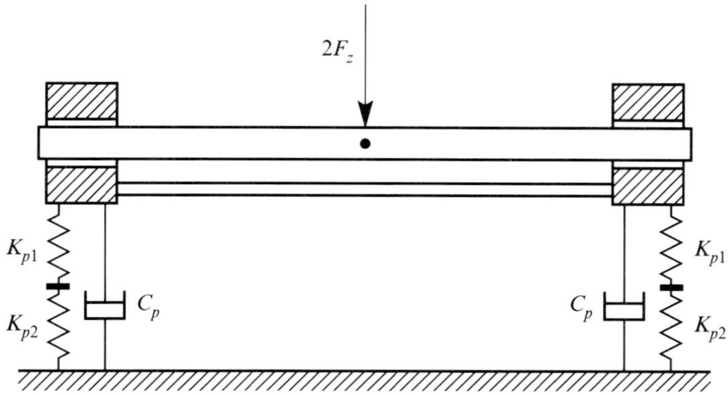

Figure 5.7 Rotor system supported in two gas bearings.

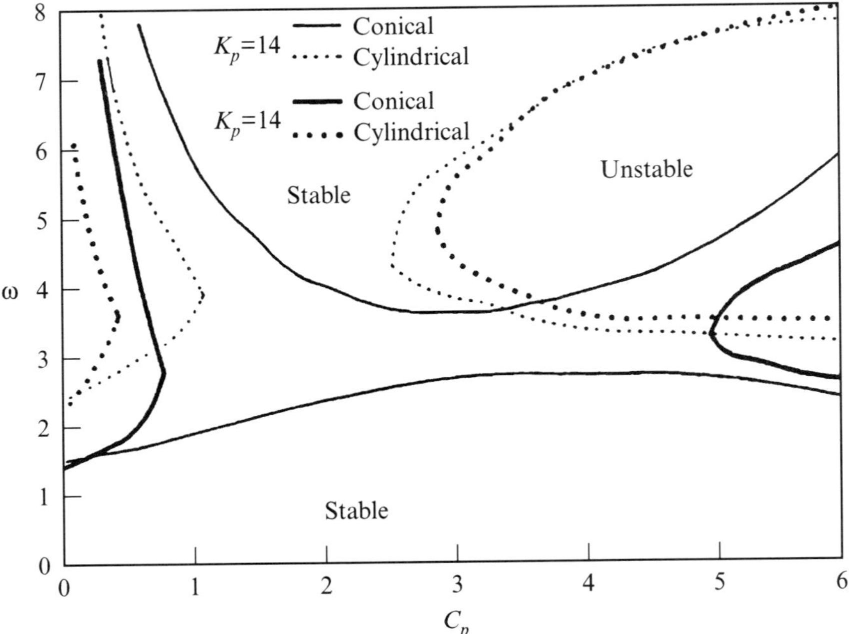

Figure 5.8 Stability map of the rotor system.

interval in which $\omega_1 \leq \omega \leq \omega_2$. When ω is near to ω_1, we may diminish K_p from 22 to 14. As we can see, the unstable regions for both the conical (thick solid lines) and cylindrical (thick dotted lines) vibrations vanish for $1.25 \leq C_p \leq 2.55$. When ω is bigger than ω_2, we may change K_p again.

The change of K_p may be realized by the use of an additional springs K_{p2}. For $0 < \omega < \omega_1$ and $\omega_2 < \omega$, the connection points of springs are fixed and in this way only the top springs are working and

$$K_p = K_{p1} = 22 \tag{5.3}$$

In the range $\omega_1 < \omega < \omega_2$, the connection points are released, and both (top and bottom) springs go into action. Thus the global stiffness coefficient is equal to

$$K_p = \frac{K_{p1}K_{p2}}{K_{p1} + K_{p2}} = \frac{22 \cdot 38.5}{22 + 38.5} = 14 \tag{5.4}$$

We have described the controlling method which allows control of Hopf bifurcation in transients. This method guarantees the passage through the unstable zone between the Hopf and reversed Hopf bifurcation points. The passage is possible due to the temporal change of one of the system parameters.

The described method can have significant practical applications as it allows the rotational velocity of rotors of many machines to be increased instead of being restricted to lower velocities, $\omega < \omega_1$. In most of the current engineering systems the small temporal change of one of the system parameters can be easily done. This and similar methods can stimulate the idea of the nonlinear design, i.e. the system can be designed in such a way to benefit from nonlinear effects.

Further reading

Abed, E.H., Wang, H.O. and Chen, R.C. (1992) Stabilization of period doubling bifurcations and implications for control of chaos. *IEEE Proceedings of Control Decision Conference*, Tucson, AZ, pp. 2119–2124.

Abed, E.H., Wang, H.O. and Chen, R.C. (1994) Stabilization of period doubling bifurcations and implications for control of chaos. *Physica D*, **70**, 154–164.

Afraimovich, V.S., Verichev, N.N. and Rabinovich, M.I. (1986) Stochastic synchronization of oscillations in dissipative systems. *Radiophys. Quant. Electron.* **29**, 795–803.

Alekseev, V.V. and Loskutov, A.Y. (1987) Control of a system with a strange attractor through periodic parametric action. *Sov. Phys. Dokl.*, **32**, 1346–1348.

Alexeyev, A.A. and Shalfeev, V.D. (1995) Chaotic synchronization of mutually-coupled generators with frequency-controlled feedback loop. *Int. J. Bifurcat. Chaos*, **5**, 551–558.

Anishchenko, V.S., Vadivasova, T.E., Postnov, D.E. and Safonova, M.A. (1991) Forced and mutual synchronization of chaos. *Radioeng. Electron.* **36**, 338–351.

Azevedo, A. and Rezende, S.M. (1991) Controlling chaos in spin-wave instabilities. *Phys. Rev. Lett.*, **66**, 1342–1345.

Baillieul, J., Brockett, R.W. and Washburn, R.B. (1980) Chaotic motion in nonlinear feedback systems. *IEEE Trans. Circ. Syst.*, **27**, 990–997.

Bandyopadhyay, J.K., Kumar, V.R. and Kulkarni, B.D. (1992) On dynamic control of chaos: a study with reference to a reacting system. *Phys. Lett. A*, **166**, 197–204.

Bernhardt, P.A. (1994) Communications using chaotic frequency modulation. *Int. J. Bifurcat. Chaos*, **4**, 427–440.

Berreby, D. (1993) Chaos hits Wall Street. *Discover*, March, 76–84.

Bielawski, S., Bouazaoui, M., Derozier, D. and Glorieux, D. (1992) Controlling laser chaos. *Proceedings of Nonlinear Dynamics in Optical Systems*, Alpbach, Austria, June, 22–26.

Boccaletti, S. and Arecchi, F.T. (1995) Adaptive control of chaos. *Europhys. Lett.*, **31**, 127–132.

Breeden, J.L. (1994) Open-loop control of nonlinear systems. *Phys. Lett. A*, **190**, 264–272.

Breeden, J.L., Dinkelacker, F. and Hübler, A.W. (1990) Using noise in the modelling and control of dynamical systems. *Phys. Rev. A*, **42**, 5827–5836.

Brockett, R.W. (1982) On conditions leading to chaos in feedback systems. *IEEE Proceedings of Control Decision Conference*, pp. 932–936.

Carroll, T.L. and Pecora, L.M. (1991) Synchronizing chaotic circuits. *IEEE Trans. Circ. Syst.*, **38**, 453–456.

Carroll, T.L. and Pecora, L.M. (1993) Using chaos to keep period-multiplied systems in phase. *Phys. Rev. E*, **48**, 2426–2436.

Carroll, T.L., Heagy, J. and Pecora L.M. (1994) Synchronization and desynchronization in pulse coupled relaxation oscillators. *Phys. Lett. A*, **186**, 225–229.

Carroll, T.L., Triandaf, I. and Pecora L. (1992) Tracking unstable orbits in an experiment. *Phys. Rev. A*, **46**, 6189–6192.

Celka, P. (1994) Experimental verification of Pyragas's chaos control method applied to Chua's circuit. *Int. J. Bifurcat. Chaos*, **4**, 1703–1706.

Chen, G. (1994) Optimal control of chaotic systems. *Int. J. Bifurc. Chaos*, **4**, 461–464.

Chen, G. and Dong, X. (1993b) Ordering chaos of Chua's circuit – A feedback control approach. *IEEE Proceedings of International Symposium on Circuits and Systems*, Chicago, IL, May, 2604–2607.

Chen, G. and Dong, X. (1993c) On feedback control of chaotic continuous-time systems. *IEEE Trans. Circ. Syst. Pt l*, **40**, 591–601.

Chezhevsky, V.N. and Glorieux, P. (1995) Targeting unstable periodic orbits. *Phys. Rev. E*, **51**, 2701–2704.

Chua, L.O., Itoh, M., Kocarev, L. and Eckert, K. (1993) Chaos synchronization in Chua's circuit. *J. Circ. Syst. Comput.*, **3**, 309–324.

Chui, C.K. and Chen, G. (1989) *Linear Systems and Optimal Control*, Springer-Verlag, New York.

Chui, C.K. and Chen, G. (1992) *Signal Processing and Systems Theory: Selected Topics*, Springer-Verlag, New York.

Coles, D. (1965) Transition in circular Couette flow. *J. Fluid Mech.*, **21**, 385–425.

Corcoran, E. (1991) Ordering chaos. *Scient. Am.*, August, 97–98.

Cortese, A. (1992) Order out of chaos. *Inform. Wk*, August, 22.

Dabrowski, A., Galias, Z. and Ogorzalek, M.J. (1993) Online identification and control of chaos in a real Chua's circuit. *Kybernetika*, Czech Acad. Sci., **31**, 321–367.

De Figueiredo, R.J.P. and Chen, G. (1993) *Nonlinear Feedback Control Systems: An Operator Theoretic Approach*, Academic Press, New York.

De Sousa Vieira, M., Lichtenberg, A.J. and Lieberman, M.A. (1994) Self-synchronization of many coupled oscillators. *Int. J. Bifurcat. Chaos*, **4**, 1563–1578.

Ding, M. and Otto, E. (1994) Enhancing synchronism of chaotic systems. *Phys. Rev. E*, **49**, 945–948.

Ding, M., Otto, E. and Grebogi, C. (1994) Crisis control: Preventing chaos-induced capsizing of a ship. *Phys. Rev. E*, **50**, 4228–4230.

Dong, X. and Chen, G. (1992a) Control of discrete-time chaotic systems. *IEEE Proceedings of American Control Conference*, June, Chicago, IL, pp. 2234–2235.

Dong, X. and Chen, G. (1992b) Controlling chaotic continuous-time systems via feedback. *IEEE Proceedings of Decision Control Conference*, Tucson, AZ, December, 2502–2503.

Duffing, G. (1918) *Erzwungene Schwingungen bei Veränderlicher Eigenfrequenz*, F. Vieweg u. Sohn, Braunschweig.

Endo, T. and Chua, L.O. (1991) Synchronizing chaos from electronic phase-locked loops. *Int. J. Bifurcat. Chaos*, **1**, 701–710.

Feng, X. and Loparo, K.A. (1992) A study of chaos in discrete time linear systems with quantized state feedback. *IEEE Proceedings of Decision Control Conference*, Tucson, AZ, December, 2107–2112.

Fenstermacher, P.R., Swinney, H.L. and Gollub, J.P. (1979) Dynamical instabilities and transition to chaotic taylor vortex flow. *J. Fluid Mech.*, **94**, 103–128.

Fowler, T.B. (1989) Application of stochastic control techniques to chaotic nonlinear systems. *IEEE Trans. Autom. Control.*, **34**, 201–205.

Freeman, W.J. (1991) The physiology of perception, *Scient. Am.* February, 78–85.

Frison, T.W. (1992) Controlling chaos with a neural network. *Proc. of Intl Conference on Neural Networks*, Baltimore, MD, June, 75–80.

Fronzoni, L. and Giocondo, M. (1991) Experimental evidence of suppression of chaos by resonant parametric perturbations. *Phys. Rev. A*, **43**, 6483–6487.

Fuh, C.C. and Tung, P.C. (1995) Controlling chaos using differential geometric method. *Phys. Rev. Lett.*, **75**, 2952–2955.

Galias, Z. (1995) New method for stabilization of unstable periodic orbits in chaotic systems. *Int. J. Bifurcat. Chaos*, **5**, 281–296.

Garfinkel, A., Spano, M.L., Ditto, W.L. and Weiss, J.N. (1992) Controlling cardiac chaos. *Science*, **257**, 1230–1235.

Gauthier, D.J., Sukow, D.W., Concannon, H.M. and Socolar, J.E.S. (1994) Stabilizing unstable periodic orbits in a fast diode resonator using continuous time-delay autosynchronization. *Phys. Rev. E*, **50**, 2343–2346.

Genesio, R. and Tesi, A. (1992) A harmonic balance approach for chaos prediction: The Chua's circuit. *Int. J. Bifurcat Chaos*, **2**, 61–79.

Genesio, R. and Tesi, A. (1993) Distortion control of chaotic systems: The Chua's circuit. *J. Circ. Syst. Comput.*, **3**, 151–171.

Genesio, R., Basso, M., Stranghini, M., Atesi, A., Ciofini, M. and Meucci, R. (1995) Limit cycle bifurcation analysis and its implification for controlling complex dynamics in a CO_2 laser. *Proceedings of the 3rd International Specialist Workshop on Nonlinear Dynamics of Electronic Systems*, Dublin, Dublin University Press, pp. 215–220.

Glass, L., Shrier, A. and Bélair, J. (1986) Chaotic cardiac rhythms, *Chaos* (ed. A.V. Holden) Princeton Univ Press, Princeton, RI, pp. 237–256.

Gligoroski, D., Dimovski, D. and Urumov V. (1995) Control in multidimensional chaotic systems by small perturbations. *Phys. Rev. E*, **51**, 1690–1694.

Glorieux, P., Bielawski, S., Derozier, D. and Szwaj, C. (1992) Control of chaos in fiber lasers. Poster paper in: *Workshop on Measures of Complexity and Chaos 2*, Bryn Mawr, PA, August 1992.

Gonzalez, G.A. (1995) Controlling chaos of an uncertain Lozi system via adaptive techniques. *Int. J. Bifurcat. Chaos*, **5**, 559–562.

Grassberger, P. and Procaccia, I. (1984) Dimensions and entropies of strange attractors from a fluctuating dynamics approach. *Physica D*, **13**, 34–54.

Hartley, T.T. and Mossayebi, F. (1992) A classical approach to controlling the Lorenz equations. *Int. J. Bifurcat. Chaos*, **2**, 881–887.

Hartley, T.T. and Mossayebi, F. (1993) Control of Chua's circuit. *J. Circ. Syst. Comput.*, **3**, 173–194.

Hayes, S. and Grebogi, C. (1993) Coding information in the natural complexity of chaos. Invited paper, in: *SPIE 2038, Chaos in Communications*, pp. 1–9.

Hayes, S., Grebogi, C. and Otto, E. (1993) Communicating with chaos. *Phys. Rev. Lett.*, **70**, 3031–3034.

Heagy, J.F., Carroll T.L. and Pecora L.M. (1994) Synchronous chaos in coupled oscillator systems. *Phys. Rev. E*, **50**, 1874–1885.

Herzel, H. and Freund, J. (1995) Chaos, noise and synchronization revisited. *Phys. Rev. E*, (in press).

Hogg, T. and Huberman, B.A. (1991) Controlling chaos in distributed systems. *IEEE Trans. Syst. Man. Cyber.*, **21**, 1325–1332.

Huberman, B.A. and Lumer, E. (1990) Dynamics of adaptive systems. *IEEE Trans. Circ. Syst.* **37**, 547–550.

Jackson, E.A. (1991b) Controls of dynamic flows with attractors. *Phys. Rev. A*, **44**, 4839–4853.

Jackson, E.A. and Grosu, J. (1995) An open-plus-closed loop (OPCL) control of complex dynamic system. *Physica D*, **85**, 1–9.

John, J.K. and Amritkar, R.E. (1994) Synchronization by feedback and adaptive control. *Int. J. Bifurcat. Chaos*, **4**, 1687–1696.

Johnson, G.A. and Hunt, E.R. (1993) Maintaining stability in Chua's circuit driven into regions of oscillation and chaos. *J. Circ. Syst. Comput.*, **3**, 119–123.

Kalman, R.E. (1956) Nonlinear aspects of sample data control systems. *Proc. of Symposium on Nonlinear Circuit Analysis*, Brooklyn, April, pp. 273–313.

Kaluyanov, E.V. (1987) Synchronous and stochastic self-excited oscillations in a transistor microwave oscillator with delayed feedback, subjected to the parametric action of an external force. *Sov. J. Commun. Tech. Electron.*, **32**, 784.

Kameda, T., Aihara, K. and Hori, Y. (1991) Application of a TDOF controller to chaotic dynamical systems. *Proceedings of Korean Automatic Control Conference*, pp. 1549–1552. (Full version in *Trans. IEEE Japan* (1993) **113-C**(1), 43–49, in Japanese.)

Lai, Y.-C. and Grebogi C. (1994) Synchronization of spatiotemporal chaotic systems by feedback control. *Phys. Rev. E*, **50**, 1894–1899.

Lai, Y.-C., Ding, M. and Grebogi, C. (1993) Controlling Hamiltonian chaos. *Phys. Rev. E*, **47**, 86–92.

Landa P.S. and Perminov, S.M. (1985) Interaction of periodic and stochastic oscillations. *Radio-phys. Quant. Electron.*, **28**, 424.

Lathrop, D.P., Kostelich, E.J. (1989) Characterization of an experimental strange attractor by periodic orbits. *Phys. Rev. A*, **40**, 4023–4031.

Lima, R. and Pettini, M. (1990) Suppression of chaos by resonant parametric perturbations. *Phys. Rev. A*, **41**, 726–733.

Liu, Y. and Barbosa L.C. (1995) Periodic locking in coupled Lorenz systems. *Phys. Lett. A*, **197**, 13–18.

Lorenz, E. (1963) Deterministic nonperiodic flow. *J. Atmos. Sci.*, **20**, 130–141.

Luce, R. and Kernévez, J.P. (1991) *Controlability of Lorenz equation. International Series on Numerical Mathematics*, Birkhäuser, Boston, MA, vol. 97, pp. 257–261.

Lugiato, L.A. (1984) Theory of optical bistability. *Prog. Optics*, **21**, 71–211.

Mareels, I.M.Y. and Bitmead, R.R. (1986) Nonlinear dynamics in adaptive control: Chaotic and periodic stabilization. *Automatica*, **22**, 641–655.

Mareels, I.M.Y. and Bitmead, R.R. (1988) Nonlinear dynamics in adaptive control: Chaotic and periodic stabilization. II. Analysis. *Automatica*, **24**, 485–497.

Matsumoto, T. (1984) A chaotic attractor from Chua's circuit. *IEEE Trans. Circ. Syst.*, **31**, 1055–1058.

Matsumoto, T., Chua, L.O. and Komuro, M. (1985) The double scroll. *IEEE Trans. Circ. Syst.*, **32**, 798–817.

Mees, A.I. (1981) *Dynamics of Feedback Systems*, Wiley, Chichester.

Mees, A.I. (1986) Chaos in feedback systems. *Chaos* (ed. A.V. Holden), Princeton University Press, Princeton, RI, pp. 99–110.

Mees, A.I. and Sparrow, C.T. (1981) A review for engineers. *Proc. Inst. Electr. Engrs*, **D128**, 201–205.

Minreh, D. (1993) Method of controlling chaos in laser equations. *Phys. Rev. E*, **47**, 714–717.

Mpitsos, G.J. and Burton, Jr. R.M. (1992) Convergence and divergence in neural networks, processing of chaos and biological analogy. *Neur. Net.*, **5**, 605–625.

Murali, K. and Lakshmanan, M. (1993) Controlling of chaos in the driven Chua's circuit. *J. Circ. Syst. Comput.*, **3**, 125–137.

Murali, K., Lakshmanan, M. and Chua, L.O. (1995) Controlling and synchronization of chaos in the simplest dissipative non-autonomous circuit. *Int. J. Bifurcat. Chaos*, **5**, 563–572.

Newell, T.C., Alsing, P.M., Gavrielides, A. and Kovanis, V. (1994a) Synchronization of chaotic diode resonators by occasional proportional feedback. *Phys. Rev. Lett.*, **72**, 1647–1650.

Newell, T.C., Alsing, P.M., Gavrielides, A. and Kovanis V. (1994b) Synchronization of chaos using proportional feedback. *Phys. Rev. E*, **49**, 313–318.

Nicolis, J.S. (1987) Chaotic dynamics in biological information processing: A heuristic outline. *Chaos in Biological Systems*, (eds H. Degn, A.V. Holden and L.F. Olsen), Plenum Press, New York, pp. 221–232.

Nitsche, G. and Dressler, U. (1992) Controlling chaotic dynamical systems using time delay coordinates. *Physica D*, **58**, 153–164.

Ogorzalek, M.J., (1993a) Taming chaos – part I: Synchronization. *IEEE Trans. Circ. Syst.*, **40**, 693–699.

Ogorzalek, M.J. (1993b) Taming chaos – Part II: Control. *IEEE Trans. Circ. Syst.*, **40**, 700–721.

Ogorzalek, M.J. and Galias, Z. (1993) Exploring chaos in Chua's circuit via unstable periodic orbits. *IEEE Proceedings of International Symposium on Circuits and Systems*, 3–6 May, Chicago, IL, pp. 2608–2611.

Ohle, F. (1995) No-feedback control and slaving principles in intermittent wake flows. *Chaos Solitons Fractals*, **5**, 1871–1880.

Parker, T.S. and Chua, L.O. (1989) *Practical Numerical Algorithms for Chaotic Systems*, Springer, New York.

Paskota, M., Mees, A.I. and Teo, K.L. (1994) Stabilizing higher periodic orbits. *Int. J. Bifurcat. Chaos*, **4**, 447–460.

Pecora, L.M. (1995) Overview of chaos and communications research. *SPIE Press*, (in press).

Pecora, L.M. and Carroll, T.L. (1991b) Pseudoperiodic driving: Eliminating multiple domains of attraction using chaos. *Phys. Rev. Lett.*, **67**, 945–948.

Peng, B., Petrov, V. and Showalter, K. (1991) Controlling chemical chaos. *J. Phys. Chem.*, **95**, 4957–4959.

Peterman, D.W., Ye, M. and Wigen, P.E. (1995) High frequency synchronization of chaos. *Phys. Rev. Lett.*, **74**, 1740–1742.

Pikovsky, A.S. (1984) Synchronization and stochastization of nonlinear oscillations by external noise. *Izv. Vyssh. Uchebn. Zaved. Radiofiz.*, **27**, 576.

Pikovsky, A.S. (1990) Synchronization of the stochastic self-excited oscillation phase using a periodic external signal. *Radiophys. Quant. Electron*, **32**, 962.

Plapp, B.B. and Hübler, A.W. (1990) Nonlinear resonance and suppression of chaos in the rf-biased Josephson junction. *Phys. Rev. Lett.*, **65**, 2302–2305.

Poddar, G., Chakrabarty, K. and Banerjee, A. (1995) Experimental control of chaotic behaviour of buck converter. *IEEE Trans. Circ. Syst.*, **CAS 42**, 502–504.

Pyragas, K. (1995) Control of chaos via extended delay feedback. *Phys. Lett.*, **206A**, 323–330.

Rajasekar, S. (1995) Controlling of chaotic motion by chaos and noise signals in a logistic map and a Bonhoeffer–van der Pol oscillator. *Phys. Rev. E*, **51**, 775–778.

Rajasekar, S. and Lakshmanan, M. (1992) Controlling of chaos in Bonhoeffer–van der Pol oscillator. *Int. J. Bifurcat. Chaos*, **2**, 201–204.

Rapp, P.E., Mees, A.I. and Sparrow, C.T. (1981) Frequency encoded biochemical regulation is more accurate than amplitude dependent control. *J. Theor. Biol.*, **90**, 531–544.

Reyl, C., Flepp, L., Badii, R. and Brun, E. (1993) Control of NMR-laser chaos in high-dimensional embedding space. *Phys. Rev. E*, **47**, 267–272.

Rollins, R.W., Parmananda, P. and Sherard, P. (1992) Controlling chaos in highly dissipative systems: A simple recursive algorithm. *Phys. Rev. E*, **47**, 780–783.

Roy, R., Murphy, T.W., Maier, T.D., Gills, Z. and Hunt, E.R. (1992) Dynamical control of a chaotic laser: Experimental stabilization of a globally coupled system. *Phys. Rev. Lett.*, **68**, 1259–1262.

Rulkov, N.F., Sushchik, M.M., Tsimring, L.S., Abarbanel H.D.I. (1995) Generalized synchronization of chaos in directionally coupled chaotic systems. *Phys. Rev. E*, **51**, 980–994.

Qu, Z., Hu, G., Yang, G. and Qin, G. (1995) Phase effect in taming nonautonomous chaos by weak harmonic perturbations. *Phys. Rev. Lett.*, **74**, 1736–1739.

Savage, H.T., Ditto, W.L., Braza, P.A., Spano, M.L., Rauseo, S.N. and Spring III, W.C. (1990) Crisis induced intermittency in a parametrically driven, gravitationally buckled, magnetoelastic amorphous ribbon experiment. *J. Appl. Phys.*, **67**, 5919–5923.

Schiff, S.J., Jerger, K., Duong, D.H., Chang, T., Spano, M.L. and Ditto, W.L. (1994) Controlling chaos in the brain. *Nature*, **370**, 615–620.

Schweizer, J., Kennedy, M.P., Hasler, M. and Dedieu, H. (1995) Synchronization theorem for a chaotic system. *Int. J. Bifurcat. Chaos*, **5**, 297–302.

Shinbrot, T., Grebogi, C., Ott, E. and Yorke, J.A. (1992a) Using chaos to target stationary states of flows. *Phys. Lett. A*, **169**, 349–354.

Shinbrot, T., Ott, E., Grebogi, C. and Yorke, J.A. (1992b) Using chaos to direct orbits to targets in systems described by a one-dimensional map. *Phys. Rev. A*, **45**, 4165–4168.

Shinbrot, T., Ott, E., Grebogi, C. and Yorke, J.A. (1992c) Using the sensitive dependence of chaos (the butterfly effect) to direct trajectories in an experimental chaotic system. *Phys. Rev. Lett.*, **68**, 2863–2866.

Shinbrot, T., Grebogi, C., Ott, E. and Yorke, J.A. (1993) Using small perturbations to control chaos. *Nature*, **363**, 411–417.

Sinha, S., Ramaswamy, R. and Rao, J.S. (1990) Adaptive control in nonlinear dynamics. *Physica D*, **43**, 118–128.

Socolar, J.E.S., Sukow, D.W. and Gauthier D.J. (1994) Stabilizing unstable periodic orbits in fast dynamical systems. *Phys. Rev. E*, **50**, 3245–3248.

Sparrow, C.T. (1981) Chaos in a three-dimensional single loop feedback system with a piecewise linear feedback function. *J. Math. Anal. Appl.*, **83**, 275–291.

Stampfe, M. (1994) Controlling chaos through iteration sequences and interpolation techniques. *Int. J. Bifurcat. Chaos*, **4**, 1697–1702.

Tai, O., Skorupka, C., Pecora, L.M. (1994) Chaotic parameter variation in maps: pseudoperiodicity, crisis and synchronization. *Phys. Lett. A*, **176**, 175–179.

Taiwar, S. and Namachchivaya, N.S. (1992) Control of chaotic systems: Application to the Lorenz equation. *Nonlinear Vibrations* (eds R.A. Ibrahim *et al.*), DE Vol. 50, AMD Vol. 144, American Society of Mechanical Engineers, pp. 47–58.

Tang, Y.S., Mees, A.I. and Chua, L.O. (1983) Synchronization and chaos. *IEEE Trans. Circ. Syst.*, **30**, 620–626.

Taylor, T. (1992) A tutorial on chaos in control systems. *IEEE Proceedings of Control Decision Conference*, Tucson, AZ, December, 2102–2106.

Tél, T. (1993) Crossover between the control of permanent and transient chaos. *Int. J. Bifurcat. Chaos*, **3**, 757–164.

Tesi, A., de Angeli, A. and Genesio, R. (1994) On system decomposition for synchronizing chaos. *Int. J. Bifurcat. Chaos*, **4**, 1675–1686.

Ueda, Y. (1991) Survey of regular and chaotic phenomena in the Duffing's oscillator. *Chaos Solitons Fractals*, **1**, 199.

Veitch, D. (1992) Windows of stability in control chaos. *IEEE Trans. Circ. and Syst.*, **39**, 808–819.

Verichev, N.N. and Maksimov, A.G. (1989) Synchronization of stochastic oscillations of parametrically excited nonlinear oscillators. *Radiophys. Quant. Elect.*, **32**, 962.

Vincent, T.L. and Yu, J. (1991) Control of a chaotic system. *Dyn. Control*, **1**, 35–52.

Wang, H. and Abed, E.H. (1992) Bifurcation control of chaotic dynamical systems. *Proceedings of 2nd IFAC Nonlinear Control and Systems Design Symposium*, Bordeaux, France, June, 57–62.

Wu, C.W. and Chua, L.O. (1995) Synchronization in an array of linearly coupled dynamical systems. *IEEE Trans. Circ. Syst.*, **42**, 430–447.

Wu, S. (1987) Chua's circuit family. *IEEE Proc.*, **75**, 1022–1032.

Yang, W., Ding, M., Mandell, A.J. and Ott, E. (1995) Preserving chaos: control strategies to preserve complex dynamics with potential relevance to biological disorders. *Phys. Rev. E*, (in press).

Yevdokimova, O.N. and Kaptsov, L.N. (1988) Amplitude synchronization threshold and Lyapunov factor of stochastic oscillations of an oscillator with an inertial nonlinearity. *Sov. J. Commun. Tech. Electron*, **33**, 2558.

References

Auerbach, D., Grebogi, C., Otto, E. and Yorke, J.A. (1992) Controlling chaos in high dimensional systems. *Phys. Rev. Lett.*, **69**, 3479–3482.

Braiman, Y. and Goldhirsch, I. (1991) Taming chaotic dynamics with weak periodic perturbations. *Phys. Rev. Lett.*, **66**, 2545–2548.

Brindley, J., Kapitaniak, T. and Kocarev, L. (1995) Controlling chaos in geophysical systems. *Geophys. Res. Lett.*, **22**, 1257–1260.

Carroll, T.L. (1994) Synchronizing chaotic systems using filtered signals. *Phys. Rev. E*, **50**, 2580–2587.

Chen, G. (1993) Controlling chaotic trajectories to unstable limit cycles – A case study. *IEEE Proceedings of American Control Conference*, San Francisco, CA, June, 2413–2414.

Chen, G. and Dong, X. (1992) On feedback control of chaotic dynamical systems. *Int. J. Bifurcat. Chaos*, **2**, 407–411.

Chen, G. and Dong, X. (1993a) Controlling Chua's circuit. *J. Circ. Syst. Comput.*, **3**, 139–149.

Chen, G. and Dong, X. (1995) Identification and control of chaotic systems: An artificial neural network approach. *Proceedings of the 1995 IEEE International Symposium on Circuits and Systems*, Seattle, pp. 1177–1182.

Chua, L.O. (1993) Global unfolding of Chua's circuit. *IEICE Trans. Fundam.*, **E76-A**, 704–734.

Chua, L.O., Komuro, M. and Matsumoto, T. (1986) The double scroll family: I and II. *IEEE Trans. Circ. Syst.*, **33**, 1072–1118.

Cuomo, K.M. and Oppenheim, A.V. (1993) Circuit implementation of synchronized chaos with applications to communications. *Phys. Rev. Lett.*, **71**, 65–68.

Czolczynski, K. and Marynowski, K. (1995) How to avoid self-excited vibrations in symmetrical rotors supported in gas journal bearings. *Mach. Dyn. Prob.* (in press).

Dabrowski, A., Galias, Z. and Ogorzalek, M.J. (1992) A study of identification and control in a real implementation of Chua's circuit. *Proceedings of the 2nd IFAC Workshop on Systems Structures and Control*, September 3–5, Prague, Czechoslovakia, 278–281.

Dedieu, H. and Ogorzalek, M.J. (1994) Controlling chaos in Chua's circuit via sampled inputs. *Int. J. Bifurcat. Chaos*, **4**, 447–456.

De Sousa Vielra, M., Lichtenberg, A.J. and Lieberman, M.A. (1992) Self-synchronization of many coupled oscillators. *Phys, Rev. A*, **46**, 7359–7363.

Ditto, W.L., Rauseo, S.N. and Spano, M.L. (1990) Experimental control of chaos. *Phys. Rev. Lett.*, **65**, 3211–3214.

Dressler, U. and Nitsche, G. (1992) Controlling chaos using time delay coordinates. *Phys. Rev. Lett.*, **68**, 1–4.

Feigenbaum, M. (1978) Qualitative universality for a class of nonlinear transformations. *J. Stat. Phys.*, **19**, 25–42.

Fujisaka, H. and Yamada, T. (1983) Stability theory of synchronized motion in coupled-oscillator systems. *Prog. Theor. Phys.*, **69**, 32–47.

Galias, Z. and Ogorzalek, M.J. (1995) Bang-bang control of chaotic systems. *Proceedings of the 3rd International Specialist Workshop on Nonlinear Dynamics of Electronic Systems*, Dublin, Dublin University Press, pp. 229–232.

Genesio, R., Tesi, A. and Viloresi, F. (1993) A frequency approach for analysing and controlling chaos in nonlinear circuits. *IEEE Trans. Circ. Syst., CAS* **40**, 819–828.

Halle, K., Wu, C.W., Itoh, M. and Chua, L.O. (1993) Spread spectrum communication through modulation of chaos. *Int. J. Bifurcat. Chaos*, **3**, 469–477.

Hayashi, C. (1964) *Nonlinear Oscillations in Physical Systems*, McGraw-Hill, New York.

Herzel, H. (1988) Stabilization of chaotic orbits by random noise. *ZAMM*, **68**, 582–583.

Hübler, A.W. (1989) Adaptive control of chaotic systems. *Helv. Phys. Acta*, **62**, 343–346.

Hunt, E.R. (1991) Stabilizing high-period orbits in a chaotic system – the diode resonator. *Phys. Rev. Lett.*, **67**, 1953–1955.

Isomaki, H.M. van Boehm, J. and Raty, R. (1987) Chaotic oscillations and fractal basin boundaries of an impacting body. *Proceedings XIth Conference on Nonlinear Oscillations*, Budapest, pp. 763–772.

Jackson, E.A. (1990) The entrainment and migration controls of multiple attractor systems. *Phys. Lett. A*, **151**, 478–484.

Jackson, E.A. (1991a) On the control of complex dynamic systems. *Physica. D*, **50**, 341–366.

Jackson, E.A. (1991b) Controls of dynamic flows with attractors. *Phys. Rev. A*, **44**, 4839–4853.

Jackson, E.A. and Hübler, A. (1990) Periodic entrainment of chaotic logistic map dynamics. *Physica D*, **44**, 407–420.

Jackson, E.A. and Kodogeorgiou, A. (1991) Entrainment and migration controls of two-dimensional maps. *Physica D*, **54**, 253–265.

Johnson, G.A., Tigner, T.E. and Hunt, E.R. (1993) Controlling chaos in Chua's circuit. *J. Circ. Syst. Comput.*, **3**, 109–117.

Kapitaniak, T. (1988) *Chaos in Systems with Noise*, second edition. World Scientific, Singapore, 1990.

Kapitaniak, T. (1991a) *Chaotic Oscillations in Mechanical Systems*, Manchester University Press, Manchester.

Kapitaniak, T. (1991b) The loss of chaos in a quasiperiodically forced nonlinear oscillator. *Int. J. Bifurcat. Chaos*, **1**, 357–362.

Kapitaniak, T. (1992) Controlling chaotic oscillators without feedback. *Chaos Solitons Fractals*, **2**, 519–527.

Kapitaniak, T. (1993) Targeting unstable stationary states of Chua's circuit. *J. Circ. Syst. Comput.*, **3**, 195–199.

Kapitaniak, T. (1994) Synchronization of chaos using continuous control. *Phys. Rev., E*, **50**, 1642–1644.

Kapitaniak, T., (1995) Distributions of transient Lyapunov exponents in quasiperiodically forced systems. *Prog. Theor. Phys.*, **93**, 831–833.

Kapitaniak, T. and Chua, L.O. (1994) Hyperchaotic attractors of unidirectionally-coupled Chua's circuits. *Int. J. Bifurcat. Chaos*, **4**, 477–482.

Kapitaniak, T. and Steeb, W.-H. (1991) Transition to hyperchaos in coupled generalized van der Pol's equations. *Phys. Lett. A*, **152**, 33–36.

Kapitaniak, T. and Thylwe, K.-E. (1996) Monotonic stability. *Chaos, Solitons Fractals*, **7**, (in press).

Kapitaniak, T., Awrejcewicz, J. and Steeb, W.-H. (1987) Chaotic behaviour of anharmonic oscillator with almost periodic excitation. *J. Phys. A*, **20**, L355–359.

Kapitaniak T., Chua, L.O. and Zhong, G.-Q. (1994) Experimental synchronization of chaos using continuous control. *Int. J. Bifurcat. Chaos*, **4**, 483–488.

Kapitaniak, T., Kocarev, L. and Chua, L.O. (1993) Controlling chaos without feedback and control signals. *Int. J. Bifurcat. Chaos*, **3**, 459–468.

Kapitaniak, T., Sekieta, M. and Ogorzalek, M.J. (1996): Monotone synchronization of chaos. *Int. J. Bifurcat. Chaos*, **6**, 211–217.

Klamka, J. (1991) *Controlability of Dynamical Systems*, Kluwer, Boston.

Kocarev, L. and Kapitaniak, T. (1995) On an equivalence of chaotic attractors. *J. Phys. A.*, **28**, 249–254.

Kocarev, L. and Parlitz, U. (1995) General approach for chaotic synchronization with applications to communication. *Phys. Rev. Lett.*, **74**, 5028–5031.

Kocarev, L., Halle, K., Eckert, K. and Chua, L.O. (1992) Experimental demonstration of secure communications via chaos synchronization. *Int. J. Bifurcat. Chaos*, **2**, 709–716.

Kostelich, E.J., Grebogi, C., Otto, E. and Yorke, J.A. (1993) Higher-dimensional targeting. *Phys. Rev. E*, **47**, 305–310.

Lai, Y.-C. and Grebogi C. (1993) Synchronization of chaotic trajectories using control. *Phys. Rev. E*, **47**, 2357–2360.

Lozi, R. and Aziz-Alaoui, A. (1995) Secure communications via chaotic synchronization in Chua's circuit: numerical analysis of the errors of the recovered signal. *Proceedings ISCAS*, Seattle, (in press).

Madan, R. (1993) *Chua's Circuit: Paradigm for Chaos*, World Scientific, Singapore.

Mehta, N.J. and Henderson, R.M. (1991) Controlling chaos to generate aperiodic orbits. *Phys. Rev. A*, **44**, 4861–4865.

Murphy, C.A. and Kennedy, M.P. (1995) Chaos controller for autonomous circuits. *Proceedings of the 3rd International Specialist Workshop on Nonlinear Dynamics of Electronic Systems*, Dublin, Dublin University Press, pp. 225–228.

Ogorzalek, M.J. (1994) Chaos control: How to avoid chaos or take advantage of it. *J. Franklin Inst.*, **331B**, 681–704.

Ott, E., Grebogi, C. and Yorke, J.A. (1990) Controlling chaos. *Phys. Rev. Lett.*, **64**, 1196–1199.

Parlitz, U., Chua, L.O., Kocarev, L., Halle, K. and Shang, A. (1992) Transmission of digital signals by chaotic synchronization. *Int. J. Bifurcat. Chaos*, **2**, 973–977.

Pecora, L.M. and Carroll, T.L. (1990) Synchronization in chaotic systems. *Phys. Rev. Lett.*, **64**, 821–824.

Pecora, L.M. and Carroll, T.L. (1991a) Driving systems with chaotic signals. *Phys. Rev. A*, **44**, 2374–2383.

Pérez, G. and Cerdeira, H. (1995) Extracting messages masked by chaos. *Phys. Rev. Lett.*, **74**, 1970–1973.

Peterson, I. (1991) Ribbon of chaos: researchers develop a lab technique for snatching order out of chaos. *Sci. News*, **139** (Jan.), 60–61.

Pettini, M. (1989) Controlling chaos through parametric excitations. *Dynamics and Stochastic Processes* (eds R. Lima, L. Streit and R.V. Mendes), Springer-Verlag, New York, pp. 242–250.

Pyragas, K. (1992) Continuous control of chaos by self-controlling feedback. *Phys. Lett.*, **170A**, 421–428.

Pyragas, K. (1993) Predictable chaos in slightly perturbed unpredictable chaotic systems. *Phys. Lett.*, **181A**, 203–210.

Pyragas, K. and Tamasevicis, A. (1993) Experimental control of chaos by delayed self-controlling feedback. *Phys. Lett.*, **180A**, 99–104.

Romeiras, F.J., Grebogi, C., Ott, E. and Dayawansa, W.P. (1992] Controlling chaotic dynamical systems. *Physica D*, **58**, 165–192.

Rossler, O. (1979) An equation for hyperchaos. *Phys. Lett. A*, **71**, 155–157.

Qu, Z., Hu, G. and Ma, B. (1993) Note on continuous chaos control. *Phys. Lett. A*, **178**, 265–272.

Sato, S., Sano, M. and Sawada, Y. (1983) Universal scaling property in bifurcation structure of Duffing's and of generalized Duffing's equations. *Phys. Rev. A*, **28**, 1654–1659.

Shinbrot, T., Ott, E., Grebogi, C. and Yorke, J.A. (1990) Using chaos to direct trajectories to targets. *Phys. Rev. Lett.*, **65**, 3215–3218.

Shinbrot, T., Grebogi, C., Ott, E. and Yorke, J.A. (1992) Using chaos to target stationary states of flows. *Phys. Lett. A*, **169**, 349–354.

Singer, J., Wang, Y. and Bau, H. (1991) Controlling a chaotic system. *Phys. Rev. Lett.*, **66**, 1123–1125.

Steeb, W.-H. and Louw, J.A. (1986) *Chaos and Quantum Chaos*, World Scientific, Singapore.

Steeb, W.-H, Louw, J.A. and Kapitaniak, T. (1986) Chaotic behaviour of an anharmonic oscillator with two external periodic forces. *J. Phys. Soc. Japn*, **55**, 3279–3281.

Stefanski, A. and Kapitaniak, T. (1995) Steady state locking in coupled chaotic systems. *Phys. Lett. A*, **210**, 279–282.

Szemplinska-Stupnicka, W. (1989) The refined approximate criterion for chaos in a two state mechanical system. *Ingen.-Archiv*, **58**, 354–365.

Tél, T. (1991) Controlling transient chaos. *J. Phys. A: Math. Gen.*, **24**, 1359–1368.

Ueda, Y. (1979) Randomly transitional phenomena in the system governed by Duffing's oscillator. *J. Statist. Phys.*, **20**, 181.

Wolf, A., Swift, J.B., Swinney, H.L. and Vastano, J.A. (1985) Determining Lyapunov exponents from time series. *Physica*, **16D**, 285–317.

PART II:
Selected Reprints

Paper 1
Controlling chaos

Edward Ott,[a],[b] Celso Grebogi,[a] and James A. Yorke[c]

University of Maryland, College Park, Maryland 20742, USA

[a] *Laboratory for Plasma Research.*
[b] *Departments of Electrical Engineering and of Physics.*
[c] *Institute for Physical Science and Technology and Department of Mathematics.*

Received 22 December 1989

It is shown that one can convert a chaotic attractor to any one of a large number of possible attracting time-periodic motions by making only *small* time-dependent perturbations of an available system parameter. The method utilizes delay coordinate embedding, and so is applicable to experimental situations in which *a priori* analytical knowledge of the system dynamics is not available. Important issues include the length of the chaotic transient preceding the periodic motion, and the effect of noise. These are illustrated with a numerical example.

The presence of chaos in physical systems has been extensively demonstrated and is very common. In practice, however, it is often desired that chaos be avoided and/or that the system performance be improved or changed in some way. Given a chaotic attractor, one approach might be to make some large and possibly costly alteration in the system which completely changes its dynamics in such a way as to achieve the desired behavior. Here we assume that this avenue is not available. Thus, we address the following question: Given a chaotic attractor, how can one obtain improved performance and a desired attracting time-periodic motion by making only *small* time-dependent perturbations in an *accessible* system parameter?

The key observation is that a chaotic attractor typically has embedded within it an infinite number of unstable periodic orbits [1]. Since we wish to make only small perturbations to the system, we do not envision creating new orbits with very different properties from the existing ones. Thus, we seek to exploit the already existing unstable periodic orbits. Our approach is as follows: We first determine some of the unstable low-period periodic orbits that are embedded in the chaotic attractor. We then examine these orbits and choose one which yields improved system performance. Finally, we tailor our small time-dependent parameter perturbations so as to stabilize this already existing orbit. In this Letter we describe how this can be done, and we illustrate the method with a numerical example. The method is very general and should be capable of yielding greatly improved performance in a wide variety of situations.

It is interesting to note that if the situation is such that the suggested method is practical, then the presence of chaos can be a great advantage. The point is that any one of a number of different orbits can be stabilized, and the choice can be made to achieve the best system performance among those orbits. If, on the other hand, the attractor is not chaotic but is, say, periodic, then small parameter perturbations can only change the orbit slightly. Basically we are then stuck with whatever system performance the stable periodic orbit gives, and we have no option for substantial improvement, short of making large alterations in the system.

Furthermore, one may want a system to be used for different purposes or under different conditions at different times. Thus, depending on the use, different requirements are made of the system. If the system is chaotic, this type of multiple-use situation might be accommodated without alteration of the gross system configuration. In particular, depending on the use desired, the system behavior could be changed by switching the temporal programming of the small parameter perturbations to stabilize different orbits. In contrast, in the absence of chaos, completely separate systems might be required for each use. Thus, when designing a system intended for multiple uses, purposely building chaotic dynamics into the system may allow for the desired flexibility. Such multipurpose flexibility is essential to higher life forms, and we, therefore, speculate that chaos may be a necessary ingredient in their regulation by the brain.

To simplify the analysis we consider continuous-time dynamical systems which are *three dimensional* and depend on one system parameter which we denote p (for example, $d\mathbf{x}/dt = \mathbf{F}(\mathbf{x},p)$, where $\mathbf{x}$ is three dimensional). We assume that the parameter p is available for external adjustment, and we wish to temporally program our adjustments of p so as to achieve improved performance. We emphasize that our restriction to a three-dimensional system is mainly for ease of presentation, and that the case of higher-dimensional (including infinite-dimensional) systems can be treated by similar methods [2].

We imagine that the dynamical equations describing the system are not known, but that experimental time series of some scalar-dependent variable $z(t)$ can be measured. Using delay coordinates [3,4] with delay T one can form a delay-coordinate vector,

$$\mathbf{X}(t) = [z(t), z(t-T), z(t-2T), \ldots, z(t-MT)].$$

We are interested in periodic orbits and their stability properties, and we shall use $\mathbf{X}$ to obtain a surface of section for this purpose. In the surface of section, a continuous-time-periodic orbit appears as a discrete-time orbit cycling through a finite set of points. We require the dynamical behavior of the surface of section map in *neighborhoods* of these points in order to study the stability of the periodic orbits. To embed a small neighborhood of a point from $\mathbf{x}$ into $\mathbf{X}$, we typically only require as many dimensions as there are coordinates of the point. Thus, for our purposes, $M = D - 1$ is generally sufficient. (This is in contrast with [3] $M + 1 = 2D + 1$, typically required for global embedding of the original phase space in the delay-coordinate space.) Hence, for the case considered ($D = 3$), our surface of section is two dimensional.

We suppose that the parameter p can be varied in a small range about some nominal value p_0. Henceforth, without loss of generality, we set $p_0 \equiv 0$. Let the range in which we are allowed to vary p be $p_* > p > -p_*$.

Using an experimental surface of section for the embedding vector $\mathbf{X}$, we imagine that we obtain many experimental points in the surface of section for $p = 0$. We denote these points $\xi_1, \xi_2, \xi_3, \ldots, \xi_k$, where ξ_n denotes the coordinates in the surface of section at the nth piercing of the surface of section by the orbit $\mathbf{X}(t)$. For example, a common choice of the surface of section would be $z(t - MT)$ equals a constant, and $\xi_n = [z(t_n), \ldots, z(t_n - (M-1)T)]$, where $t = t_n$ denotes the time at the nth piercing. From such experimentally determined sequences it has been demonstrated that a large number of distinct unstable periodic orbits on a chaotic attractor can be determined [5,6]. We then examine these unstable periodic orbits and select the one which gives the best performance. Again using an experimentally determined sequence, we obtain the stability properties of the chosen periodic orbit (cf. Refs. 5 and 6 for discussion of how this can be done and for descriptions of its implementation in concrete experimental cases). For the purposes of simplicity, let us assume in what follows that this orbit is a fixed point of the surface of section map (i.e., period one; the case of higher period is a straightforward extension). Let λ_s and λ_u be the experimentally determined stable and unstable eigenvalues of the surface of section map at the chosen fixed point of the map ($|\lambda_u| > 1 > |\lambda_s|$). Let $\mathbf{e}_s$ and $\mathbf{e}_u$ be the experimentally determined unit vectors in the stable and unstable directions. Let $\xi = \xi_F \equiv 0$ be the desired fixed point. We then change p slightly from $p = 0$ to some other value $p = \bar{p}$. The fixed-point coordinates in the experimental surface of

section will shift from 0 to some nearby point $\xi_F(\bar{p})$ and we determine this new position. For small $\bar{p}$ we approximate $\mathbf{g} \equiv \partial \xi_F(p)/\partial p \big|_{p=0} \cong \bar{p}^{-1} \xi_F(\bar{p})$, which allows an experimental determination of the vector $\mathbf{g}$.

Thus, in the surface of section, near $\xi = 0$, we can use a linear approximation for the map, $\xi_{n+1} - \xi_F(p) \cong \mathbf{M} \cdot [\xi_n - \xi_F(p)]$, where $\mathbf{M}$ is a 2×2 matrix. Using $\xi_F(p) \cong p\mathbf{g}$ we have

$$\xi_{n+1} \cong p_n \mathbf{g} + [\lambda_u \mathbf{e}_u \mathbf{f}_u + \lambda_s \mathbf{e}_s \mathbf{f}_s] \cdot [\xi_n + p_n \mathbf{g}]. \tag{1}$$

[In the linearization (1), we have considered p_n to be small and of the same order as ξ_n.] We emphasize that $\mathbf{g}$, $\mathbf{e}_u$, $\mathbf{e}_s$, λ_u, and λ_s are all experimentally accessible by the embedding technique just discussed. In (1) $\mathbf{f}_u$ and $\mathbf{f}_s$ are contravariant basis vectors defined by $\mathbf{f}_s \cdot \mathbf{e}_s = \mathbf{f}_u \cdot \mathbf{e}_u = 1$, $\mathbf{f}_s \cdot \mathbf{e}_u = \mathbf{f}_u \cdot \mathbf{e}_s = 0$. Note that we have written the location of the fixed point as $p_n \mathbf{g}$ because we imagine that we adjust p to a new value p_n after each piercing of the surface of section. That is, we observe ξ_n and then adjust p to the value p_n. Thus p_n depends on ξ_n. Further, we only envision making this adjustment when the orbit falls near the desired fixed point for $p = 0$.

Assume that ξ_n falls near the desired fixed point at $\xi = 0$ so that (1) applies. We then attempt to pick p_n so that ξ_{n+1} falls on the stable manifold of $\xi = 0$. That is, we choose p_n so that $\mathbf{f}_u \cdot \xi_{n+1} = 0$. If ξ_{n+1} falls on the stable manifold of $\xi = 0$, we can then set the parameter perturbations to zero, and the orbit for subsequent time will approach the fixed point at the geometrical rate λ_s. Thus, for sufficiently small ξ_n, we can dot (1) with $\mathbf{f}_u$ to obtain

$$p_n = \lambda_u(\lambda_u - 1)^{-1}(\xi_n \cdot \mathbf{f}_u)/(\mathbf{g} \cdot \mathbf{f}_u), \tag{2}$$

which we use when the magnitude of the right-hand side of (2) is less than p_*. When it is greater than p_*, we set $p_n = 0$. We assume in (2) that the generic condition $\mathbf{g} \cdot \mathbf{f}_u \neq 0$ is satisfied. Thus, the parameter perturbations are activated (i.e., $p_n \neq 0$) only if ξ_n falls in a narrow strip $|\xi_n^u| < \xi_*$, where $\xi_n^u = \mathbf{f}_u \cdot \xi_n$, and from (2) $\xi_* = p_* |(1 - \lambda_u^{-1}) \mathbf{g} \cdot \mathbf{f}_u|$. Thus, for small p_*, a typical initial condition will execute a chaotic orbit, unchanged from the uncontrolled case, until ξ_n falls in the strip. Even then, because of nonlinearity not included in (1), the control may not be able to bring the orbit to the fixed point. In this case the orbit will leave the strip and continue to wander chaotically as if there was no control. Since the orbit on the uncontrolled chaotic attractor is ergodic, at some time it will eventually satisfy $|\xi_n^u| < \xi_*$ and also be sufficiently close to the desired fixed point that attraction to $\xi = 0$ is achieved. (In rare cases applying Eq. (2) when the trajectory enters the strip, but is still far from 0, may result in stabilizing the wrong periodic orbit which visits the strip.)

Thus, we create a stable orbit, but, for a typical initial condition, it is preceded in time by a chaotic transient in which the orbit is similar to orbits on the uncontrolled chaotic attractor. The length τ of such a chaotic transient depends sensitively on the initial condition, and, for randomly chosen initial conditions, has an exponential probability distribution [7] $P(\tau) \sim \exp[-(\tau/\langle\tau\rangle)]$ for large τ. The average length of the chaotic transient $\langle\tau\rangle$ increases with decreasing p_* and follows a power-law relation[7] for small p_*, $\langle\tau\rangle \sim p_*^{-\gamma}$.

We will now derive a formula for the exponent γ. Dotting the linearized map for ξ_{n+1}, Eq. (1), with $\mathbf{f}_u$, we obtain $\xi^u_{n+1} \cong 0$. In obtaining this result from (1) we have substituted p_n appropriate for $|\xi^u_n| < \xi_*$. We note that the result $\xi^u_{n+1} \cong 0$ is a linearization, and typically has a lowest-order nonlinear correction that is quadratic. In particular, $\xi^s_n = \mathbf{f}_s \cdot \xi_n$ is not restricted by $|\xi^u_n| < \xi_*$, and thus may not be small when the condition $|\xi^u_n| < \xi_*$ is satisfied. Hence the correction quadratic in ξ^s_n is most significant. Including such a correction we have $\xi^u_{n+1} \cong \kappa(\xi^s_n)^2$, where κ is a constant. Thus, if $|\kappa|(\xi^s_n)^2 > \xi_*$, then $|\xi^u_{n+1}| > \xi_*$, and attraction to $\xi = 0$ is not achieved, even though $|\xi^u_n| < \xi_*$. Attraction to $\xi = 0$ is achieved when the orbit falls in the small parallelogram P_c given by $|\xi^u_n| < \xi_*$, $|\xi^s_n| < (\xi_*/|\kappa|)^{1/2}$. For very small ξ_*, an initial condition will bounce around on the set comprising the uncontrolled chaotic attractor for a long time before it falls in the parallelogram P_c. At any given iterate the probability of falling in P_c is $\mu(P_c)$, the measure of the uncontrolled attractor contained in P_c. Thus, $\langle\tau\rangle^{-1} = \mu(P_c)$. The scaling of $\mu(P_c)$ with ξ_* is

$$\mu(P_c) \sim (\xi_*)^{d_u}[(\xi_*/|\kappa|)^{1/2}]^{d_s} \sim \xi_*^{d_u+(1/2)d_s},$$

where d_u and d_s are the partial pointwise dimensions for the uncontrolled chaotic attractor at $\xi = 0$ in the unstable direction and the stable direction, respectively. Thus, $\mu(P_c) = \xi_*^\gamma$, where $\gamma = d_u + d_s/2$. Since we assume the attractor to be effectively smooth in the unstable direction, $d_u = 1$. The partial pointwise dimension in the stable direction is given in terms of the eigenvalues[7] at $\xi = 0$, $d_s = \ln|\lambda_u|/\ln|\lambda_s|^{-1}$. Thus,

$$\gamma = 1 + \tfrac{1}{2}\ln|\lambda_u|/\ln|\lambda_s|^{-1}. \tag{3}$$

To study the effect of noise we add a term $\varepsilon\delta_n$ to the right-hand side of the linearized equations for ξ_{n+1}, Eq. (1), where δ_n is a random variable and ε is a small parameter specifying the intensity of the noise. The quantities δ_n are taken to have zero mean ($\langle\delta_n\rangle = 0$), be independent ($\langle\delta_n\delta_m\rangle = 0$ for $m \neq n$), and have a probability density independent of n. Dotting (1) with noise included with $\mathbf{f}_u$ we obtain $\xi^u_{n+1} = \varepsilon\delta^u_n$, where $\delta^u_n \equiv \mathbf{f}_u \cdot \delta_n$. Thus, if the noise is bounded, $|\delta^u_n| < \delta_{max}$, then the stability of $\xi = 0$ will not be affected by the noise if the bound is small enough, $\varepsilon\delta_{max} < \xi_*$. If this condition is not satisfied, then the noise can kick an orbit which is initially in the parallelogram P_c into the region outside P_c. We are particularly interested in the case where such kickouts are caused by low-probability tails on the probability density and are thus rare. (If they are frequent, then our procedure is ineffective.) In such a case the average time to be kicked out $\langle\tau'\rangle$ will be long. Thus, an orbit will typically alternate between epochs of chaotic motion of average duration $\langle\tau\rangle$ in which it is far from $\xi = 0$, and epochs of average length $\langle\tau'\rangle$ in which the orbit lies in the parallelogram P_c. For small enough noise the orbit spends most of its time in P_c, $\langle\tau'\rangle \gg \langle\tau\rangle$, and one might then regard the procedure as being effective.

We now consider a specific numerical example. Our purpose is to illustrate and test our analyses of the average time to achieve control and the effect of noise. To do this we shall utilize the Henon map, $x_{n+1} = A - x^2_n + By_n$, $y_{n+1} = x_n$, where we take $B = 0.3$. We assume that the quantity A can

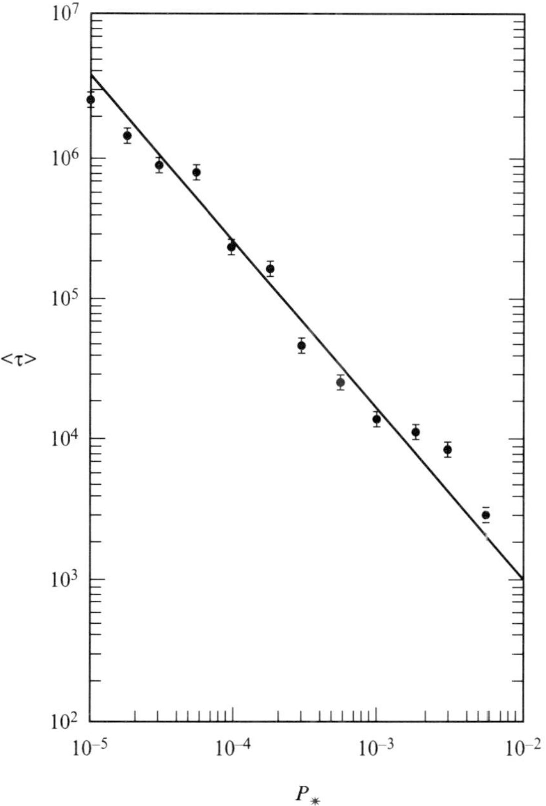

Figure 1 $\langle\tau\rangle$ vs p_*. Points were computed using 128 randomly selected initial conditions. $A_0 = 1.4$.

be varied by a small amount about some value A_0. Accordingly, we write A as $A = A_0 + p$, where p is the control parameter. For the values of A_0 which we investigate, the attractor for the map is chaotic and contains an unstable period-one (fixed-point) orbit. The coordinates (x_F, y_F) of the fixed point which is in the attractor for $p = 0$ along with the associated parameters and vectors appearing in Eq. (1) may be explicitly calculated. The quantity ξ_n appearing in (1) is $\xi_n = (x_n - x_F)\mathbf{x}_0 + (y_n - y_F)\mathbf{y}_0$. To test our prediction for the dependence of $\langle\tau\rangle$, the average time to approach $\xi = 0$, on the maximum allowed size of the parameter perturbation p_*, we proceed as follows. We iterate the map with $p = 0$ using a large number of randomly chosen initial conditions until all these initial conditions are distributed over the attractor (500 iterates were typically used). We then turn on the parameter perturbations and determine for each orbit how many further iterates τ are necessary before the orbit falls within a circle of radius $\tfrac{1}{2}\xi_*$ centered at the fixed point. We then calculate the average of these times. We do this for many different values of p_* and plot the results as a function of p_*. This is shown on the log-log plot in Figure 1 along with the theoretical straight line of slope given by the exponent (3). We see that the agreement is good although there are significant variations about the general power-law trend. These are to

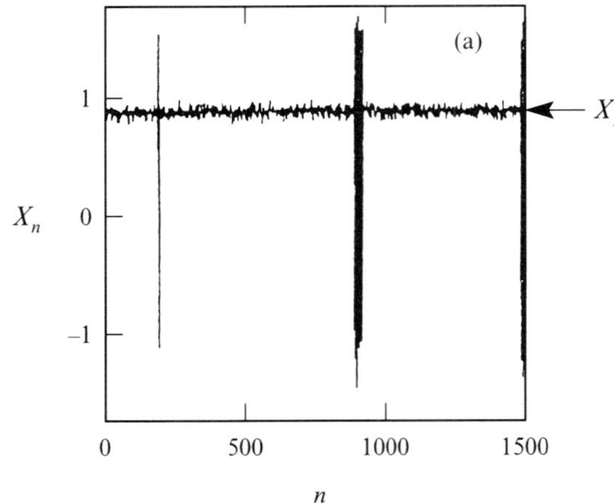

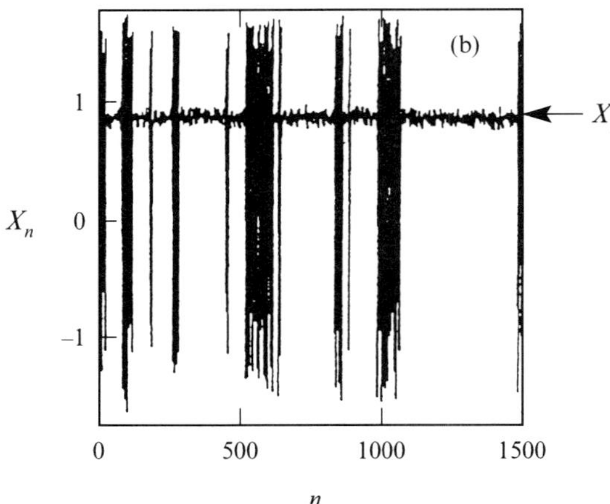

Figure 2 x_n vs n for two cases with the same realization of the random vector δ. $p_* = 0.2$ and $A_0 = 1.29$ for both cases. (a) $\varepsilon = 3.5 \times 10^{-2}$; (b) $\varepsilon = 3.8 \times 10^{-2}$.

be expected due to the fractal nature of the attractor and have also been seen in numerical calculations of the pointwise dimension for points on chaotic attractors (cf. Grebogi, Ott, and Yorke [1]).

Next, we consider the issue of noise. We add terms $\varepsilon\delta_{xn}$ and $\varepsilon\delta_{yn}$ to the right-hand sides of the Henon map equations. The random quantities δ_{xn} and δ_{yn} are independent of each other, have mean value 0 and mean-squared value 1 ($\langle\delta_x^2\rangle = \langle\delta_y^2\rangle = 1$), and have a Gaussian probability density. Figure 2 shows orbit plots, x_n vs n for 1500 iterates of the noisy map with parameter perturbations given by (2), for two different noise levels and p_* held fixed at $p_* = 0.2$. As predicted the

orbit stays near the fixed point with occasional bursts into the region far from $\xi = 0$, and these bursts are less frequent for small noise levels.

In conclusion, we have shown that there is great inherent flexibility in situations in which the dynamical motion is on a chaotic attractor. In particular, by using only small (carefully chosen) parameter perturbations it is possible to create a large variety of attracting periodic motions and to choose amongst these periodic motions the most desirable [8].

This research was supported by the U.S. Department of Energy (Scientific Computing Staff Office of Energy Research). The computation was done at the National Energy Research Supercomputer Center.

Notes

[1] The periodic orbits are dense in the attractor (i.e., periodic orbits pass through any neighborhood, however small, of any point on the attractor). For discussions of the relation of ergodic properties of an attractor to its dense set of unstable periodic orbits, see, for example, C. Grebogi, E. Ott, and J.A. Yorke, *Phys. Rev. A* **37**, 1711 (1988); **36**, 3522 (1987); D. Auerbach *et al.*, *Phys. Rev. Lett.* **58**, 2387 (1987); H. Hata *et al.*, *Prog. Theor. Phys.* **78**, 511 (1987); A. Katok, *Publ. Math. IHES* **51**, 137 (1980); R. Bowen, *Trans. Am. Math. Soc.* **154**, 377 (1971).

[2] E. Ott, C. Grebogi, and J.A. Yorke, in *Chaos: Proceedings of a Soviet-American Conference* (American Institute of Physics, New York, 1990).

[3] F. Takens, in *Dynamical Systems and Turbulence*, edited by D. Rand and L.S. Young (Springer-Verlag, Berlin, 1981), p. 230.

[4] N.H. Packard *et al.*, *Phys. Rev. Lett.* **45**, 712 (1980).

[5] G.H. Gunaratne, P.S. Linsay, and M.J. Vinson, *Phys. Rev. Lett.* **63**, 1 (1989).

[6] D.P. Lathrop and E.J. Kostelich, 'The Characterization of an Experimental Strange Attractor by Periodic Orbits' (to be published).

[7] C. Grebogi, E. Ott, and J.A. Yorke, *Phys. Rev. Lett.* **57**, 1284 (1986); P. Romeiras, C. Grebogi, E. Ott, and J.A. Yorke, *Phys. Rev. A* **36**, 5365 (1987).

[8] The general problem of controlling chaotic systems, while clearly very important, has, so far, received almost no attention. Two exceptions (which are quite different from our approach) are the papers of Hubler (who typically requires large controlling signals) and Fowler [A. Hubler, *Helv. Phys. Acta* **62**, 343 (1989); T.B. Fowler, *IEEE Trans. Autom. Control* **34**, 201 (1989)].

Paper 2
Controlling chaotic dynamical systems

Filipe J. Romeiras[a,b], Celso Grebogi[a,c,d], Edward Ott[a,e,f] and W.P. Dayawansa[f,g]

[a]*Laboratory for Plasma Research, University of Maryland, College Park, MD 20742, USA*
[b]*Centro de Electrodinâmica (INIC) and Departamento de Matemática, Instituto Superior Técnico, 1096 Lisbon Codex, Portugal*
[c]*Department of Mathematics, University of Maryland, College Park, MD 20742, USA*
[d]*Institute for Physical Science and Technology, University of Maryland, College Park, MD 20742, USA*
[e]*Department of Physics, University of Maryland, College Park, MD 20742, USA*
[f]*Department of Electrical Engineering, University of Maryland, College Park, MD 20742, USA*
[g]*Systems Research Center, University of Maryland, College Park, MD 20742, USA*

Received 18 November 1991
Revised manuscript received 16 January 1992
Accepted 16 January 1992

We describe a method that converts the motion on a chaotic attractor to a desired attracting time periodic motion by making only small time dependent perturbations of a control parameter. The time periodic motion results from the stabilization of one of the infinite number of previously unstable periodic orbits embedded in the attractor. The present paper extends that of Ott, Grebogi and Yorke [*Phys. Rev. Lett.* **64** (1990) 1196], allowing for a more general choice of the feedback matrix and implementation to higher-dimensional systems. The method is illustrated by an application to the control of a periodically impulsively kicked dissipative mechanical system with two degrees of freedom resulting in a four-dimensional map (the 'double rotor map'). A key issue addressed is that of the dependence of the average time to achieve control on the size of the perturbations and on the choice of the feedback matrix.

1. Introduction

It is common for systems to evolve with time in a chaotic way. In practice, however, it is often desired that chaos be avoided and/or that the system be optimized with respect to some performance criterion. Given a system which behaves chaotically, one approach might be to make some large (and possibly costly) alteration in the system which completely changes its dynamics in such a way as to achieve the desired objectives. Here we assume that this avenue is not available. Thus we address the following question: Given a chaotic system, how can we obtain improved performance and achieve a desired attracting time-periodic motion by making only *small* controlling temporal perturbations in an accessible system parameter?

The key observation is that a chaotic attractor typically has embedded densely within it an infinite number of unstable periodic orbits [1–5]. In addition, chaotic attractors can also sometimes contain unstable steady states (e.g., the Lorenz attractor has such an embedded steady state). Since we wish to make only small controlling perturbations to the system, we do not envision creating new orbits with very different properties from the already existing orbits. Thus we seek to exploit the already existing unstable periodic orbits and unstable steady states. Our approach is as follows: We first determine some of the unstable low-period periodic orbits and unstable steady states that are embedded in the chaotic attractor. We then examine these orbits and choose one which yields improved system performance. Finally, we apply small controls so as to stabilize this already existing orbit.

Some comments concerning this method are the following:

(1) Before settling into the desired controlled orbit the trajectory experiences a chaotic transient whose expected duration diverges as the maximum allowed size of the control approaches zero.
(2) Small noise can result in occasional bursts in which the orbit wanders far from the controlled orbit.
(3) Controlled chaotic systems offer an advantage in flexibility in that any one of a number of different orbits can be stabilized by the small control, and the choice can be switched from one to another depending on the current desired system performance.

Although we describe the details only in the case of discrete time systems, this method is applicable in the continuous time case as well by considering the discrete time system obtained from the induced dynamics on a Poincaré section.

In order to illustrate the method we apply it to a periodically forced mechanical system (the kicked double rotor), which results in a four-dimensional map. Amongst the examples considered, we study cases where the unstable orbit of the uncontrolled system has two unstable eigenvalues and two stable eigenvalues, and the stabilization is achieved by variation of one control parameter characterizing the strength of the periodic forcing. The present paper generalizes our previous work [6] to the case of higher-dimensional systems [7] and also includes new material illustrating the effect of the choice of stabilization on the length of the chaotic transient experienced by the orbit before control is achieved. Other relevant references on the feedback stabilization of periodic or steady orbits embedded in chaotic attractors are the experiments of Ditto *et al.* [8], Singer *et al.* [9], and the paper of Fowler [10]. (Other works in the general field are listed in Ref. [11].)

The plan of the paper is as follows. In Section 2, we give an implementation of the method, initially developed in Ref. [6], by using the 'pole placement technique' [7, 12]. In particular, we address the problem of stabilization of periodic orbits with more than one unstable eigenvalue. We also discuss experimental implementation in the absence of an *a priori* mathematical system model and generalization of the method to deal with cases where delay coordinates embedding is used. In Section 3 we present some results for the control of the Hénon map [13], a two-dimensional system that is used as a paradigm in the study of dynamical systems; these results extend those given in Ref. [6] in directions relevant to our present study. In Section 4 we present results for the control of the double rotor map [14], a four-dimensional system that describes a particular impulsively periodically forced mechanical system. Finally, in Section 5 we present the main conclusions of the work.

2. Description of the method

2.1. Formulation

For the sake of simplicity we consider a discrete time dynamical system,

$$Z_{i+1} = F(Z_i, p), \qquad (2.1)$$

where $Z_i \in \mathbb{R}^n$, $p \in \mathbb{R}$ *and* F is sufficiently smooth in both variables. Here, p is considered a real parameter which is available for external adjustment but is restricted to lie in some small interval,

$$|p - \bar{p}| < \delta, \qquad (2.2)$$

around a nominal value $\bar{p}$. We assume that the nominal system (i.e. for $p = \bar{p}$) contains a chaotic attractor. Our objective is to vary the parameter p with time i in such a way that for almost all initial conditions in the basin of the chaotic attractor, the dynamics of the system converge onto a desired time periodic orbit contained in the attractor. The control strategy is the following. We will find a stabilizing local feedback control law which is defined on a neighborhood of the desired periodic orbit. This is done by considering the first order approximation of the system at the chosen unstable periodic orbit. Here we assume that this

approximation is stabilizable. Since stabilizability is a generic property of linear systems, this assumption is quite reasonable. The ergodic nature of the chaotic dynamics ensures that the state trajectory eventually enters into the neighborhood. Once inside, we apply the stabilizing feedback control law in order to steer the trajectory towards the desired orbit.

For simplicity we shall describe the method as applied to the stabilization of fixed points (i.e., period one orbits) of the map F. The consideration of periodic orbits of period larger than one is straightforward and is discussed in Section 2.5. Let $Z_*(p)$ denote an unstable fixed point on the attractor. For values of p close to $\bar{p}$ and in the neighborhood of the fixed point $Z_*(\bar{p})$ the map (2.1) can be approximated by the linear map

$$Z_{i+1} - Z_*(\bar{p}) = \mathbf{A}[Z_i - Z_*(\bar{p})] + B(p - \bar{p}), \qquad (2.3)$$

where $\mathbf{A}$ is an $n \times n$ Jacobian matrix and B is an n-dimensional column vector,

$$\mathbf{A} = \mathbf{D}_Z F(Z, p), \qquad (2.4)$$

$$B = \mathbf{D}_p F(Z, p), \qquad (2.5)$$

and these partial derivatives are evaluated at $Z = Z_*(\bar{p})$ and $p = \bar{p}$. We now introduce the time-dependence of the parameter p by assuming that it is a linear function of the variable Z_i of the form

$$p - \bar{p} = -K^{\mathrm{T}}[Z_i - Z_*(\bar{p})], \qquad (2.6)$$

The $1 \times n$ matrix K^{T} is to be determined so that the fixed point $Z_*(\bar{p})$ becomes stable. Substituting (2.6) into (2.3) we obtain

$$Z_{i+1} - Z_*(\bar{p}) = \left(\mathbf{A} - BK^{\mathrm{T}}\right)[Z_i - Z_*(\bar{p})], \qquad (2.7)$$

which shows that the fixed point will be stable provided the matrix $\mathbf{A} - BK^{\mathrm{T}}$ is asymptotically stable; that is, all its eigenvalues have modulus smaller than unity.

The solution to the problem of the determination of K^{T}, such that the eigenvalues of the matrix $\mathbf{A} - BK^{\mathrm{T}}$ have specified values, is well known from control systems theory and is called 'pole placement technique' (see, for example, Ogata [12]). We summarize the relevant results.

2.2. Review of the pole placement technique

The eigenvalues of the matrix $\mathbf{A} - BK^{\mathrm{T}}$ are called the 'regulator poles', and the problem of placing these poles at the desired locations by choosing K^{T} with $\mathbf{A}$ and B given is the 'pole placement problem'.

Pole placement problem. Determine the matrix K^{T} in such a way that the eigenvalues of the matrix $\mathbf{A} - BK^{\mathrm{T}}$ have specified (complex) values $\{\mu_1, \ldots, \mu_n\}$.

The following results [12] give a necessary and sufficient condition for a unique solution of the pole placement problem to exist, and also a method for obtaining it (Ackermann's method).

(1) The pole placement problem has a unique solution if and only if the $n \times n$ matrix

$$\mathbf{C} = (B : AB : A^2B : \cdots : A^{n-1}B),$$

is of rank n. ($\mathbf{C}$ is called the controllability matrix).
(2) The solution of the pole placement problem is given by

$$K^{\mathsf{T}} = (\alpha_n - a_n \ldots \alpha_1 - a_1)\mathbf{T}^{-1},$$

where $\mathbf{T} = \mathbf{CW}$, and

$$\mathbf{W} = \begin{pmatrix} a_{n-1} & a_{n-2} & \ldots & a_1 & 1 \\ a_{n-2} & a_{n-3} & \ldots & 1 & 0 \\ \vdots & \vdots & & \vdots & \vdots \\ a_1 & 1 & \ldots & 0 & 0 \\ 1 & 0 & \ldots & 0 & 0 \end{pmatrix}.$$

Here $\{a_1, \ldots, a_n\}$ are the coefficients of the characteristic polynomial of $\mathbf{A}$,

$$|s\mathbf{I} - \mathbf{A}| = s^n + a_1 s^{n-1} + \cdots + a_n,$$

and $\{\alpha_1, \ldots, \alpha_n\}$ are the coefficients of the desired characteristic polynomial of $\mathbf{A} - BK^{\mathsf{T}}$,

$$\Pi_{j=1}^{n}(s - \mu_j) = s^n + \alpha_1 s^{n-1} + \cdots + \alpha_n.$$

2.3. Control parameter

Our considerations so far are based on the linear equation (2.7) and therefore only apply in the local region near $Z_*(\bar{p})$. On the other hand, the limitation in the size of the parameter perturbations given by (2.2), when combined with (2.6), yields

$$\left| K^{\mathsf{T}}[Z_i - Z_*(\bar{p})] \right| < \delta. \tag{2.8}$$

This defines a slab of width $2\delta/|K^{\mathsf{T}}|$. We choose to activate the control according to (2.6) only for values of Z_i inside this slab, and we choose to leave the control parameter at its nominal value (i.e., $p = \bar{p}$) when Z_i is outside this slab. Other choices are possible.

In summary, the control is determined by

$$p - \bar{p} = -K^{\mathsf{T}}[Z_i - Z_*(\bar{p})]\, u(\delta - |K^{\mathsf{T}}[Z_i - Z_*(\bar{p})]|), \tag{2.9}$$

for arbitrary Z_i [not necessarily close to $Z_*(\bar{p})$], where u is the unit step function defined by

$$u(\alpha) = \begin{cases} 0, & \alpha < 0, \\ 1, & \alpha > 0. \end{cases}$$

At this stage it should be pointed out that the matrix K^{T} can be chosen in many different ways. In principle, any choice of regulator poles inside the unit circle serves our purpose. In Ref. [6], the authors made a very special, though quite natural, choice of the gain matrix K^{T}: the resulting value of $p - \bar{p}$ forces the orbit onto the (linear) stable manifold of the fixed point at each iteration. In terms of regulator poles this choice corresponds to setting n_s of these poles equal to the n_s stable eigenvalues of matrix $\mathbf{A}$ and the remaining $n - n_s$ to 0. In terms of the slab (2.8) this

choice corresponds not only to orientating it parallel to the stable manifold but also taking an appropriate width.

The choice of the matrix K^{T} will be discussed at some length in our applications of the method in Sections 3 and 4.

2.4. Time to achieve control

The control is activated (i.e., $p \neq \bar{p}$) only if Z_i falls in the narrow slab (2.8). Thus, for small δ, a typical initial condition will execute a chaotic orbit, unchanged from the uncontrolled case, until Z_i falls in this slab. Even then, because of nonlinearity not included in the linearized equation (2.7), the control may not be able to bring the orbit to the fixed point. In this case the orbit will leave the slab and continue to wander chaotically as if there was no control. Since the orbit on the uncontrolled chaotic attractor is ergodic, at some time it will eventually satisfy (2.8) and also be sufficiently close to the desired fixed point so that control is achieved.

Thus, we create a stable orbit, which, for a typical initial condition, is preceded by a chaotic transient [15–18] in which the orbit is similar to orbits on the uncontrolled chaotic attractor. The length τ of such chaotic transient depends sensitively on the initial condition of the particular orbit. For initial conditions randomly chosen in the basin of attraction the distribution of chaotic transient lengths is exponential [15, 16],

$$\phi(\tau) \simeq \frac{1}{\langle \tau \rangle} \exp\!\left(-\frac{\tau}{\langle \tau \rangle}\right), \tag{2.10}$$

for large τ. The quantity $\langle \tau \rangle$ is the characteristic length of the chaotic transient, called in the present case the *average time to achieve control*. Estimates of the scaling of $\langle \tau \rangle$ with δ for small δ are given in Appendix A for the case of two-dimensional maps.

2.5. Control of periodic orbits of period greater than one

The analysis of periodic orbits given in Sections 2.1–2.3 can be extended to nontrivial periodic orbits (i.e., orbits with period greater than one). The most direct way is to take the Tth iterate of the map, where T denotes the period of the orbit to be stabilized. For the T times iterated map, any point on the periodic orbit is a fixed point, and we can then apply the discussion Sections 2.1–2.3. This method is, however, overly sensitive to noise, especially when long period periodic orbits are involved. Next we outline another method which we believe should, in general, be better. In terms of the treatment of Section 2.2, the prescription we give below corresponds to placing the unstable eigenvalues of the uncontrolled problem at zero, while leaving the stable eigenvalues unchanged. (This is only one of many possibilities that could be given.)

We denote the periodic orbit by $Z_{i^*}(p)$, where $Z_{(i+T)^*}(p) = Z_{i^*}(p)$. In addition, we introduce the set of T matrices $\mathbf{A}_i$

which are $n \times n$ and the set of T column vectors $\boldsymbol{B}_i$ which are of dimension n, where

$$\boldsymbol{A}_i = \boldsymbol{A}_{i+T} = \boldsymbol{D}_Z \boldsymbol{F}(\boldsymbol{Z}, p),$$

$$\boldsymbol{B}_i = \boldsymbol{B}_{i+T} = \boldsymbol{D}_p \boldsymbol{F}(\boldsymbol{Z}, p),$$

and the partial derivatives are evaluated at $\boldsymbol{Z} = \boldsymbol{Z}_{i^*}(\bar{p})$ and $p = \bar{p}$.

Linearizing as in Eq. (2.3), we have

$$\boldsymbol{Z}_{i+1} - \boldsymbol{Z}_{(i+1)^*}(\bar{p})$$
$$= \boldsymbol{A}_i[\boldsymbol{Z}_i - \boldsymbol{Z}_{i^*}(\bar{p})] + \boldsymbol{B}_i(p + \bar{p}). \tag{2.11}$$

Say that the periodic orbit has u unstable eigenvalues (i.e., u eigenvalues with magnitude greater than one) and s stable eigenvalues, where $u + s = n$. At each point $\boldsymbol{Z}_{i^*}(\bar{p})$ on the $p = \bar{p}$ periodic orbit, determine vectors $\{\boldsymbol{v}_{i,1}, \boldsymbol{v}_{i,2}, \ldots, \boldsymbol{v}_{i,s}\}$ which span the linearized stable subspace. Now let

$$\boldsymbol{\phi}_{i,j} = \boldsymbol{A}_{i+u-1} \boldsymbol{A}_{i+u-2} \cdots \boldsymbol{A}_{i+j+1} \boldsymbol{A}_{i+j},$$

for $j = 1, 2, \ldots, (u-1)$ and

$$\boldsymbol{C}_i = \left(\boldsymbol{\phi}_{i,1} \boldsymbol{B}_i \vdots \boldsymbol{\phi}_{i,2} \boldsymbol{B}_{i+1} \vdots \cdots \vdots \boldsymbol{\phi}_{i,u-1} \boldsymbol{B}_{i+u-2} \vdots \boldsymbol{B}_{i+u-1} \right.$$
$$\left. \vdots \boldsymbol{v}_{i+u,1} \vdots \boldsymbol{v}_{i+u,2} \vdots \cdots \vdots \boldsymbol{v}_{i+u,s} \right)$$

(One choice of the vectors $\{\boldsymbol{v}_{i,1}, \boldsymbol{v}_{i,2}, \ldots, \boldsymbol{v}_{i,s}\}$ is the stable eigenvectors of $\boldsymbol{A}_i \boldsymbol{A}_{i-1} \cdots \boldsymbol{A}_{i-T+1}$.) The controllability condition (analogous to that in Section 2.2) is that $\boldsymbol{C}_i$ be nonsingular. The desired result for the control is then specified by

$$p_i - \bar{p} = -\boldsymbol{K}_i^{\mathrm{T}}[\boldsymbol{Z}_i - \boldsymbol{Z}_{i^*}(\bar{p})], \tag{2.12a}$$

where

$$\boldsymbol{K}_i^{\mathrm{T}} = \kappa \boldsymbol{C}_i^{-1} \boldsymbol{\phi}_{i,0}, \tag{2.12b}$$

and κ denotes an n-dimensional row vector whose first entry is one and all of whose remaining entries are zeros.

To derive Eqs. (2.12) we iterate (2.11) u times,

$$\boldsymbol{Z}_{i+u} - \boldsymbol{Z}_{(i+u)^*}(\bar{p}) = \boldsymbol{\phi}_{i,0}[\boldsymbol{Z}_i - \boldsymbol{Z}_{i^*}(\bar{p})]$$
$$+ \boldsymbol{\phi}_{i,1} \boldsymbol{B}_i(p_i - \bar{p}) + \boldsymbol{\phi}_{i,2} \boldsymbol{B}_{i+1}(p_{i+1} - \bar{p})$$
$$+ \cdots + \boldsymbol{B}_{i+u-1}(p_{i+u-1} - \bar{p}). \tag{2.13a}$$

We then demand that $\boldsymbol{Z}_{i+u}$ land on the linearized stable manifold of the periodic orbit through the point $\boldsymbol{Z}_{(i+u)^*}(\bar{p})$. That is, we choose the p's such that there exist s coefficients $\alpha_1, \alpha_2, \ldots, \alpha_s$ such that

$$\boldsymbol{Z}_{i+u} - \boldsymbol{Z}_{(i+u)^*}(\bar{p}) = \alpha_1 \boldsymbol{v}_{i+u,1} + \alpha_2 \boldsymbol{v}_{i+u,2} + \cdots + \alpha_s \boldsymbol{v}_{i+u,s} \tag{2.13b}$$

Regarding (2.13a) and (2.13b) as $n = u + s$ equations in the n unknowns, $p_i, p_{i+1}, \ldots, p_{i+u-1}, \alpha_1, \alpha_2, \ldots, \alpha_s$, we then solve for p_i to obtain (2.12).

(Note from the above that at time i we could, once and for all, calculate all the control parameter values to be applied in the next u iterates, $p_i, p_{i+1}, \ldots, p_{i+u-1}$. In the presence of noise, however, this is not a good idea (assuming $u > 1$), since it does not take advantage of the opportunity to correct for the noise on each iterate. Therefore, we believe that, in the presence of noise, it is best to perform the calculation of p_i via Eq. (2.12) on each iterate.)

2.6. *Use of delay coordinates*

In experimental studies of chaotic dynamical systems, delay coordinates are often used to represent the system state. This is sometimes useful because it only requires measurement of the time series of a *single* scalar state variable which, we denote $\xi(t)$. A delay coordinate vector can be formed as follows:

$$\boldsymbol{Z}(t) = (\xi(t), \xi(t - T_{\mathrm{D}}), \xi(t - 2T_{\mathrm{D}}), \ldots, \xi(t - MT_{\mathrm{D}})),$$

where T_{D} is some conveniently chosen delay time, and the time variable t is assumed continuous. Embedding theorems guarantee that for $M \geq 2n$, where n is the system dimensionality, the vector $\boldsymbol{Z}$ is generically a global one-to-one representation of the system state. (Actually, for our purposes, we do not require a global embedding; we only require $\boldsymbol{Z}$ to be one-to-one in the small region near the periodic orbit, and this can typically be achieved with $M = n - 1$.) To obtain a map, one can take a Poincaré surface of section. For the often encountered case of a system which is periodically forced at a period T_{F}, one can define a 'stroboscopic surface of section' by sampling the state at discrete times $t_i = iT_{\mathrm{F}} + t_0$. In this case we have the discrete state variable

$$\boldsymbol{Z}_i = \boldsymbol{Z}(t_i).$$

As pointed out by Dressler and Nitsche [19], in the presence of parameter variation, delay coordinates lead to a map of a different form than

$$\boldsymbol{Z}_{i+1} = \boldsymbol{F}(\boldsymbol{Z}_i, p_i),$$

which is the form assumed in Sections 2.1–2.5. For example, in the periodically forced case, since the components of $\boldsymbol{Z}_i$ are $\xi(t_i - mT_{\mathrm{D}})$ for $m = 0, 1, \ldots, M$, the vector $\boldsymbol{Z}_{i+1}$ must depend not only on p_i, but also on all previous values of the parameter that were in effect during the time interval $t_i \leq t \leq t_i - MT_{\mathrm{D}}$. In particular, let r be the smallest integer such that $MT_{\mathrm{D}} < rT_{\mathrm{F}}$. Then the relevant map is in general of the form

$$\boldsymbol{Z}_{i+1} = \boldsymbol{G}(\boldsymbol{Z}_i, p_i, p_{i-1}, \ldots, p_{i-r}). \tag{2.14a}$$

For $r = 1$ we have

$$\boldsymbol{Z}_{i+1} = \boldsymbol{G}(\boldsymbol{Z}_i, p_i, p_{i-1}). \tag{2.14b}$$

We now discuss how the technique of Section 2.2 can be applied in the case of delay coordinates, and, for simplicity, we limit the discussion to $r = 1$, Eq. (2.14b). Linearizing as in Eq. (2.3) and again restricting our attention to the case of a fixed point orbit, we have

$$\boldsymbol{Z}_{i+1} - \boldsymbol{Z}_*(\bar{p}) = \boldsymbol{A}[\boldsymbol{Z}_i - \boldsymbol{Z}_*(\bar{p})] + \boldsymbol{B}_a(p_i - \bar{p}) + \boldsymbol{B}_b(p_{i-1} - \bar{p}), \tag{2.15}$$

where $\boldsymbol{A} = \boldsymbol{D}_Z \boldsymbol{G}(\boldsymbol{Z}, p, p')$, $\boldsymbol{B}_a = \boldsymbol{D}_p \boldsymbol{G}(\boldsymbol{Z}, p, p')$, and $\boldsymbol{B}_b = \boldsymbol{D}_{p'} \boldsymbol{G}(\boldsymbol{Z}, p, p')$, and all partial derivatives are evaluated at $\boldsymbol{Z} = \boldsymbol{Z}_*(\bar{p})$ and $p = \bar{p} = p'$.

Now define a new state variable with one extra component by

$$\tilde{\boldsymbol{Z}}_{i+1} = \begin{pmatrix} \boldsymbol{Z}_{i+1} \\ p_i \end{pmatrix}, \tag{2.16}$$

and introduce the linear control law,

$$p_i - \bar{p} = -\boldsymbol{K}^{\mathrm{T}}[\boldsymbol{Z}_i - \boldsymbol{Z}_*(\bar{p})] - k(p_{i-1} - \bar{p}), \qquad (2.17)$$

Combining these equations, we obtain

$$\tilde{\boldsymbol{Z}}_{i+1} - \tilde{\boldsymbol{Z}}_*(\bar{p}) = (\tilde{\boldsymbol{A}} - \tilde{\boldsymbol{B}}\tilde{\boldsymbol{K}}^{\mathrm{T}})[\tilde{\boldsymbol{Z}}_i - \tilde{\boldsymbol{Z}}_*(\bar{p})], \qquad (2.18)$$

where

$$\tilde{\boldsymbol{Z}}_*(\bar{p}) = \begin{pmatrix} \boldsymbol{Z}_*(\bar{p}) \\ \bar{p} \end{pmatrix}, \quad \tilde{\boldsymbol{A}} = \begin{pmatrix} \boldsymbol{A} & \boldsymbol{B}_\mathrm{b} \\ \boldsymbol{0} & 0 \end{pmatrix}, \quad \tilde{\boldsymbol{B}} = \begin{pmatrix} \boldsymbol{B}_\mathrm{a} \\ 1 \end{pmatrix}, \quad \tilde{\boldsymbol{K}} = \begin{pmatrix} \boldsymbol{K} \\ k \end{pmatrix}.$$

Since (2.18) is now of the same form as (2.3), the method of Section 2.2 can be applied. (A similar result for any $r > 1$ also clearly holds.)

Another method of control for delay coordinates is to reduce (2.14b) directly to the form $\boldsymbol{Z}_{i+1} = \boldsymbol{F}(\boldsymbol{Z}_i, p_i)$ and then proceed as in Sections 2.1 and 2.2. This reduction can be done by setting $p_i \equiv \bar{p}$ for every other time step. For example, say $p_i = 0$ for i odd, and $j = \tfrac{1}{2}i$ for even i. Then making the replacements $\boldsymbol{Z}_i \to \hat{\boldsymbol{Z}}_j$, $p_i \to \hat{p}_j$ for even i, and iterating (2.14b) twice we have

$$\hat{\boldsymbol{Z}}_{j+1} = \boldsymbol{G}[\boldsymbol{G}(\hat{\boldsymbol{Z}}_j, \bar{p}, \hat{p}_j), \hat{p}_j, \bar{p}]$$
$$\equiv \hat{\boldsymbol{F}}(\hat{\boldsymbol{Z}}_j, \hat{p}_j), \qquad (2.19)$$

which is of the required form. We believe, however, that the first method we have given (i.e., that based on Eq. (2.18)) should usually be capable of yielding superior results to the method based on (2.19) with respect to noise sensitivity and time to achieve control. This is because our second method does not take advantage of the opportunity to control on each time iterate while our first method does.

3. Controlling the Hénon map

In Ref. [6], the authors used the Hénon map to illustrate the control method and, in particular, to test their theoretical predictions concerning the average time to achieve control. As already pointed out, their work is based on a particular choice of the gain matrix $\boldsymbol{K}^{\mathrm{T}}$. In this section we consider how different choices of $\boldsymbol{K}^{\mathrm{T}}$ affect the average time to achieve control for the Hénon map.

The Hénon map [13] is the two-dimensional map

$$\boldsymbol{Z} \mapsto \boldsymbol{Z}' = \boldsymbol{F}(\boldsymbol{Z}),$$

defined by

$$\begin{pmatrix} x \\ y \end{pmatrix} \mapsto \begin{pmatrix} x' \\ y' \end{pmatrix} = \begin{pmatrix} a - x^2 + by \\ x \end{pmatrix},$$

where $(x, y) \in \mathbb{R} \times \mathbb{R}$. We keep the parameter b fixed throughout ($b = 0.3$) and allow the control parameter a to vary around a nominal value $\bar{a}$ ($\bar{a} = 1.4$) for which the map has a chaotic attractor.

For $a = \bar{a} = 1.4$ there is an unstable saddle fixed point contained in the chaotic attractor. This fixed point is located at

$$\boldsymbol{Z}_*(a) = x_*(a)\begin{pmatrix} 1 \\ 1 \end{pmatrix},$$

$$x_*(a) = -c + (c^2 + a)^{1/2}, \quad c = \tfrac{1}{2}(1 - b),$$

for $a \geq -c^2$. Noting that the Jacobian matrix of partial derivatives of the map is

$$\boldsymbol{D}_z\boldsymbol{F}(\boldsymbol{Z}) = \begin{pmatrix} -2x & b \\ 1 & 0 \end{pmatrix},$$

and that the stability of the fixed point is determined by the roots of the characteristic equation

$$|\boldsymbol{D}_z\boldsymbol{F}[\boldsymbol{Z}_*(a)] - s\boldsymbol{1}| = 0,$$

one can easily check that the fixed point is stable for $-c^2 < a < 3c^2$ and unstable for $a > 3c^2$ (Hence the fixed point is unstable for $b = 0.3$, $a = \bar{a} = 1.4$ since $c = 0.35$.)

The quantities that appear in Section 2.2 are as follows:

$$\boldsymbol{A} = \begin{pmatrix} -2\bar{x}_* & b \\ 1 & 0 \end{pmatrix}, \quad \boldsymbol{B} = \begin{pmatrix} 1 \\ 0 \end{pmatrix}, \quad \boldsymbol{C} = (\boldsymbol{B} : \boldsymbol{AB}) = \begin{pmatrix} 1 & -2\bar{x}_* \\ 0 & 0 \end{pmatrix},$$

$$\boldsymbol{W} = \begin{pmatrix} 2\bar{x}_* & 1 \\ 1 & 0 \end{pmatrix}, \quad \boldsymbol{T} = \begin{pmatrix} 0 & 1 \\ 1 & 0 \end{pmatrix}, \quad \boldsymbol{K}^{\mathrm{T}} = (\alpha_1 - a_1 \quad \alpha_2 - a_2),$$

where

$$a_1 = 2\bar{x}_* = -(\lambda_\mathrm{u} + \lambda_\mathrm{s}), \quad a_2 = -b = \lambda_\mathrm{u}\lambda_\mathrm{s},$$

and

$$\alpha_1 = -(\mu_1 + \mu_2), \quad \alpha_2 = \mu_1\mu_2.$$

Here $\bar{x}_* = \bar{x}_*(\bar{a})$, and λ_u and λ_s are the eigenvalues of matrix $\boldsymbol{A}$,

$$\left.\begin{matrix} \lambda_\mathrm{s} \\ \lambda_\mathrm{u} \end{matrix}\right\} = -\bar{x}_* \pm (\bar{x}_*^2 + b)^{1/2}.$$

The quantities μ_1 and μ_2 are the regulator poles (i.e., the eigenvalues of $(\boldsymbol{A} - \boldsymbol{B}\boldsymbol{K}^{\mathrm{T}})$).

In order to better illustrate the different choices of regulator poles or, equivalently, of the matrix $\boldsymbol{K}^{\mathrm{T}}$, we have used the plane (α_1, α_2) (cf. Figure 1). In this plane we have plotted the lines of marginal stability $\mu_i = \pm 1$ ($1 \pm \alpha_1 + \alpha_2 = 0$) and $\mu_1\mu_2 = 1$ ($\alpha_2 = 1$); the bounded triangular region delimited by these lines (shown shaded in the figure) is the region where the regulator poles are stable. In addition, we

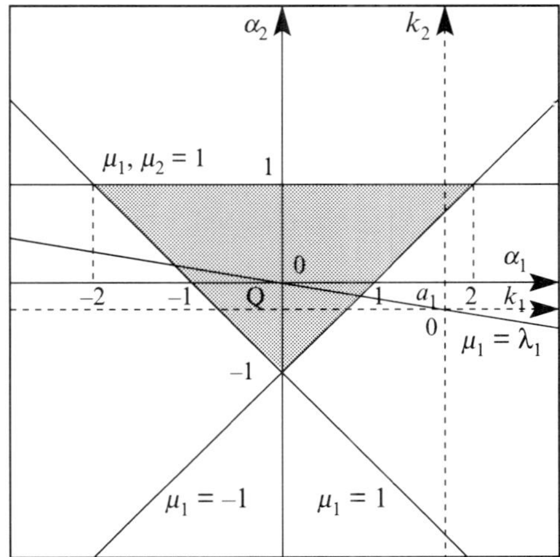

Figure 1 Hénon map: choice of regulator poles.

have plotted as dashed lines the axes $(k_1, k_2) = \boldsymbol{K}^\mathrm{T}$ which are related to (α_1, α_2) by the translations

$$k_1 = \alpha_1 - a_1, \quad k_2 = \alpha_2 - a_2.$$

The straight solid line in the figure going through the origin of the (k_1, k_2) plane has slope $-\lambda_s$ and intersects the line $\alpha_2 = 0$ at the point Q with coordinates $(\alpha_1, \alpha_2) = (-\lambda_s, 0)$. To this point corresponds the regulator poles

$$\mu_1 = 0, \quad \mu_2 = \lambda_s,$$

and the matrix

$$\boldsymbol{K}^\mathrm{T} = \lambda_u(1 \quad -\lambda_s) \equiv \boldsymbol{K}_Q^\mathrm{T}.$$

$\boldsymbol{K}_Q^\mathrm{T}$ is the special choice of matrix $\boldsymbol{K}^\mathrm{T}$ made in Ref. [6].

Before proceeding with the discussion, it is convenient to express the vector $\boldsymbol{K}^\mathrm{T}$ in polar coordinates

$$\boldsymbol{K}^\mathrm{T} = |\boldsymbol{K}^\mathrm{T}|(\cos\theta, \sin\theta).$$

We consider the following two ways of varying the vector $\boldsymbol{K}^\mathrm{T}$ (inside the triangular region of stability):

(I) θ fixed, $|\boldsymbol{K}^\mathrm{T}|$ variable;
(II) $|\boldsymbol{K}^\mathrm{T}|$ fixed, θ variable.

In terms of the control slab defined by Eq. (2.8) we have that in situation (I) the slab is kept oriented in a fixed direction while its width $w = 2\delta/|\boldsymbol{K}^\mathrm{T}|$ varies, whereas in situation (II) the direction of the slab is rotated while its width is kept fixed at $w = 2\delta/|\boldsymbol{K}_Q^\mathrm{T}|$. The choice of the $\boldsymbol{K}^\mathrm{T}$ in Ref. [6] has $\theta = \theta_Q \equiv \tan^{-1}(-\lambda_s)$ and, as we shall see, this choice is optimal from the point of view of the time to achieve control. (To see that the choice of Ref. [6] corresponds to $\theta = \theta_Q$, we note that with this choice one obtains a convergence rate to the periodic orbit of μ_s as in Ref. [6].)

In the numerical experiments we calculated the average time to achieve control by the method described in Appendix B. We also allowed for different values of the maximum amplitude of the parameter perturbations, δ.

First we consider the case where θ is fixed (case I) at the value

$$\theta = \theta_Q.$$

This case has a simple interpretation in terms of regulator poles: $\mu_2 = \lambda_s$ is kept fixed while μ_1 is allowed to vary between -1 and $+1$. μ_1 and $|\boldsymbol{K}^\mathrm{T}|$ are related by

$$|\boldsymbol{K}^\mathrm{T}| = |\mu_1 - \lambda_u|(1 + \lambda_s^2)^{1/2}.$$

Figure 2 shows results for $\langle\tau\rangle$ for this case. We see that the average time to achieve control increases with μ_1, although only moderately.

Figure 3 shows results for $\langle\tau\rangle$ versus θ for $|\boldsymbol{K}|^\mathrm{T}$ held fixed (case II) at

$$|\boldsymbol{K}^\mathrm{T}| = |\boldsymbol{K}_Q^\mathrm{T}| = |\lambda_u|(1 + \lambda_s^2)^{1/2}.$$

We see that the average time to achieve control has a strong minimum at $\theta = \theta_Q$.

Figure 4 shows $\langle\tau\rangle$ versus $|\boldsymbol{K}^\mathrm{T}|$ for three values of θ, $\theta = \theta_0$, $\theta = \theta_Q$, and $\theta = \theta_1$, where $\theta_0 < \theta_Q < \theta_1$ and θ_0 and θ_1 are close to θ_Q ($\theta_0 = 170.4°$, $\theta_Q = 171.1°$, $\theta_1 = 172.0°$). We observe that the $\theta = \theta_Q$ result is always below the results for $\theta = \theta_0$ and $\theta = \theta_1$ indicating that the average time to achieve

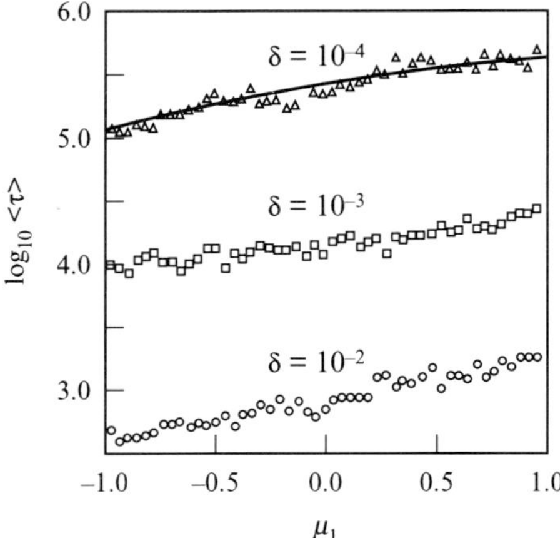

Figure 2 Hénon map: $\log_{10}\langle\tau\rangle$ versus μ_1, with $\mu_2 = \lambda_s$, for ($\bigcirc$) $\delta = 10^{-2}$, ($\square$) $\delta = 10^{-3}$, ($\triangle$) $\delta = 10^{-4}$. The theoretical curve was calculated using Eq. (A.9) of Appendix A. ($\bar{a} = 1.4$, $b = 0.3$).

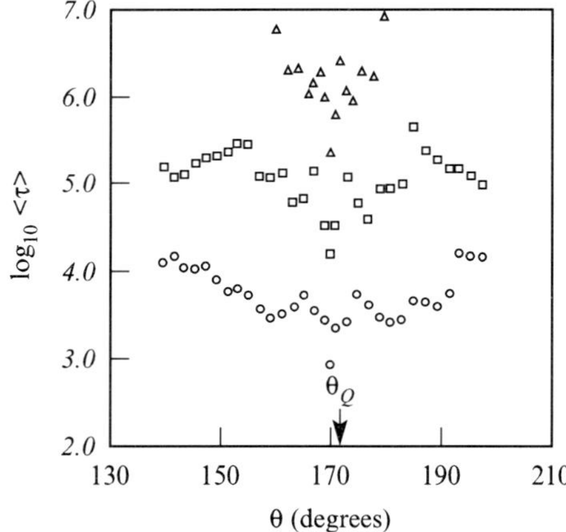

Figure 3 Hénon map: $\log_{10}\langle\tau\rangle$ versus θ, with $|\boldsymbol{K}^\mathrm{T}| = |\boldsymbol{K}_Q^\mathrm{T}|$, for ($\bigcirc$) $\delta = 10^{-2}$, ($\square$) $\delta = 10^{-3}$, ($\triangle$) $\delta = 10^{-4}$ ($\bar{a} = 1.4$, $b = 0.3$).

control has a strong minimum at $\theta = \theta_Q$ not only for $|\boldsymbol{K}^\mathrm{T}| = |\boldsymbol{K}_Q^\mathrm{T}|$ but for all values of $|\boldsymbol{K}^\mathrm{T}|$. Thus the condition $\theta = \theta_Q$ is optimal.

In Appendix A we show how the average time to achieve control can be obtained theoretically in the case of two-dimensional maps and verify that there is excellent agreement between the theoretical and experimental results in the case of the Hénon map.

4. Controlling the double rotor

In this section we apply the control method described in Section 2 to a dynamical system known as the double rotor map. We start by deriving the map (Section 4.1 and

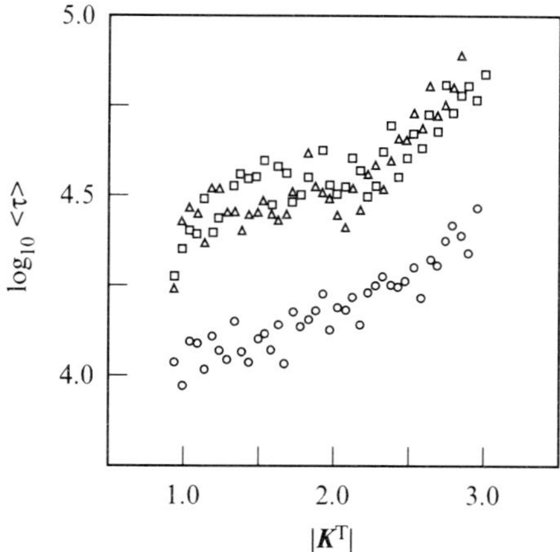

Figure 4 Hénon map: $\log_{10}\langle\tau\rangle$ versus $|K^{\mathrm{T}}|$ for ($\bigcirc$) $\theta = \theta_Q$, ($\square$) $\theta = \theta_0$, ($\triangle$) $\theta = \theta_1$ ($\theta_0 < \theta_Q < \theta_1$; $\theta_0 = \tan^{-1}[-\lambda_u\lambda_s/(\lambda_u + \lambda_s)]$) $= 170.4°$, $\theta_Q = \tan^{-1}(-\lambda_s) = 171.1°$, $\theta_1 = 172.0°$), and $\delta = 10^{-3}$.

Appendix B), then study its fixed points (Section 4.2) and its attractors (Section 4.3), including chaotic ones, and finally proceed to control some of the fixed points embedded in one of the chaotic attractors (Sections 4.4 and 4.5).

4.1. The double rotor map

The double rotor map is a four-dimensional map which describes the time evolution of a mechanical system known as the kicked double rotor [14]. This system is a four-dimensional extension of the kicked (single) rotor, a two-dimensional system that is described by the well-known dissipative standard map [20].

The double rotor is composed of two thin, massless rods connected as shown in Figure 5. The first rod, of length l_1, pivots about P_1 (which is fixed), and the second rod, of length $2l_2$, pivots about P_2 (which moves). The angles $\theta_1(t)$, $\theta_2(t)$ specify the orientations at time t of the first and second rods, respectively. A mass m_1 is attached at P_2, and masses $\frac{1}{2}m_2$ are attached to each end of the second rod (P_3 and P_4). Friction at P_1 (with coefficient v_1) slows the first rod at a rate proportional to its angular velocity $\dot\theta_1(t) \equiv d\theta_1(t)/dt$; friction at P_2 (with coefficient v_2) slows the second rod (and simultaneously accelerates the first rod) at a rate proportional to $\dot\theta_2(t) - \dot\theta_1(t)$. The end of the second rod marked P_3 receives periodic impulse kicks at times $t = T$, $2T, \ldots$, always from the same direction and with constant strength f_0. There is no gravity.

In Appendix C we write the differential equations that describe the kicked double rotor and proceed to derive from them the *double rotor map* relating the state of the system just after consecutive kicks. We obtain the four-dimensional map

$$Z \mapsto Z' = F(Z),$$

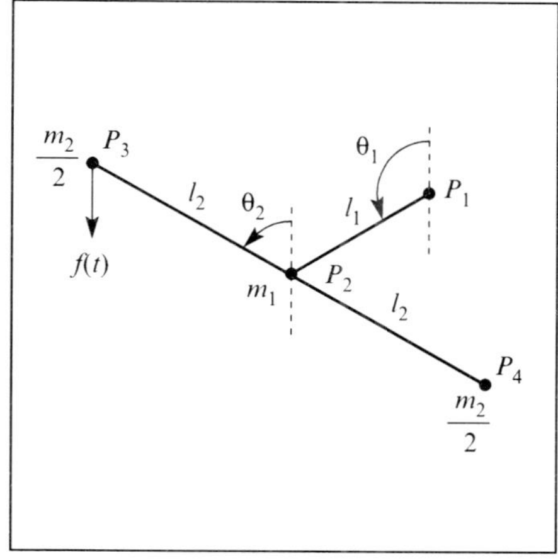

Figure 5 The double rotor.

defined by

$$\begin{pmatrix} X \\ Y \end{pmatrix} \mapsto \begin{pmatrix} X' \\ Y' \end{pmatrix} = \begin{pmatrix} \mathbf{M}Y + X \\ \mathbf{L}Y + G(X') \end{pmatrix}, \tag{4.1}$$

where

$$X = \begin{pmatrix} x_1 \\ x_2 \end{pmatrix} \in S^1 \times S^1, \quad Y = \begin{pmatrix} y_1 \\ y_2 \end{pmatrix} \in \mathbb{R} \times \mathbb{R},$$

and

$$G(X') = \begin{pmatrix} c_1 \sin x'_1 \\ c_2 \sin x'_2 \end{pmatrix}. \tag{4.2}$$

x_1, x_2 are the angular positions of the rods at the instant of the kth kick, $x_j = \theta_j(kT)$, while y_1, y_2 are the angular velocities of the rods immediately after the kth kick, $y_j = \dot\theta_j$ (kT^+). S^1 is the circle $\mathbb{R}(\mathrm{mod}\ 2\pi)$. $\mathbf{L}$ and $\mathbf{M}$ are constant 2×2 matrices defined by

$$\mathbf{L} = \sum_{j=1}^{2} \mathbf{W}_j e^{\lambda_j T}, \quad \mathbf{M} = \sum_{j=1}^{2} \mathbf{W}_j \frac{e^{\lambda_j T} - 1}{\lambda_j},$$

$$\mathbf{W}_1 = \begin{pmatrix} a & b \\ b & d \end{pmatrix}, \quad \mathbf{W}_2 = \begin{pmatrix} d & -b \\ -b & a \end{pmatrix},$$

$$a = \frac{1}{2}\left(1 + \frac{v_1}{\Delta}\right), \quad d = \frac{1}{2}\left(1 - \frac{v_1}{\Delta}\right), \quad b = -\frac{v_2}{\Delta},$$

$$\left.\begin{array}{c} \lambda_1 \\ \lambda_2 \end{array}\right\} = -\tfrac{1}{2}(v_1 + 2v_2 \pm \Delta), \quad \Delta = (v_1^2 + 4v_2^2)^{1/2}.$$

Finally, c_1 and c_2 are given by

$$c_j = \frac{f_0}{I}\, l_j, \quad j = 1, 2,$$

where

$$I = (m_1 + m_2)l_1^2 = m_2 l_2^2.$$

The following relation between matrices $\mathbf{L}$ and $\mathbf{M}$ will be useful below:

$$\mathbf{L} = \mathbf{I} + \mathbf{A}_v\mathbf{M}, \tag{4.3}$$

where

$$\mathbf{A}_v = \begin{pmatrix} -(v_1 + v_2) & v_2 \\ v_2 & -v_2 \end{pmatrix}.$$

(λ_1, λ_2 are precisely the eigenvalues of $\mathbf{A}_v$.) Note also that

$$|\mathbf{L}| = e^{(\lambda_1 + \lambda_2)T},$$

$$|\mathbf{M}| = \frac{e^{\lambda_1 T} - 1}{\lambda_1} \frac{e^{\lambda_2 T} - 1}{\lambda_2},$$

$$|\mathbf{A}_v| = v_1 v_2.$$

From now on we assume that $v_1 = v_2 \equiv v$. This leads to

$$\left.\begin{array}{c}\lambda_1 \\ \lambda_2\end{array}\right\} = -\tfrac{1}{2}v\left(3 \pm \sqrt{5}\right),$$

$$\left.\begin{array}{c}a \\ d\end{array}\right\} = \tfrac{1}{2}\left(1 \pm \tfrac{1}{5}\sqrt{5}\right),$$

$$b = -\tfrac{1}{5}\sqrt{5}.$$

In all the numerical work described in the rest of this section the parameters v, T, I, m_1, m_2, l_1, and l_2 were kept fixed at the values

$$v = T = I = m_1 = m_2 = l_2 = 1,$$
$$l_1 = 1/\sqrt{2}. \tag{4.4}$$

The only parameter which we shall vary is the forcing f_0 used as the control parameter.

4.2. Fixed points of the double rotor map

The fixed points $\mathbf{Z}_* = (\mathbf{X}_*, \mathbf{Y}_*)$ of the map (4.1) are solutions of the system

$$\mathbf{X}_* = \mathbf{M}\mathbf{Y}_* + \mathbf{X}_* - 2\pi\mathbf{N},$$
$$\mathbf{Y}_* = \mathbf{L}\mathbf{Y}_* + \mathbf{G}(\mathbf{X}_*), \tag{4.5}$$

where the components of the vector $\mathbf{N} = (n_1, n_2)$ are integer and are the rotation numbers in the x_1, x_2 variables. The rotation numbers n_1, n_2 are defined as the multiples of 2π by which x_{1*}, x_{2*} are increased in one iteration of the map before being brought to the interval $[0, 2\pi]$. From Eqs. (4.5) we obtain, using (4.3).

$$\mathbf{Y}_* = 2\pi\mathbf{M}^{-1}\mathbf{N},$$
$$\mathbf{G}(\mathbf{X}_*) = -2\pi\mathbf{A}_v\mathbf{N}. \tag{4.6}$$

Using the definitions of the matrices $\mathbf{G}$ and $\mathbf{A}_v$ we rewrite the second of the Eqs. (4.6) in the form

$$\begin{pmatrix}\sin x_{1*} \\ \sin x_{2*}\end{pmatrix} = -\frac{2\pi v l}{f_0}\begin{pmatrix}(1/l_1)(-2n_1 + n_2) \\ (1/l_2)(n_1 - n_2)\end{pmatrix}$$

$$\equiv \frac{1}{f_0}\begin{pmatrix}f_{01} \\ f_{02}\end{pmatrix}, \tag{4.7}$$

where the identity on the right defines the two new quantities f_{01} and f_{02}. These equations show that for each pair of rotation numbers (n_1, n_2) a set of four possible solutions for (x_{1*}, x_{2*}) exists if $|f_0| \geq |f_{0c}|$, where $|f_{0c}| = \max(|f_{01}|, |f_{02}|)$. The four fixed points correspond to the four combinations of values of (x_{1*}, x_{2*}) that have the same pair of values of $(\sin x_{1*}, \sin x_{2*})$. When necessary we will use the notation

$$\mathbf{Z}_*^{[N;q]} = \left(\mathbf{X}_*^{[N;q]}, \mathbf{Y}_*^{[N]}\right),$$

or more simply $[N; q]$, to identify the fixed points, where the index q labels the four possible solutions of (4.7) ($q = 1, 2, 3, 4$) and, as shown in Figure 6, corresponds to the ordering

$$x_{1*}^{[N;1]} = x_{1*}^{[N;2]} < x_{1*}^{[N;3]} = x_{1*}^{[N;4]},$$

$$x_{2*}^{[N;1]} = x_{2*}^{[N;3]} < x_{2*}^{[N;2]} = x_{2*}^{[N;4]}.$$

Note that $\mathbf{Y}_*^{[N]} = (y_{1*}^{[N]}, y_{2*}^{[N]})$ is the same for the four fixed points (i.e., it does not depend on q). Equations (4.7) also show that for $|f_0| \geq |f_{0c}|$, $(n_1, n_2) \neq (0, 0)$, another set of four fixed points exists with rotation numbers $(-n_1, -n_2)$. It is easy to see that to each point $(x_{1*}, x_{2*}, y_{1*}, y_{2*})$ of the first set corresponds a point of the second set given by $(2\pi - x_{1*}, 2\pi - x_{2*}, -y_{1*}, -y_{2*})$. This is a reflection of the fact that the double rotor map (4.1) itself is invariant under the change of variables $(x_1, x_2, y_1, y_2) \mapsto (2\pi - x_1, 2\pi - x_2, -y_1, -y_2)$.

In Table 1 we summarize the properties of the five sets of fixed points (36 fixed points) with smaller values of f_{0c} (when the other parameters of the map take the values

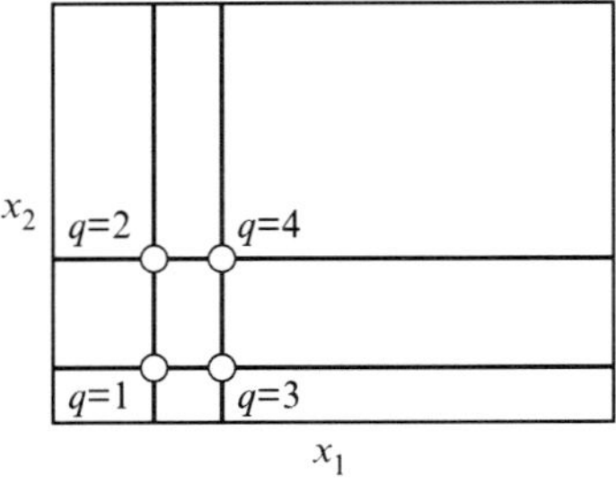

Figure 6 Double rotor map: labeling of fixed points.

Table 1. Double rotor map: fixed points. The only stable fixed points are $[(0, 0); 4]$ in the interval $0 < f_0 < 4.27\ldots$; $[(1, 2); 4]$ and $[(-1, -2); 1]$ in the interval $2\pi < f < 7.01\ldots$ (the other parameters are given by Eq. (4.4))

(n_1, n_2)	f_{01}	f_{02}	f_{0c}	h_{11c}	h_{22c}
$(0, 0)$	0	0	0	0	0
$\pm(1, 2)$	0	$2\pi v l / l_2$	$2\pi v l / l_2$	$\pm 2\pi v l_1 / l_2$	0
$\pm(0, 1)$	$2\pi v l / l_1$	$2\pi v l / l_2$	$2\pi v l / l_1$	0	$\pm 2\pi v (l_2^2/l_1^2 - 1)^{1/2}$
$\pm(1, 1)$	$2\pi v l / l_1$	0	$2\pi v l / l_1$	0	$\pm 2\pi v l_2 / l_1$
$\pm(2, 3)$	$2\pi v l / l_1$	$2\pi v l / l_2$	$2\pi v l / l_1$	0	$\pm 2\pi v (l_2^2/l_1^2 - 1)^{1/2}$

specified by Eqs. (4.4)), with rotation numbers $N = (0,0)$, $\pm(1,2)$, $\pm(0,1)$, $\pm(1,1)$, $\pm(2,3)$. Note that the last three sets have the same value of f_{0c}. In Figure 7 we have plotted these fixed points in the plane (x_1, x_2). Their (y_1, y_2) coordinates are given by the first of Eqs. (4.6).

Let us now turn our attention to the stability of the fixed points. The basic element of the analysis is the Jacobian (4×4) matrix of partial derivatives of the map (4.1),

$$\mathbf{D}_z F(Z) = \begin{pmatrix} \mathbf{I}_2 & \mathbf{M} \\ \mathbf{H}(X') & \mathbf{L} + \mathbf{H}(X')\mathbf{M} \end{pmatrix},$$

where

$$\mathbf{H}(X') = \mathbf{D}_x G(X') = \begin{pmatrix} c_1 \cos x_1' & 0 \\ 0 & c_2 \cos x_2' \end{pmatrix},$$

and $\mathbf{I}_n$ denotes the $n \times n$ identity matrix. The characteristic polynomial of $\mathbf{D}_z F(Z_*)$ is

$$P(s) = \left| \mathbf{D}_z F(Z_*) - s\mathbf{I}_4 \right|$$

$$= \left| s^2 \mathbf{I}_2 - s(\mathbf{I}_2 + \mathbf{L} + \mathbf{HM}) + \mathbf{L} \right|, \qquad (4.8a)$$

where, for simplicity, we have set $\mathbf{H} \equiv \mathbf{H}(X_*)$. The characteristic equation

$$P(s) = 0, \qquad (4.8b)$$

determines the stability of the fixed points: if all the four roots have modulus smaller than one, the fixed point is stable. The stability of the fixed points as determined from Eqs. (4.8) is discussed in Appendix D.

For $f_0 = 9.0$, the nominal value in the control experiments of Sections 4.4 and 4.5, all the fixed points are unstable. We have indicated in Figure 7 the number of unstable eigendirections at each fixed point.

We observe, from Eq. (4.7), that as the forcing f_0 increases, the number of fixed points increases without bound. Not all these fixed points are necessarily embedded

in the chaotic attractor, but those that are embedded in it are necessarily unstable. Furthermore, we find that the fixed points are roughly spread throughout the attractor, suggesting that there can be substantial flexibility to select among a variety of asymptotic behaviors by selecting different fixed points for control. (Even more flexibility can be achieved if we also consider periodic orbits of period greater than one.)

4.3. Bifurcation diagram

A bifurcation diagram shows how the attractors of a dynamical system change with a system parameter.

In Figures 8(a) and (b) we present a bifurcation diagram for the double rotor map, which was obtained in the following way. For each value of the parameter we took a large number of initial angles (x_1, x_2) with both x_1 and x_2

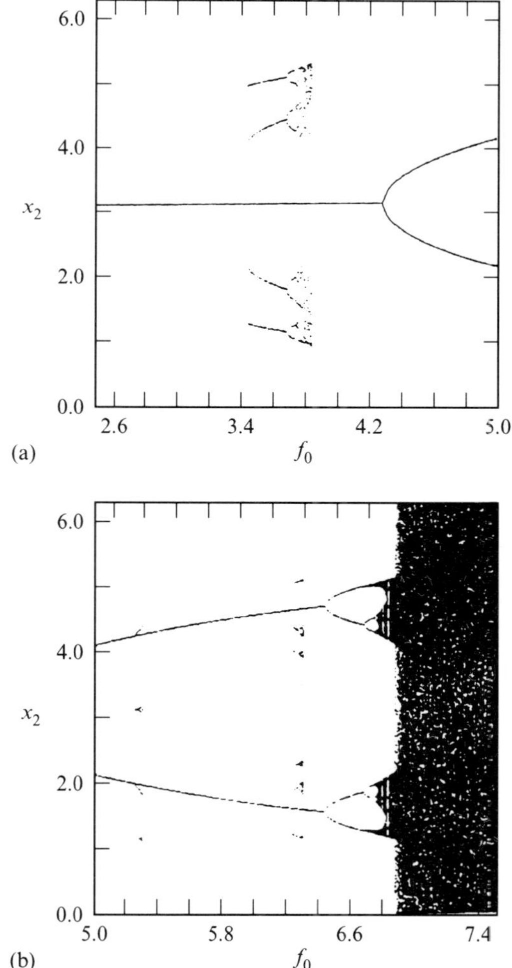

(a)

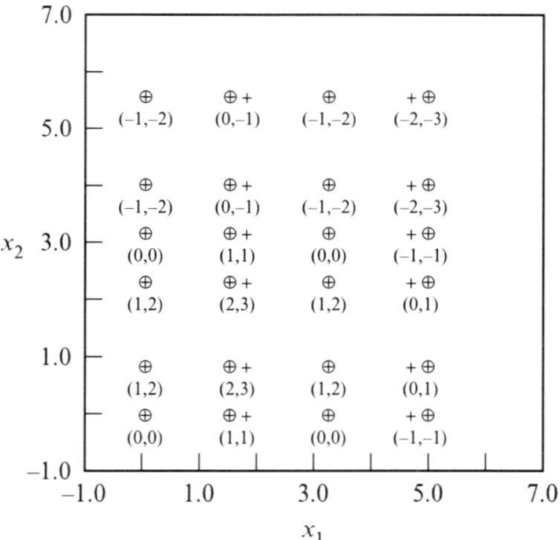

Figure 7 Double rotor map: fixed points with rotation numbers (n_1, n_2). The symbol $(+)$ denotes fixed points with one unstable eigendirection, while the symbol $\oplus$ denotes fixed points with two unstable eigendirections. ($f_0 = 9.0$, other parameters given by Eq. (4.4).)

(b)

Figure 8 (a), (b) Double rotor map: bifurcation diagram. (Parameters given by Eq. (4.4): number of values of f_0 in each figure: 251; number of initial conditions for each f_0: 625: snapshot taken after 6000 iterations.)

distributed uniformly in $[0, 2\pi]$ and iterated them starting with zero angular velocity [i.e., $(y_1, y_2) = (0, 0)$]. After iterating a sufficient number of times so that the orbits are essentially on the attractor, we plotted the x_2 component of all the orbits.

The diagram clearly exhibits a main branch that develops continuously for all values of the parameter. This main branch illustrates a period-doubling bifurcation sequence to chaos: a period-1 periodic orbit bifurcates (at $f_0 \simeq 4.27$) to a period-2 periodic orbit which then bifurcates (at $f_0 \simeq 6.42$) to a period-4 periodic orbit which then bifurcates (at $f_0 \simeq 6.67$) to a period-8 periodic orbit, and so on, with an accumulation point of period doublings at $f_0 \simeq 6.75$ beyond which chaos appears. The period-2 periodic orbit in the sequence results from the bifurcation of the stable orbit $Z_*^{[(0,0);4]} = (\pi, \pi, 0, 0)$ discussed in Section 4.2 which exists for $f_0 \geq 0$; at the value $f_{0u}^{[(0,0);4]}$ at which this orbit becomes unstable the stable period-2 periodic orbit is born.

Although it cannot be seen in the diagram, this period doubling sequence is peculiar in the following sense: what appears to be a period-2^m periodic orbit, $m \geq 2$, is in fact 2 period-2^{m-1} periodic orbits. This is a consequence of the symmetry of the double rotor map that forces the period-1 orbit to become unstable (at $f_0 \cong 4.2$) through an eigenvalue 1 instead of through -1 as occurs in the normal period doubling bifurcation (an example of which is the bifurcation of the period-1 periodic orbit).

Besides this main branch, there are other period doubling sequences, one of which starts with a period-4 periodic orbit (at $f_0 \simeq 3.42$) and ends with a crisis (at $f_0 \simeq 3.84$). (A *crisis* is the sudden disappearance of a chaotic attractor by collision with an unstable periodic orbit [15, 16].)

It is convenient to have some quantitative characterization of the chaotic attractors revealed by the bifurcation diagram. For this purpose we introduce the spectrum of *Lyapunov exponents*, defined as follows [21, 22].

Consider an n-dimensional map $Z \mapsto F(Z)$ and its Jacobian matrix of partial derivatives $J(Z) = D_Z F(Z)$. Consider also the sequence $\{Z_0, Z_1 \ldots, Z_{k-1}\}$ generated by successive iteration of the initial condition Z_0. For this sequence introduce the matrix

$$J_k = J(Z_{k-1})J(Z_{k-2}) \ldots J(Z_1)J(Z_0)$$

Now let

$$\zeta_1(k) \geq \zeta_2(k) \geq \cdots \geq \zeta_n(k),$$

denote the n eigenvalues of $(J_k^T J_k)^{1/2}$, where J_k^T is the transpose of J_k. The Lyapunov numbers of the map are then defined by

$$\eta_j = \lim_{k \to \infty} [\zeta_j(k)]^{1/k}, \quad j = 1, \ldots, n,$$

where the positive real kth root is taken. They satisfy the same ordering as the $\zeta_j(k)$, $j = 1, \ldots, n$. The Lyapunov exponents are the logarithms of the Lyapunov numbers,

$$L_j = \log_e \eta_j, \quad j = 1, \ldots, n,$$

satisfying the same ordering

$$L_1 \geq L_2 \geq \cdots \geq L_n.$$

Hence, for chaotic attractors of an n-dimensional map there are n Lyapunov exponents, L_j, $j = 1, \ldots, n$. A chaotic attractor is defined to be one which possesses a positive Lyapunov exponent, $L_1 > 0$.

For typical dynamical systems the Lyapunov exponents are the same for almost all initial conditions on the basin of attraction of the attractor. (This is true in particular for the chaotic attractors of the double rotor map for which we calculated Lyapunov exponents; these results are reported below.) Thus the spectrum of Lyapunov exponents may be indeed considered to be a property of the attractor. For maps such that the determinant of the Jacobian matrix is independent of the variable Z the Lyapunov exponents satisfy the identity

$$\sum_{j=1}^{n} L_j = \log_e |J|.$$

This is true in the case of the double rotor map for which we have

$$\sum_{j=1}^{4} L_j = \log_e |L| = (\lambda_1 + \lambda_2)T = -3\nu,$$

the last equality applying when $\nu_1 = \nu_2 \equiv \nu$ and $T = 1$.

From the spectrum of Lyapunov exponents define the *Lyapunov dimension*.

$$d_L = k_L + \frac{\sum_{j=1}^{k_L} L_j}{|L_{k_L+1}|}$$

where $1 \leq k_L \leq n - 1$ is the largest integer for which $\sum_{j=1}^{k_L} L_j \geq 0$. If $L_1 < 0$, we define $d_L = 0$; if $\sum_{j=1}^{n} L_j \geq 0$, we define $d_L = n$. (Note that $d_L = n$ is not possible in the case of dissipative systems for which $\log_e |J| < 0$.) Kaplan and Yorke [23, 24] conjecture that d_L, as given above in terms of the Lyapunov exponents, is typically equal to the fractal dimension of the support of the measure of the attractor (the information dimension).

We have numerically calculated the Lyapunov exponents and the Lyapunov dimension of the chaotic attractor in the main branch of the bifurcation diagram as a function of the forcing f_0. We used the method described in Refs. [21, 22] to calculate the exponents of a large number of orbits in the basin of attraction and then took the average of these values. The results of the calculation at evenly spaced values along the f_0 axis are shown in Figure 9. The Lyapunov dimension first becomes positive at the onset of chaos ($f_0 \simeq 6.75$). The attractor dimension goes through the integer values $d_L = 2$ and 3 at $f_0 \simeq 6.88$ and 12.7, respectively.

In the numerical experiments on control that we describe in Sections 4.4 and 4.5 we took $\bar{f}_0 = 9.0$ as the nominal value of the control parameter. In Table 2 we list the corresponding values of the four Lyapunov exponents and the Lyapunov dimension. In order to illustrate the point made above regarding the fact that the Lyapunov exponents are the same for almost all initial conditions on the basin of attraction of the attractor, we also give some details on the numerical calculation of these exponents.

We have now described in sufficient detail the two ingredients necessary to the application of the control method to the double rotor map: chaotic attractors and fixed points. It remains to be checked if the fixed points determined in Section 4.2 are embedded in the chaotic

Table 2. Double rotor map: calculation of Lyapunov exponents and Lyapunov dimension of chaotic attractor ($f_0 = 9.0$, other parameters given by Eq. (4.4); number of initial conditions = $N_0 = 256$; number of iterations = 10 000). $d_L = 2 + (L_1 + L_2)/|L_3| = 2.838$

	j			
	1	2	3	4
$L_j = (1/N_0)\sum_{i=1}^{N_0} L_j^{(i)}$	1.205	0.256	−1.744	−2.717
$\min_{i=1,N_0} L_j^{(i)}$	1.185	0.228	−1.771	−2.734
$\max_{i=1,N_0} L_j^{(i)}$	1.229	0.284	−1.719	−2.693
$[(1/N_0)\sum_{i=1}^{N_0} (L_j^{(i)} - L_j)^2]^{1/2}$	0.00816	0.0102	0.00910	0.00724

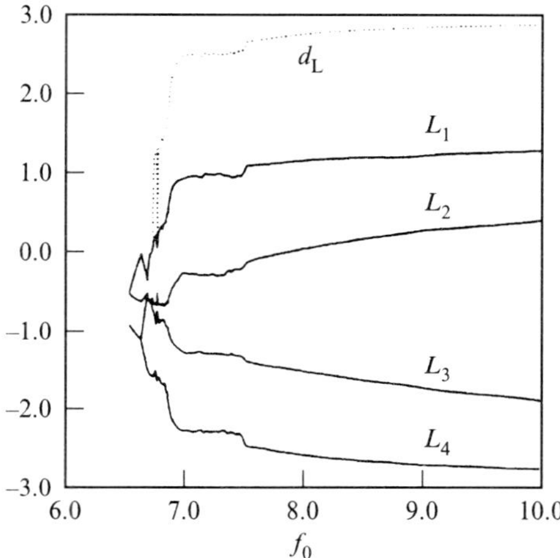

Figure 9 Double rotor map: spectrum of Lyapunov exponents and Lyapunov dimension of chaotic attractors versus f_0 (Eq. (4.4)).

attractor. By this we mean that any neighborhood of the fixed point contains an infinite number of points of the chaotic attractor. In order to check this, we consider the intersection of the attractor in its four-dimensional phase space with a three-dimensional hyperplane containing the fixed points $\mathbf{Z}_*$, that we wish to check. Numerically we approximate the hyperplane by a very narrow slab through each fixed point of the form

$$|\hat{K}^T(\mathbf{Z} - \mathbf{Z}_*)| < w. \qquad (4.9)$$

Actually we took the slabs parallel to the plane (x_1, x_2) which implies that each slab contains the four fixed points with the same rotation number. We then examine a very long orbit and plot only those points satisfying (4.9). The intersection of our 2.8-dimensional attractor with a three dimensional hyperplane is a 1.8-dimensional cross-section. The small scale structure of this 1.8-dimensional intersection is somewhat fuzzed out due to the finite slab thickness. The results, for $f_0 = 9.0$, are given in Figures 10(a)–(e), which refer to the rotation numbers $N = (0, 0)$, $(1, 2)$, $(0, 1)$. $(1, 1)$ and $(2, 3)$. respectively. In these figures the relevant fixed points are denoted by a + symbol. The results indicate, with different degrees of certitude, that the first four sets of fixed points are indeed embedded in the attractor while the fifth is not. Note that Figure 10(a) nicely

reveals the symmetry of the map with respect to the point $(\pi, \pi, 0, 0)$. Note also the fractal-like structure in this figure.

We conclude this discussion by mentioning what seems to be an interesting issue: the loss of hyperbolicity due to the existence of fixed points embedded in the attractor that have a number of unstable directions (that is, eigenvalues with magnitude bigger than one) different from the number of unstable directions of the attractor (that is, positive Lyapunov exponents). In fact, from the observation of Figure 7 and Table 2, we see that while the chaotic attractor for $f_0 = 9.0$ has two positive Lyapunov exponents some of the unstable fixed points embedded in the attractor have only one unstable eigenvalue.

4.4. Control

We now proceed to apply the method developed in Section 2 to control the fixed points of the double rotor map with control parameter f_0. Let us denote by $\bar{\mathbf{Z}}_*$ the fixed point to be controlled at the nominal value $\bar{f}_0$ of the parameter. The quantities that were introduced in Section 2 now take the following particular form:

$$\mathbf{A} = \begin{pmatrix} \mathbf{I}_2 & \mathbf{M} \\ \mathbf{H}(\bar{X}_*) & \mathbf{L} + \mathbf{H}(\bar{X}_*)\mathbf{M} \end{pmatrix},$$

$$\mathbf{H}(\bar{X}_*) = \frac{\bar{f}_0}{I}\begin{pmatrix} l_1\cos \bar{x}_{1*} & 0 \\ 0 & l_2\cos \bar{x}_{2*} \end{pmatrix},$$

$$\mathbf{B}^T = \begin{pmatrix} 0 & 0 & \frac{l_1}{I}\sin \bar{x}_{1*} & \frac{l_2}{I}\sin \bar{x}_{2*} \end{pmatrix},$$

$$\mathbf{C} = (\mathbf{B} : \mathbf{AB} : \mathbf{A}^2\mathbf{B} : \mathbf{A}^3\mathbf{B}),$$

$$\mathbf{T} = \mathbf{CW},$$

$$\mathbf{W} = \begin{vmatrix} a_3 & a_2 & a_1 & 1 \\ a_2 & a_1 & 1 & 0 \\ a_1 & 1 & 0 & 0 \\ 1 & 0 & 0 & 0 \end{vmatrix},$$

$$\mathbf{K}^T = (\alpha_4 - a_4 \quad \alpha_3 - a_3 \quad \alpha_2 - a_2 \quad \alpha_1 - a_1)\mathbf{T}^{-1}.$$

One immediate conclusion that can be drawn from these results is that the controllability matrix $\mathbf{C}$ is identically zero in the case of the fixed points with rotation numbers $N = (0, 0)$ for which $\sin \bar{x}_{1*} = \sin \bar{x}_{2*} = 0$. Hence these points are uncontrollable, at least when the control parameter is f_0. We will show in the next subsection that this set of fixed points can be controlled if we modify the double rotor map to

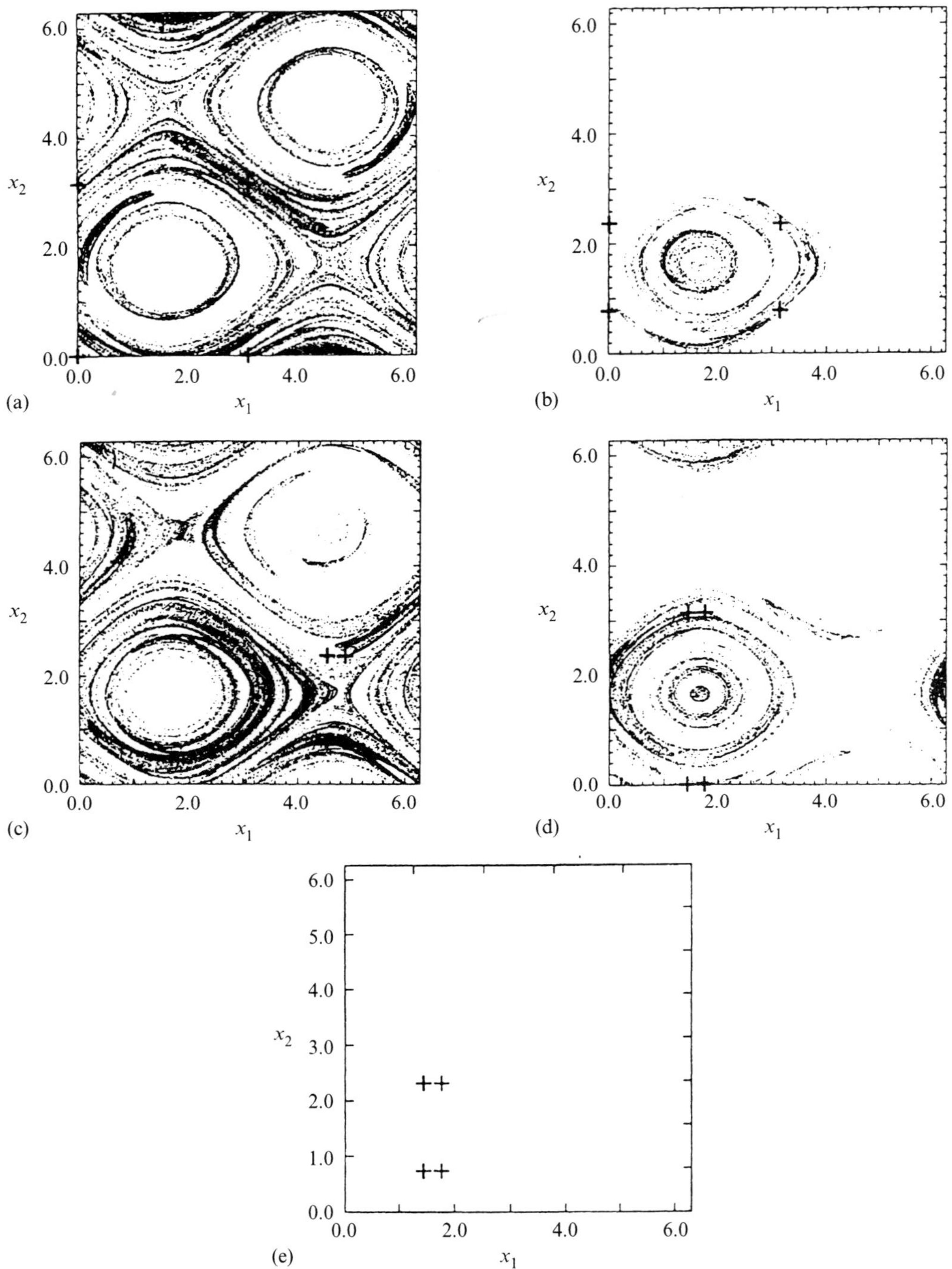

Figure 10 Double rotor map: sections of chaotic attractor by slab $|\hat{K}^{\mathrm{T}}(Z - Z_*)| < w$, $\hat{K}^{\mathrm{T}} = (0, 0, 1, 1)$, $w = 10^{-2}$, through the fixed points (+) with rotation numbers (a) $N = (0, 0)$, (b) $N = (1, 2)$, (c) $N = (0, 1)$, (d) $N = (1, 1)$, (e) $N = (2, 3)$. The map was iterated 10^8 times ($f_0 = 9.0$, Eq. (4.4).)

allow for kicks with variable direction and then take as control parameter the angle the kicks make with the vertical direction in Figure 5.

The method is illustrated in Figures 11(a) and (b). The control of the first fixed point was turned on at $i = 0$ with switches to control other fixed points occurring at later times. We plot the x_1 and x_2 coordinates of an orbit as a function of (discrete) time. The parameter perturbations

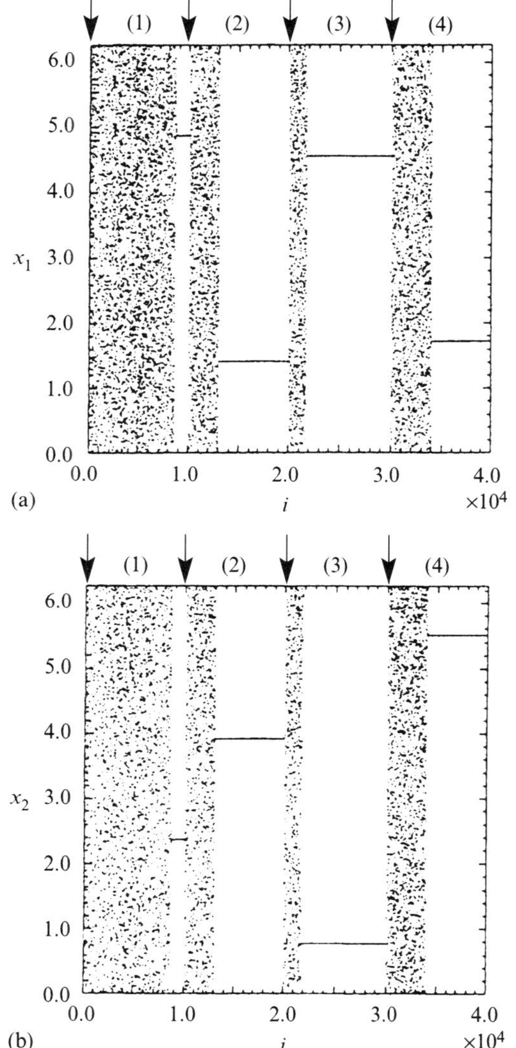

(a)

(b)

Figure 11 Double rotor map: successive control of fixed points (1) $[(0, 1); 4]$, (2) $[(0, -1); 1]$, (3) $[(0, 1); 1]$, (4) $[(0, -1); 4]$. The arrows indicate the times of switching. The regulator poles correspond to projection onto the stable manifold ($\delta = 1.0$, $\bar{f}_0 = 9.0$, Eq. (4.4)).

were programmed to control successively four different fixed points of the set with rotation numbers $N = \pm(0, 1)$. The times at which we switched the control from stabilizing one fixed point to stabilizing another are labeled by the arrows in the figure. The figure clearly illustrates the flexibility offered by the method in controlling different periodic motions embedded in the attractor. The figure also shows that the time to achieve control is different from case to case.

We now report the results of several numerical experiments that were carried out with the purpose of understanding the behavior of the time to achieve control.

The first experiment was intended to confirm that the

time to achieve control indeed follows an exponential probability distribution as indicated in Section 2.4. We proceeded to control the fixed point $[(0, 1); 4]$ by starting at a large number of different points on the attractor and measuring the time each orbit took to reach the fixed point. We then obtained the distribution function of the time to achieve control $\phi(\tau)$ by plotting a histogram of τ using bins of constant size. The results are presented as a semilog plot in Figure 12 and show excellent agreement with the predicted fit to a straight line.

In our next experiment we looked at the dependence of the average time to achieve control on the size of the parameter perturbations, δ. The results are shown in Figure 13, where we have used logarithmic scales in both axes. The two fixed points $[(0, 1); 4]$ and $[(0, 1); 1]$ were controlled. (The first of these points has two unstable eigenvalues while the second has only one unstable eigenvalue.) We see that for the smaller values of δ the results closely follow straight lines indicating a power law dependence,

$$\langle \tau \rangle \sim \delta^{-\gamma},$$

in accord with the theoretical predictions of Ref. [5] for two-dimensional maps.

In the experiments described until this point the choice of the regulator poles (eigenvalues of $\mathbf{A} - \mathbf{B}\mathbf{K}^\mathrm{T}$) corresponded to projection onto the stable manifold of the fixed points. That is, the stable eigenvalues of matrix $\mathbf{A}$ were left unchanged, and the unstable eigenvalues were shifted to zero.

In our next set of experiments we looked at how different choices of regulator poles affect the average time to achieve control. We considered the fixed point $[(0, 1); 4]$ with two unstable eigendirections and kept two of the regulator poles equal to the two stable eigenvalues of the fixed point. As

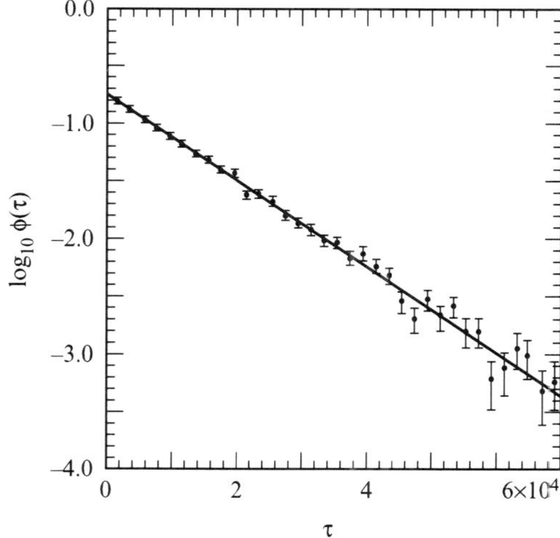

Figure 12 Double rotor map: histogram of the time to achieve control τ of a sample of 8192 orbits. The fixed point controlled was $[(0, 1); 4]$. The regulator poles correspond to projection onto the stable manifold. ($\delta = 1.0$, $\bar{f}_0 = 9.0$, Eq. (4.4).)

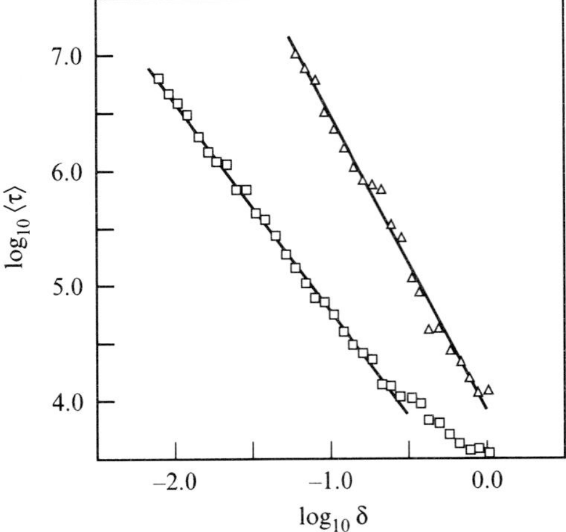

Figure 13 Double rotor map: $\log_{10}\langle\tau\rangle$ versus $\log_{10}\delta$ for control of the fixed points ($\square$) [(0, 1); 1], ($\triangle$) [(0, 1); 4]. The regulator poles correspond to projection onto the stable manifold. The straight lines are least square fits to the data [excluding the last nine data points in the case of ($\square$)]. ($\bar{f}_0 = 9.0$, Eq. (4.4).)

regards the other two regulator poles, μ_1 and μ_2, three cases were considered:

  (I) $\mu_2 = 0$,
 (II) $\mu_2 = \mu_1$,
(III) $\mu_2 = -\mu_1$.

μ_1 was then allowed to vary in the interval $(-1, 1)$. The results of the experiments are shown in Figure 14. In cases (I) and (II) the average time to achieve control essentially

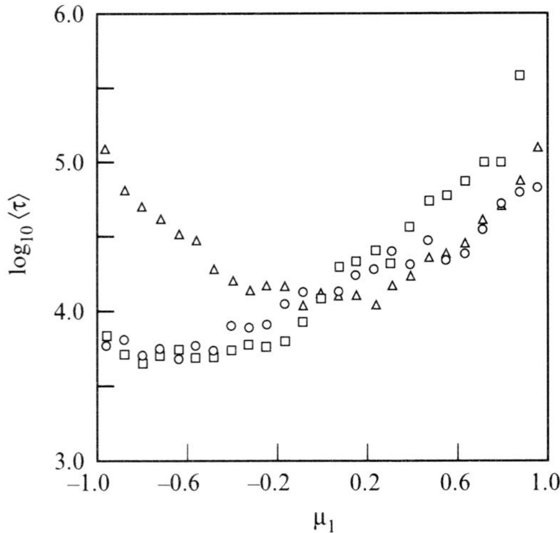

Figure 14 Double rotor map: $\log_{10}\langle\tau\rangle$ versus μ_1 for ($\bigcirc$) $\mu_2 = 0$, ($\square$) $\mu_2 = \mu_1$, ($\triangle$) $\mu_2 = -\mu_1$. The other two regulator poles were kept equal to the stable eigenvalues of the uncontrolled map. The fixed point controlled was [(0, 1); 4]. ($\delta = 1.0$, $\bar{f}_0 = 9.0$, Eq. (4.4).)

increases with μ_1, indicating behavior similar to that found for the Hénon map in Figure 2. In case (III) the average time to achieve control passes through a broad minimum. (Note that the point $\mu_1 = \mu_2 = 0$, which is common to the three cases, corresponds to projection onto the stable manifold.)

4.5. f_0-uncontrollable fixed points

We saw that the set of four fixed points with rotation numbers $N = (0, 0)$ could not be controlled by changes in the parameter f_0 because the controllability matrix at these points is identically zero.

We show now that these fixed points can be controlled by modifying the double rotor map to allow for kicks with variable direction and then taking the direction of the kicks to be the control parameter, with the nominal value corresponding to the previously fixed direction.

Let us assume that the direction of the kicks makes an angle ψ with the vertical downward direction. Going back to the derivation of the double rotor map in Appendix B, it is easy to verify that the introduction of kicks with variable direction can be taken into account by simply replacing the function G used in the definition of the map and given by Eq. (4.2) by the new function

$$\tilde{G}(X) = \begin{pmatrix} c_1 \sin(x_1 - \psi) \\ c_2 \sin(x_2 - \psi) \end{pmatrix}.$$

Taking ψ to be the control parameter with variations around the nominal value $\bar{\psi} = 0$, the application of the method now involves the following quantities:

$$\mathbf{A} = \begin{pmatrix} \mathbf{I} & \mathbf{M} \\ \tilde{\mathbf{H}}(\bar{X}_*) & \mathbf{L} + \tilde{\mathbf{H}}(\bar{X}_*)\mathbf{M} \end{pmatrix},$$

$$\tilde{\mathbf{H}}(\bar{X}_*) = \frac{f_0}{I}\begin{pmatrix} l_1 \cos x_{1*} & 0 \\ 0 & l_2 \cos \bar{x}_{2*} \end{pmatrix},$$

$$\mathbf{B}^{\mathrm{T}} = \begin{pmatrix} 0 & 0 & -\dfrac{f_0 l_1}{I}\cos \bar{x}_{1*} & -\dfrac{f_0 l_2}{I}\cos \bar{x}_{2*} \end{pmatrix}.$$

The fixed points are now all controllable by small perturbations of the parameter ψ around the nominal value $\bar{\psi} = 0$.

Figures 15(a) and (b) illustrate the control of the fixed points [(0, 0); 3] and [(0, 0); 4] by kicks of variable direction. The parameter perturbations were programmed to control the first of these points from $i = 0$ to $i = 10^4$ and the second from $i = 10^4$ to $i = 2 \times 10^4$.

5. Discussion

The transient phase where the orbit wanders chaotically before locking in to a controlled orbit can be greatly shortened by applying the technique discussed by Shinbrot et al. [25]. In the latter paper it was pointed out that orbits can be rapidly brought to a target region on the attractor (in the present case the neighborhood of the periodic orbit which we wish to stabilize) by using small control perturbations when the orbit is far from the neighborhood of the periodic orbit to be stabilized. The idea was that,

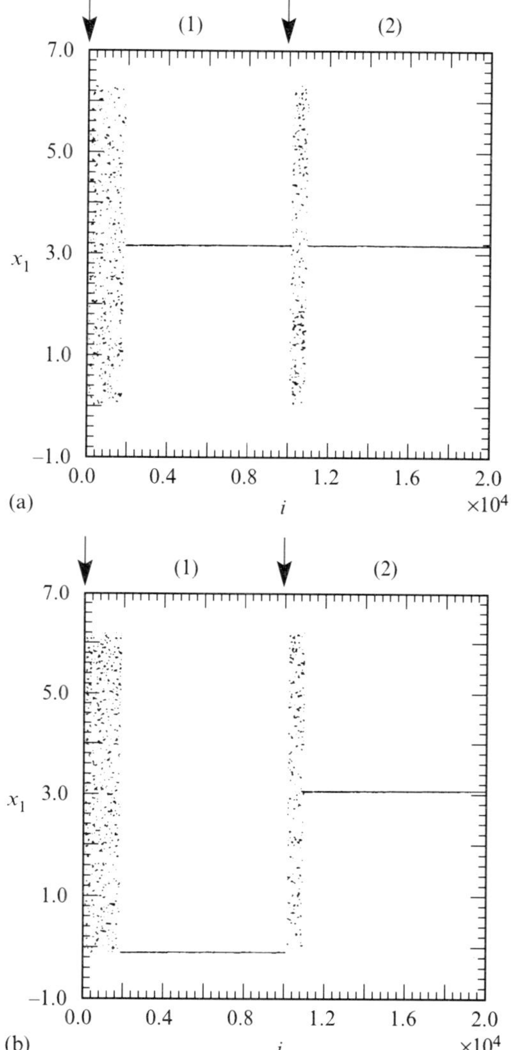

Figure 15 Modified double rotor map: successive control of fixed points (1) [(0, 0); 3], (2) [(0, 0); 4] by kicks of variable direction. The arrows indicate the times of switching. The regulator poles correspond to projection onto the stable manifold ($\delta = 0.05$, $f_0 = 9.0$, Eq. (4.4).)

since chaotic systems are exponentially sensitive to perturbations, careful choice of even small control perturbations can, after some time, have a large effect on the orbit location and can be used to guide it. Thus the time to achieve control can, in principle, be greatly shortened by properly applying small controls when the orbit is far from the neighborhood of the desired periodic orbit.

One issue which we have not addressed is the effect of noise. If the noise remains small, it may not be sufficient to kick the orbit out of the neighborhood of the chosen periodic orbit where the control is activated. In this case, the orbit remains near the desired periodic orbit indefinitely. However, it may be that the random noise is such that it may

occasionally kick the orbit far enough away frcm the periodic orbit that the orbit falls outside the small controlled phase space region. In this case, after the orbit is kicked out of the controlled phase space region, it wanders chaotically over the attractor until it falls in the controlled region again. Thus there are epochs where the orbit is kept near the desired orbit interspersed with epochs wherein the orbit wanders chaotically far from the desired orbit. If the latter are, on average, relatively much shorter than the former, then one might still regard the control as being effective. See Ref. [6] for numerical experiments on this effect using the Hénon map. We also remark that the procedure discussed in the previous paragraph [25] can be used to greatly reduce the duration of the noise induced epochs where the orbit bursts out of the controlled phase space region.

In this paper we have considered the case where there is only a single control parameter available for adjustment. While generically a single parameter is sufficient for stabilization of a desired periodic orbit, there may be some advantage to utilizing several control variables. Therefore, the single control parameter p becomes a vector (e.g., Ref. [26] discusses the case where the number of control parameters is equal to the number of unstable eigenvalues). In particular, the added freedom in having several control parameters might allow better means of choosing the control so as to minimize the time to achieve control, as well as the effects of noise.

Finally we wish to point out that full knowledge of the system dynamics is not necessary in order to apply our technique (see also Ref. [6]). In particular, we only require the location of the desired periodic orbit, the linearized dynamics about the periodic orbit, and the dependence of the location of the periodic orbit on small variation of the control parameter. Recently, delay coordinate embedding [19, 27] has been utilized in several experimental studies (Refs. [8, 28–31]) to extract such information purely from observations of experimental chaotic orbits on the atractor without any *a priori* knowledge of the system of equations governing the dynamics, and such information has been utilized to control periodic orbits [9]. Hence, application of our method is not limited to cases where a complete knowledge of the system is available.

In conclusion, we have demonstrated that chaotic dynamics can often be converted, by using only a small feedback control, to motion on a desired periodic orbit. Furthermore, by switching the small control, one can switch the time asymptotic behavior from one periodic orbit to another. In some situations, where the flexibility offered by the ability to do such switching is desirable, it may be advantageous to design the system so that it is chactic. In other situations, where one is presented with a chaotic system, the method may allow one to eliminate the chaos and achieve greatly improved behavior at relatively low cost.

Acknowledgements

This work was supported by the US Department of Energy (Scientific Computing Staff, Office of Energy

Research), the Portuguese Junta Nacional de Investigação Científica e Tecnológica, and the National Science Foundation (Engineering Research Center Program). The computation was done at the National Energy Research Supercomputer Center.

Appendix A. Time to achieve control in the case of two-dimensional maps

We assume that control is achieved if the orbit remains in the slab (2.8) for two consecutive iterations of the map. The two conditions

$$\left|K^{\mathrm{T}}(Z - Z_*(\bar{p}))\right| < \delta, \quad \left|K^{\mathrm{T}}(Z' - Z_*(\bar{p}))\right| < \delta,$$
(A.1)

define a control 'parallelepiped' P_c, where $Z' = F(Z, p)$. For small δ, an initial condition will bounce around on the set comprising the uncontrolled chaotic attractor for a long time before it falls in the control parallelepiped P_c. At any given iterate the probability of falling in P_c is approximately the natural measure (see, for example, [17, 18, 23]) of the uncontrolled chaotic attractor contained in P_c. If we follow many orbits this probability $\mu(P_c)$ also gives the rate at which these orbits fall into P_c. Thus $\mu(P_c)$ is the inverse of the average time for a typical orbit to first fall in P_c,

$$\langle \tau \rangle^{-1} = \mu(P_c).$$
(A.2)

An estimate for $\mu(P_c)$ can be given in the two-dimensional case [23]:

$$\mu(P_c) \sim \int_{P_c} \rho \left|v_s\right|^{d_s-1}\left|v_u\right|^{d_u-1} dv_s\, dv_u.$$
(A.3)

where v_s and v_u denote linear coordinates in the stable and unstable directions. In here d_u and d_s are the pointwise dimensions [1] for the uncontrolled chaotic attractor at the fixed point in the unstable and the stable directions, respectively; ρ is a normalizing constant. Assuming that the attractor is smooth in the unstable direction we have $d_u = 1$, while d_s is given in terms of the eigenvalues at the fixed point [1, 17, 18] by

$$d_s = \frac{\log_e\left|\lambda_u\right|}{\log_e(1/\left|\lambda_s\right|)}.$$

In order to determine the control parallelepiped, we need to obtain Z' in the neighborhood of the fixed point $Z_*(\bar{p})$ with a better approximation than that provided by the linear map (2.3). We therefore take

$$V' = \mathbf{A}V + B(p - \bar{p}) + \tfrac{1}{2}Q(V, V) + \tfrac{1}{2}D(p - \bar{p})^2,$$
(A.4)

where

$$V = Z - Z_*(\bar{p}), \quad V' = Z' - Z_*(\bar{p}).$$

$\mathbf{A}$, B were defined by (2.4), (2.5) and Q, D are two vectors with components q_k, d_k ($k = 1, 2$) defined by

$$q_k = \sum_{i,j=1}^{2} \left.\frac{\partial^2 f_k}{\partial x_i \partial x_j}\right|_{(Z_*(\bar{p}),\bar{p})} v_i v_j,$$

$$d_k = \left.\frac{\partial^2 f_k}{\partial p^2}\right|_{(Z_*(\bar{p}),\bar{p})}.$$

Here x_k, f_k, and v_k ($k = 1, 2$) denote the components of the vectors X, F and V. Using (2.6) to eliminate $p - \bar{p}$ from Eq. (A.4), we obtain

$$V' = \mathbf{A}V + \tfrac{1}{2}Q(V, V) - B(K^{\mathrm{T}}V) + \tfrac{1}{2}D(K^{\mathrm{T}}V)^2.$$
(A.5)

The control 'parallelogram' P_c will therefore be defined by the two equations

$$\left|K^{\mathrm{T}}V\right| < \delta, \quad \left|K^{\mathrm{T}}V'\right| < \delta.$$
(A.6)

In order to compare with the numerical experimental results described in Section 3 we have carried out the calculation of (A.3) in the case of the Hénon map. Writing this map in the form

$$\begin{pmatrix} x_1 \\ x_2 \end{pmatrix} \mapsto \begin{pmatrix} f_1(x_1, x_2) \\ f_2(x_1, x_2) \end{pmatrix} = \begin{pmatrix} a - x_1^2 + bx_2 \\ x_1 \end{pmatrix},$$

and taking a to be the control parameter while b is kept fixed, we obtain

$$\mathbf{A} = \begin{pmatrix} -2\bar{x}_* & b \\ 1 & 0 \end{pmatrix}, \quad B = \begin{pmatrix} 1 \\ 0 \end{pmatrix},$$

$$D = \begin{pmatrix} 0 \\ 0 \end{pmatrix}, \quad Q(V,V) = \begin{pmatrix} -2v_1^2 \\ 0 \end{pmatrix},$$

$$K^{\mathrm{T}}V = k_1 v_1 + k_2 v_2,$$

$$K^{\mathrm{T}}V' = -k_1 v_1^2 + (k_2 - k_1^2 - 2\bar{x}_* k_1)v_1 + k_1(b - k_2)v_2.$$

Also we note that for the Hénon map the variables (v_1, v_2) and (v_s, v_u) are related by

$$\begin{pmatrix} v_1 \\ v_2 \end{pmatrix} = \begin{pmatrix} \lambda_s \gamma_s & \lambda_u \gamma_u \\ \gamma_s & \gamma_u \end{pmatrix} \begin{pmatrix} v_s \\ v_u \end{pmatrix},$$
(A.7)

where

$$\gamma_s = (1 + \lambda_s^2)^{-1/2}, \quad \gamma_u = (1 + \lambda_u^2)^{-1/2}.$$

Letting

$$\xi = v_1, \quad \eta = v_1 + t v_2, \quad t = k_2/k_1,$$

and using (A.7) to change the variables of integration, Eq. (A.3) can be written in the form

$$\mu(P_c) \sim \rho_0 \frac{|r|}{\left|1 - r\right|^{d_s}} \int_{P_c} \left|\xi - r\eta\right|^{d_s-1} d\xi\, d\eta,$$
(A.8)

where

$$\frac{1}{r} = 1 + \frac{t}{\lambda_u}$$

and

$$\frac{1}{\rho_0} = \frac{1}{\rho}\,\gamma_u\left|\lambda_u\right|\left[\gamma_s(\lambda_s - \lambda_u)\right]^{d_s}.$$

The integration in the variable in the direction of the straight lines $K^{\mathrm{T}}V = \pm\delta$ can be done exactly. On the contrary, except in the case $k_1 = 0$, the integration in the other variable does not seem to be possible in closed form. We have therefore resorted to numerical integration to obtain the results presented in Figure 16 (see next page).

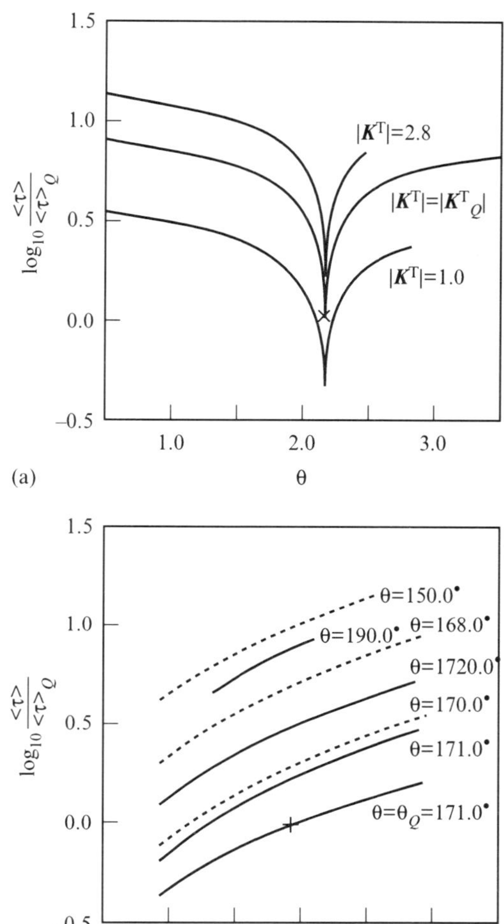

Figure 16 Hénon map (theoretical) curves of $\log_{10}$ $(\langle\tau\rangle/\langle\tau\rangle_Q)$ versus $\theta = \arg(K^{\mathrm{T}})$ with $|K^{\mathrm{T}}|$ fixed (a) and of $\log_{10}(\langle\tau\rangle/\langle\tau\rangle_Q)$ versus $|K^{\mathrm{T}}|$ with θ fixed (b). The × (+) denotes the reference value.

An accurate analytical approximation can be obtained in the case $t = -\lambda_s$ (slab parallel to the stable manifold) in the limit $\delta \to 0$:

$$\mu(P_c) \sim \rho_1 g(\mu_1)\delta^\gamma[1 + 0(\delta)], \qquad (A.9)$$

where g is the function defined by

$$g(\mu_1) = \frac{2}{\gamma} \frac{(1 + \mu_1)^\gamma - (1 - \mu_1)^\gamma}{\mu_1(|\lambda_u| + \mu_1)^\gamma},$$

$$\mu_{1m} < \mu_1 < 0 \quad 0 < \mu_1 < \mu_{1M}$$

$$g(0) = 4|\lambda_u|^{-\gamma},$$

and

$$\gamma = \tfrac{1}{2}d_s + 1,$$

$$\rho_1 = \frac{\rho_0}{d_s} \frac{|r|}{|1 - r|^{d_s}}, \quad \frac{1}{r} = 1 - \frac{\lambda_s}{\lambda_u},$$

$$\mu_{1m} = -1 + \frac{r^2}{|\lambda_u| - 1}\,\delta + 0(\delta^1),$$

$$\mu_{1M} = 1 - \frac{r^2}{|\lambda_u| + 1}\,\delta + 0(\delta^2).$$

The dependence of $\mu(P_c)$ on δ given by (A.9) is precisely that predicted in Ref. [6]. The dependence of $\mu(P_c)$ on μ_1 shows very good agreement with the experimental results – see Figure 2. Note that in plotting the theoretical curve we used as normalizing constant ρ_1 that obtained by least square fitting the theoretical curve to the experimental points.

We have used Eq. (A.8) to study the dependence of $\langle\tau\rangle = 1/\mu(P_c)$ on the gain vector K^{T}. In Figure 16(a) we have plotted curves of $\langle\tau\rangle$ versus $\theta = \arg(K^{\mathrm{T}})$ with $|K^{\mathrm{T}}|$ kept fixed and in Figure 16(b) curves of $\langle\tau\rangle$ versus $|K^{\mathrm{T}}|$ with θ kept fixed. $\langle\tau\rangle$ was normalized to its value at the point Q (see Section 3 and Figure 1). The results show that $\langle\tau\rangle$ exhibits a strong minimum at $\theta = \theta_Q$ for all values of $|K^{\mathrm{T}}|$ and increases slowly with $|K^{\mathrm{T}}|$ for all values of θ, in agreement with the experimental results of Figures 3 and 4.

Appendix B. Numerical method for calculating $\langle\tau\rangle$

In this appendix we describe the procedure used in Sections 3 and 4 to numerically obtain the average time to achieve control, $\langle\tau\rangle$.

From (2.10) we obtain the fraction of chaotic transients with length smaller than some value τ_{max},

$$\rho_{\tau_{\mathrm{max}}} = \int_0^{\tau_{\mathrm{max}}} \phi(\tau)\, d\tau = 1 - \exp\left(-\frac{\tau_{\mathrm{max}}}{\langle\tau\rangle}\right),$$

and the average length of the chaotic transients with length smaller than τ_{max},

$$\langle\tau\rangle_{\tau_{\mathrm{max}}} = \int_0^{\tau_{\mathrm{max}}} \tau\phi(\tau)\, d\tau$$

$$= \langle\tau\rangle\left[1 - \left(1 + \frac{\tau_{\mathrm{max}}}{\langle\tau\rangle}\right)\exp\left(-\frac{\tau_{\mathrm{max}}}{\langle\tau\rangle}\right)\right].$$

Combining these two equations we obtain

$$\langle\tau\rangle = \frac{\langle\tau\rangle_{\tau_{\mathrm{max}}}}{1 - (1 - \rho_{\tau_{\mathrm{max}}})[1 - \log_e(1 - \rho_{\tau_{\mathrm{max}}})]}, \qquad (B.1)$$

which is the required formula. Note that $\rho_\infty = 1$, $\langle\tau\rangle_\infty = \langle\tau\rangle$.

The numerical procedure to calculate the average time to achieve control is as follows. Take a large number N_0 of randomly chosen initial conditions and iterate each of them with the uncontrolled map (i.e., with $Z \mapsto F(Z,\bar{p})$) a sufficient number of times until they are all distributed over the attractor according to its natural measure. Then switch on the control as specified by (2.9) and determine how many further iterates are necessary for $N_f \le N_0$ orbits to fall within a circle of small radius centered at the fixed point. Letting τ_{max} be this number of iterates and $\{\tau_j\}$ with $j = 1, \ldots, N_f$ be the times required for the N_f orbits to fall within the small circle, we have

$$\rho_{\tau_{max}} = \frac{N_f}{N_0}, \quad \langle \tau \rangle_{\tau_{max}} = \frac{1}{N_f} \sum_{j=1}^{N_f} \tau_j.$$

Finally we use Eq. (B.1) to obtain $\langle \tau \rangle$. In our numerical experiments described in Sections 3 and 4, we took $N_0 = 192$, $N_f = 121$, values that led to a good compromise between accuracy and computation time.

Appendix C. Derivation of the double rotor map

The equations of motion of the kicked double rotor are

$$\frac{d}{dt}\left(\frac{\partial L}{\partial \dot\theta_j}\right) - \frac{\partial L}{\partial \theta_j} = -\frac{\partial F}{\partial \dot\theta_j}, \quad j = 1, 2, \tag{C.1}$$

where the Lagrangian function L is the difference between the kinetic energy,

$$K(\dot\theta_1, \dot\theta_2) = \tfrac{1}{2} I_1 \dot\theta_1^2 + \tfrac{1}{2} I_2 \dot\theta_2^2,$$

and the potential energy,

$$V(\theta_1, \theta_2, t) = (l_1 \cos\theta_1 + l_2 \cos\theta_2) f(t),$$

i.e., $L = T - V$, and Rayleigh's dissipation function F is

$$F(\dot\theta_1, \dot\theta_2) = \tfrac{1}{2} v_1 I_1 \dot\theta_1^2 + \tfrac{1}{2} v_2 I_2 (\dot\theta_2 - \dot\theta_1)^2.$$

The sequence of forcing kicks is given by the semi-infinite comb of delta functions of period T and strength f_0.

$$f(t) = f_0 \sum_{k=1}^{\infty} \delta(t - kT). \tag{C.2}$$

Here I_1 and I_2 are the moments of inertia.

$$I_1 = (m_1 + m_2) l_1^2, \quad I_2 = m_2 l_2^2,$$

and v_1, v_2 are the coefficients of friction.

Elimination of L and F from (C.1) yields

$$\frac{d}{dt}\begin{pmatrix}\dot\theta_1\\ \dot\theta_2\end{pmatrix} = \begin{pmatrix}-(v_1 + v_2 I_2/I_1) & v_2 I_2/I_1\\ v_2 & -v_2\end{pmatrix}\begin{pmatrix}\dot\theta_1\\ \dot\theta_2\end{pmatrix} + f(t)\begin{pmatrix}(l_1/I_1)\sin\theta_1\\ (l_2/I_2)\sin\theta_2\end{pmatrix}. \tag{C.3}$$

We now proceed to integrate Eq. (C.3). For simplicity we take $I_1 = I_2 \equiv I$.

Since the effect of the kicks is instantaneous (i.e., $f(t) = 0$, for $t \ne kT$, $k = 1, 2, \ldots$) Eqs. (C.3) are linear between successive kicks. In particular, for $0 < t < T$, Eqs. (C.3) reduce to

$$\frac{d}{dt}\begin{pmatrix}\dot\theta_1\\ \dot\theta_2\end{pmatrix} = \mathbf{A}_v \begin{pmatrix}\dot\theta_1\\ \dot\theta_2\end{pmatrix}, \quad \mathbf{A}_v = \begin{pmatrix}-(v_1 + v_2) & v_2\\ v_2 & -v_2\end{pmatrix}. \tag{C.4}$$

This system can be easily solved by the usual methods for linear differential equations with constant coefficients. Denoting by $\dot\theta_1(0)$, $\dot\theta_2(0)$ the initial angular velocities this solution is

$$\begin{pmatrix}\dot\theta_1(t)\\ \dot\theta_2(t)\end{pmatrix} = \mathbf{L}(t)\begin{pmatrix}\dot\theta_1(0)\\ \dot\theta_2(0)\end{pmatrix} \tag{C.5}$$

where

$$\mathbf{L}(t) = \sum_{j=1}^{2} \mathbf{W}_j e^{\lambda_j t}.$$

λ_1, λ_2 are the eigenvalues of matrix $\mathbf{A}_v$,

$$\left.\begin{array}{c}\lambda_1\\ \lambda_2\end{array}\right\} = -\tfrac{1}{2}(v_1 + 2v_2 \pm \Delta), \quad \Delta = (v_1^2 + 4v_2^2)^{1/2},$$

and $\mathbf{W}_1$, $\mathbf{W}_2$ are the constant matrices

$$\mathbf{W}_1 = \begin{pmatrix}a & b\\ b & d\end{pmatrix}, \quad \mathbf{W}_2 = \begin{pmatrix}d & -b\\ -b & a\end{pmatrix},$$

where

$$\left.\begin{array}{c}a\\ d\end{array}\right\} = \frac{1}{2}\left(1 \pm \frac{v_1}{\Delta}\right), \quad b = -\frac{v_2}{\Delta}.$$

The position of the rods is obtained by integration of Eq. (C.5). Denoting by $\theta_1(0)$, $\theta_2(0)$ the initial positions one obtains

$$\begin{pmatrix}\theta_1(t)\\ \theta_2(t)\end{pmatrix} = \mathbf{M}(t)\begin{pmatrix}\dot\theta_1(0)\\ \dot\theta_2(0)\end{pmatrix} + \begin{pmatrix}\theta_1(0)\\ \theta_2(0)\end{pmatrix}, \tag{C.6}$$

where

$$\mathbf{M}(t) = \int_0^t \mathbf{L}(\xi)\,d\xi = \sum_{j=1}^{2} \mathbf{W}_j \frac{e^{\lambda_j t} - 1}{\lambda_j}.$$

Equation (C.5), (C.6) completely describe the motion of the rotor for $0 < t < T$ (before the first kick).

At $t = T$ the kick instantaneously changes the angular velocity of each rod but not its position; that is, the angular velocity of each rod is discontinuous at $t = T$, while the position is continuous. Denoting by $\theta_j T^{\pm}, \dot\theta_j(T^{\pm}), j = 1, 2$ the values of $\theta_j(t), \dot\theta_j(t)$ just before and just after the kick at $t = T$, we therefore have

$$\theta_j(T^-) - \theta_j(T^+) = \theta_j(T), \tag{C.7}$$

$$\dot\theta_j(T^+) - \dot\theta_j(T^-) = \frac{f_0}{I} l_j \sin\theta_j(T), \tag{C.8}$$

for $j = 1, 2$.

The solution of Eqs. (C.3) for $T < t < 2T$ is identical to the solution of the linear system Eq. (C.4) for $0 < t < T$ except that the initial conditions $\theta_j(0), \dot\theta_j(0), j = 1, 2$ are replaced by $\dot\theta_j(T), \theta_j(T^+), j = 1, 2$.

The solution of eqs. (C.3) is a composition of the solution of Eqs. (C.4) with the effect of the kicks at $t = T, 2T, \ldots$. To study the dynamics of the rotor it is natural to consider only the state of the system immediately after each kick. Thus we obtain from (C.5)–(C.8) the *double rotor map*,

$$\begin{pmatrix}\theta_1^{(k+1)}\\ \theta_2^{(k+1)}\end{pmatrix} = \mathbf{M}(T)\begin{pmatrix}\dot\theta_1^{(k)}\\ \dot\theta_2^{(k)}\end{pmatrix} + \begin{pmatrix}\theta_1^{(k)}\\ \theta_2^{(k)}\end{pmatrix}, \tag{C.9a}$$

$$\begin{pmatrix}\dot\theta_1^{(k+1)}\\ \dot\theta_2^{(k+1)}\end{pmatrix} = \mathbf{L}(T)\begin{pmatrix}\dot\theta_1^{(k)}\\ \dot\theta_2^{(k)}\end{pmatrix} + \frac{f_0}{I}\begin{pmatrix}l_1 \sin\theta_1^{(k+1)}\\ l_2 \sin\theta_2^{(k+1)}\end{pmatrix}. \tag{C.9b}$$

where

$$\theta_j^{(k)} = \theta_j(kT), \quad j = 1, 2$$

are the positions of the rods at the instant of the kth kick, and

$$\dot{\theta}_j^{(k)} = \dot{\theta}_j(kT^+), \quad j = 1, 2$$

are the angular velocities of the rods immediately after the kth kick.

Appendix D. Stability of fixed points for the double rotor map

The coefficients of the characteristic Eq. (4.8) depend on the fixed point and on the forcing f_0 only through the two non-zero elements of the matrix $\mathbf{H}$, which we are going to denote by h_{11} and h_{22}. The discussion of the stability of the fixed points is conveniently carried out in the plane (h_{11}, h_{22}) by considering the intersections between the lines of marginal stability of the characteristic equation (where one of the roots has modulus unity) and the 'orbits' described by the paths followed by the fixed points as the forcing f_0 is varied.

The 'orbits' of the fixed points can be obtained by first eliminating x_{j^*} ($j = 1, 2$) between the two equations

$$f_0 \sin x_{j^*} = f_{0j}, \quad h_{jj} = \frac{f_0}{l} l_j \cos x_{j^*},$$

with the result

$$h_{jj} = \pm \frac{l_j}{l}(f_0^2 - f_{0j}^2)^{1/2}, \quad j = 1, 2,$$

and then eliminating f_0 between these two equations, with the result

$$\left(\frac{h_{22}}{l_2}\right)^2 - \left(\frac{h_{11}}{l_1}\right)^2 = \frac{1}{l^2}(f_{01}^2 - f_{02}^2),$$

which is the equation of the hyperbola described by each fixed point in the plane (h_{11}, h_{22}) when f_0 is varied. It should be pointed out that symmetric fixed points with respect to $(x_{1^*}, x_{2^*}, y_{1^*}, y_{2^*}) \to (2\pi - x_{1^*}, 2\pi - x_{2^*}, -y_{1^*}, -y_{2^*})$ describe the same 'orbit'. The lines of marginal stability are defined by the equation (see Eq. (4.8b))

$$P(e^{i\alpha}) = 0, \tag{D.1}$$

where α can take values in the interval $[0, 2\pi)$. When $\alpha = 0$ or $\alpha = \pi$ this equation simplifies considerably. We obtain:

(i) $P(1) = 0 = |\mathbf{HM}| = |\mathbf{H}||\mathbf{M}|$; as $|\mathbf{M}| \neq 0$
     this implies

$$h_{11}h_{22} = 0.$$

(ii) $P(-1) = 0 = |2(\mathbf{I} + \mathbf{L}) + \mathbf{HM}| = |\mathbf{H} + \mathbf{R}||\mathbf{M}|,$
      where

$$\mathbf{R} = 2(\mathbf{A}_v + 2\mathbf{M}^{-1});$$

writing $\mathbf{R} = \{r_{ij}\}_{i,j=1,2}$ this leads to

$$h_{11}h_{22} + r_{22}h_{11} + r_{11}h_{22} + (r_{11}r_{22} - r_{12}r_{21}) = 0.$$

When $\alpha \neq 0, \pi$ it can be shown that the Eq. (D.1) has no solutions in the (real) plane (h_{11}, h_{22}); that is, there are no lines of marginal stability with $\alpha \neq 0, \pi$.

In Figure 17 we have plotted in the plane (h_{11}, h_{22}) the

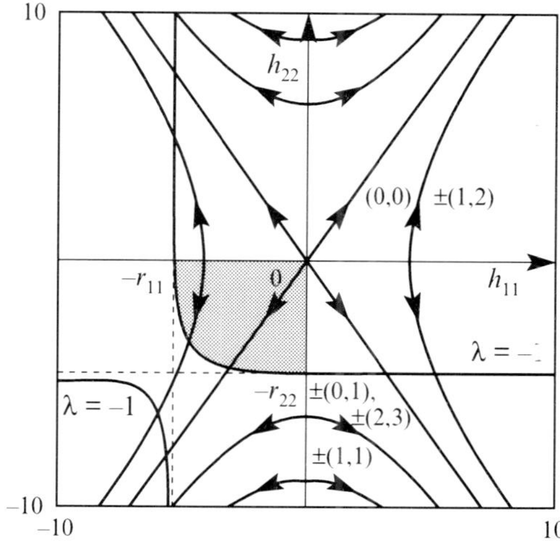

Figure 17 Double rotor map: stability diagram of the fixed points with rotation numbers (n_1, n_2). ($f_0 = 9.0$, Eq. (4.4).)

lines of marginal stability $P(1) = P(-1) = 0$ (for the parameter values given by Eq. (4.4)). The bounded region between these lines, which is the shaded region in the figure, is the only region of the plane where all the roots of the characteristic equation have modulus smaller than unity. We have also plotted the 'orbits' of the first five sets of fixed points, the arrows indicating the direction the forcing increases; the critical values of (h_{11}, h_{12}) at $f_0 = f_{0c}$, which occur on the lines $P(1) = 0$, are given in Table 1. We see that of all these orbits only two cross the shaded region: one corresponds to the fixed point $[(0, 0);4]$; the other to the fixed points $[(1, 2); 4]$ and $[(-1, -2);1]$. These fixed points are therefore stable while their 'orbits' remain in the shaded region and become unstable when the 'orbits' cross the line $P(-1) = 0$; that is, they are stable in a finite interval of values of f_0, $f_{0c} < f_0 < f_{0u}$. The other fixed points are unstable for all values of f_0. The values of f_{0u} are also given in Table 2.

Figure 17 only applies for the particular values of the parameters given by Eqs. (4.4). For other values the relative positions of the lines of marginal stability and 'orbits' of fixed points are different, and fixed points with other rotation numbers may be stable. In general, we can make the following statements regarding the stability of the fixed points: those with rotation numbers $(0, 0)$ are stable over an interval $0 < f_0 < f_{0u}^{(0,0)}$; all the others are either stable over an interval $f_{0c}^{(n_1, n_2)} < f_0 < f_{0u}^{(n_1, n_2)}$ or are always unstable.

References

[1] C. Grebogi, E. Ott and J.A. Yorke, *Phys. Rev. A* **37** (1988) 1711.

[2] D. Auerbach, P. Cvitanovich, J.P. Eckmann, G. Gunaratne and I. Procaccia, *Phys. Rev. Lett.* **58** (1987) 2387.

[3] T. Morita, H. Hata, H. Mori, T. Horita and K. Tomita, *Prog. Theor. Phys.* **78** (1987) 511.

[4] A. Katok, *Publ. Math. IHES* **51** (1980) 137.

[5] R. Bowen, *Trans. Am. Math. Soc.* **154** (1971) 377.

[6] E. Ott, C. Grebogi and J.A. Yorke, *Phys. Rev. Lett.* **64** (1990) 1196.

[7] F.J. Romeiras, E. Ott. C. Grebogi and W.P. Dayawansa. Proc. 1991 Am. Control Conf. (American Automatic Control Council, IEEE Service Center, Piscataway, NJ, 1991), pp. 1112–1119.

[8] J. Singer, Y.-Z. Wang and H.H. Bau, *Phys. Rev. Lett.* **66** (1991) 1123.

[9] W.L. Ditto, S.N. Rauseo and M.L. Spano, *Phys. Rev. Lett.* **65** (1990) 3211.

[10] T.B. Fowler, *IEEE Trans. on Automatic Control* **34** (1989) 201.

[11] A.M. Block and J.E. Marsden, *Theor. Comput. Fluid Dyn.* **1** (1989) 179;
A. Hübler and E. Lüscher, *Naturwissenshaften* **76** (1989) 67;
E.A. Jackson, *Phys. Lett. A* **151** (1990) 478;
B. Huberman and E. Lumer, *IEEE Trans. Circuits Syst.* **37** (1990) 547;
S. Sinha, R. Ramaswamy and J.S. Rao, *Physica D* **43** (1990) 118;
B. Peng, V. Petrov and K. Showalter, *J. Phys. Chem.* **95** (1991) 4957;
T. Tél, Controlling transient chaos, preprint;
A. Azevedo and S.M. Rezende, *Phys. Rev. Lett.* **66** (1991) 1342;
Y. Braiman and J. Goldhirsch, *Phys. Rev. Lett.* **66** (1991) 2545.

[12] K. Ogata. Control engineering, Second Ed. (Prentice-Hall, Englewood Cliffs, NJ, 1990), pp. 782–784.

[13] M. Hénon, *Commun. Math. Phys.* **50** (1976) 69.

[14] C. Grebogi, E. Kostelich, E. Ott and J.A. Yorke, *Physica D* **25** (1987) 347; *Phys. Lett. A* **118** (1986) 448; A 120 (1987) 497 (E).

[15] C. Grebogi, E. Ott and J.A. Yorke, Chaotic attractors in crisis, *Phys. Rev. Lett.* **48** (1982) 1507.

[16] C. Grebogi, E. Ott and J.A. Yorke, *Physica D* **7** (1983) 181.

[17] C. Grebogi, E. Ott and J.A. Yorke, *Phys. Rev. Lett.* **57** (1986) 1284.

[18] C. Grebogi, E. Ott, F.J. Romeiras and J.A. Yorke, *Phys. Rev. A* **36** (1987) 5365.

[19] G. Nitsche and U. Dressler, Controlling chaotic dynamical systems using time delay coordinates, preprint.

[20] G.M. Zaslavsky, *Phys. Lett. A* **69** (1978) 145.

[21] G. Benettin, L. Galgani, A. Giorgilli and J. Strelcyn, *Meccanica* **15** (1980) 9.

[22] G. Benettin, L. Galgani, A. Giorgilli and J. Strelcyn, *Meccanica* **15** (1980) 9.

[23] J.D. Farmer, E. Ott and J.A. Yorke, *Physica D* **7** (1983) 153.

[24] J. Kaplan and J.A. Yorke, Chaotic behavior of multidimensional difference equations, in: Functional differential equations and the approximation of fixed points, Lecture Notes in Mathematics, eds. H.O. Peitgen and H.O. Walther, Vol. 730 (Springer, Berlin, 1978), pp. 228–237.

[25] T. Shinbrot, E. Ott, C. Grebogi and J.A. Yorke, *Phys. Rev. Lett.* **65** (1990) 3215.

[26] E. Ott, C. Grebogi and J.A. Yorke, Controlling chaotic dynamical systems, in CHAOS/XAOC, Soviet–American perspective on nonlinear science, ed. D. Campbell (American Institute of Physics, New York, 1990), pp. 153–172.

[27] F. Takens, Detecting strange attractors in turbulence, in: Dynamical systems and turbulence, eds. D.A. Rand and L.-S. Young, Lecture Notes in Mathematics, Vol. 898 (Springer, New York, 1980), pp. 366–381.

[28] J.C. Sommerer. W.L. Ditto, C. Grebogi, E. Ott and M. Spano, *Phys. Lett. A* **153** (1991) 105.

[29] D.P. Lathrop and E.J. Kostelich, *Phys. Rev. A* **40** (1989) 4028.

[30] G.H. Gunaratne, P.S. Linsay and M.J. Vinson, *Phys. Rev. Lett.* **63** (1989) 1.

[31] J.C. Sommerer, W.L. Ditto, C. Grebogi, E. Ott and M.L. Spano, *Phys. Rev. Lett.* **66** (1991) 1947.

Paper 3
Controlling chaos using time delay coordinates

Ute Dressler and Gregor Nitsche
Daimler-Benz Research Institute, Goldsteinstrasse 235, 6000 Frankfurt/Main 71, Germany

Received 29 August 1991

The Ott–Grebogi–Yorke control method is analyzed in the case that the attractor is reconstructed from a time series using time delay coordinates. It turns out that the control formula of Ott, Grebogi and Yorke should be modified in order to apply to experimental systems if time delay coordinates are used. We reveal that the experimental surface of section map depends not only on the actual parameter but also on the preceding one. In order to meet this dependence two modifications are introduced which lead to a better performance of the control. To compare their control abilities they are applied to simulations of a Duffing oscillator.

In 1990 Ott, Grebogi, and Yorke (OGY) [1] proposed a new method of controlling a chaotic dynamical system by stabilizing one of the many unstable periodic orbits embedded in a chaotic attractor, through only small time-dependent perturbations in some accessible system parameter. This makes OGY's approach quite different from other previously published methods on controlling chaos [2].

OGY's method has attracted the attention of many physicists interested in applications of nonlinear dynamics. One reason for this is that OGY stress that all values needed to achieve control can be obtained from an experimental signal starting with the well-known embedding technique [3,4]. Therefore the control method can in principle be applied to experimental systems where the dynamical equations are not known. Indeed, Ditto, Rauseo, and Spano demonstrated recently [5] a first control of a physical system using the method of Ott, Grebogi, and Yorke.

With regard to possible applications we investigate the OGY control method in the case that the attractor is reconstructed from a time series using time delay coordinates. It turns out that the control formula of OGY should be modified in order to apply to experimental systems if time delay coordinates are used. The main argument will be that during the control process one switches the control parameter p from p_{i-1} to p_i at times t_i (t_i is the time of the ith piercing of the surface of section by the trajectory). But, if one uses delay coordinates, the experimental surface of section map P does not only depend on the new actual parameter p_i (as OGY implicity assume) but also on the old one p_{i-1}. In order to meet this dependence two modifications of the control algorithm are proposed. Their control abilities are compared with the original OGY formula by applying them to a time series obtained from simulations of a Duffing oscillator.

Let us briefly recall the OGY control idea. For simplicity we restrict ourselves to a two-dimensional discrete dynamical system (e.g., the surface of section map P of a three-dimensional continuous system). There also exist extensions of the method to higher-dimensional dynamical systems [6,7]. Let the system depend on some accessible parameter $p \in (p_0 - \delta p_{max}, p_0 + \delta p_{max})$ with maximal possible perturbation δp_{max}, $\xi_{i+1} = P(\xi_i, p)$. Let $\xi_F = P(\xi_F, p_0)$ denote the unstable fixed point on the attractor which one wants to stabilize. The control idea is to monitor the system until it comes close to the desired fixed point and then change p by a small amount such that the next state ξ_{i+1} will fall into the stable direction of the fixed point. To do this one uses the first-order approximation of P near ξ_F and p_0,

$$\delta\xi_{i+1} \cong A\delta\xi_i + \mathbf{w}\delta p_i,$$

with $\delta\xi_i = \xi_i - \xi_F$, $\delta p_i = p_i - p_0$, $A = D_\xi P(\xi_F, p_0)$, and $\mathbf{w} = \partial P/\partial p$ (ξ_F, p_0). Writing the linearization A as $A = \lambda_u \mathbf{e}_u \mathbf{f}_u + \lambda_s \mathbf{e}_s \mathbf{f}_s$, with $\mathbf{e}_u$ ($\mathbf{e}_s$) the unstable (stable) eigendirections of A with eigenvalues λ_u (λ_s) and $\mathbf{f}_u$ ($\mathbf{f}_s$) their contravariant basis vectors, i.e., $\mathbf{f}_s \cdot \mathbf{e}_u = \mathbf{f}_u \cdot \mathbf{e}_s = 0$ and $\mathbf{f}_s \cdot \mathbf{e}_s = \mathbf{f}_u \cdot \mathbf{e}_u = 1$, the condition that ξ_{i+1} falls on the local stable manifold of the fixed point can be formulated as $\mathbf{f}_u \cdot \delta\xi_{i+1} = 0$, which yields the control formula [8] for the new value of the control parameter $p_i = p_0 + \delta p_i$,

$$\delta p_i = -(\lambda_u/\mathbf{f}_u \cdot \mathbf{w})\mathbf{f}_u \cdot \delta\xi_i. \tag{1}$$

The control is only activated if the resulting change in the parameter δp_i is less than the maximal allowed disturbance δp_{max}; otherwise δp_i is set to zero.

Let us now consider the case that the only information about the system is obtained by some measurement process which is mathematically realized by some scalar function Z on the state space M. If $Y(t) \in M$ is the state of the system at time t, the experimental time series $z(t) = Z(Y(t))$ is obtained. Using time delay coordinates with delay τ and embedding dimension d, a d-dimensional delay coordinate vector is formed, $\mathbf{X}(t) = (z(t), z(t - \tau), \ldots, z(t -(d - 1)\tau)) \in \mathbb{R}^d$. The experimental surface of section is obtained by the common choice that one component of $\mathbf{X}(t)$ equals a constant, e.g., $[\mathbf{X}(t_i)]_1 \equiv z(t_i) = c$. This procedure gives the successive points $\xi_i \in \mathbb{R}^{d-1}$ in the surface of section and the surface of section map $\xi_{i+1} = P(\xi_i)$.

In what follows we focus our interest on the so obtained experimental surface of section map P. For the sake of simplicity let us assume that one wants to stabilize an unstable fixed point ξ_F of P which has been localized by the well-known technique of recurrent points [9–11]. Applying the OGY control algorithm implies that one (instantaneously) changes at the times t_i the parameter p from p_{i-1} to an appropriately chosen parameter p_i using Eq. (1). Let us now assume that the time between successive piercings of the surface of section is bigger than the lag window, i.e., $t_{i+1} - t_i > (d -1)\tau$. The reason that one hopes to be able to control the original system $Y(t)$ by observing $\mathbf{X}(t)$ is that for appropriately chosen embedding parameters d and τ [4] there exists a bijective relation Φ between the states $\mathbf{X}(t)$ and $Y(t)$, i.e., $\mathbf{X}(t) = \Phi(Y(t))$. The mapping Φ is, however, closely related to the dynamical equations of the system and thus, in general, dependent on the actual value of the control parameter p_i. This will be taken into account by writing Φ_{p_i} instead of Φ.

Our argumentation is now as follows. The point ξ_i at time t_i in the surface of section is related to the original state by $Y(t_i) = \Phi_{p_{i-1}}^{-1} (c, z(t_i - \tau), \ldots, z(t_i - (d - 1)\tau))$. Here we make use of our assumption that $(d - 1)\tau < t_i - t_{i-1}$ which assures that p_{i-1} is the actual value of p during the whole time interval (t_{i-1}, t_i). The time development of the original system from time t_i to the time t_{i+1} is, in case of activated control, given by $\varphi_{p_i}^{t_{i+1}-t_i}$ with φ_p^t the flow map of the dynamical system depending on p. Thus the state of the system at time t_{i+1} is obtained by $Y(t_{i+1}) = \varphi_{p_i}^{t_{i+1}-t_i} (Y(t_i))$ and the corresponding state in the embedding space by $\mathbf{X}(t_{i+1}) = \Phi_{p_i} (Y(t_{i+1}))$. Therefore we obtain $\mathbf{X}(t_{i+1}) = (\Phi_{p_i} \circ \varphi_{p_i}^{t_{i+1}-t_i} \circ \Phi_{p_{i-1}}^{-1}) (\mathbf{X}(t_i))$. This gives our main conclusion. In the case of activated control (i.e., switching the parameter from p_{i-1} to p_i at time t_i) the experimental surface of section map P depends not only on the new actual value p_i but also on the preceding value p_{i-1}, i.e.,

$$\xi_{i+1} = P(\xi_i, p_{i-1}, p_i).$$

Taking this as the starting point the algorithm of OGY is straightforwardly extended. The linearization which one has to consider now is given by

$$\delta\xi_{i+1} \cong A\delta\xi_i + \mathbf{v}\delta p_{i-1} + \mathbf{u}\delta p_i,$$

with $A = D_\xi P(\xi_F, p_0, p_0)$, $\mathbf{v} = \partial P/\partial p_{i-1}(\xi_F, p_0, p_0)$, and $\mathbf{u} = \partial P/\partial p_i(\xi_F, p_0, p_0)$. Demanding $\mathbf{f}_u \cdot \delta\xi_{i+1} = 0$ one obtains as a new control law

$$\delta p_i = -\frac{\lambda_u}{\mathbf{f}_u \cdot \mathbf{u}}\mathbf{f}_u \cdot \delta\xi_i - \frac{\mathbf{f}_u \cdot \mathbf{v}}{\mathbf{f}_u \cdot \mathbf{u}}\delta p_{i-1}. \qquad (2)$$

When P is not influenced by the preceding perturbation δp_{i-1}, i.e., $\mathbf{v} = 0$, the original OGY control formula (1) is reobtained. To see this we note that the vector $\mathbf{w}$ in the control formula (1) is related to $\mathbf{u}$ and $\mathbf{v}$ by $\mathbf{w} = \mathbf{u} + \mathbf{v}$.

The new control formula Eq. (2) contains one possible

instability. In the case that $|(\mathbf{f}_u \cdot \mathbf{v})/(\mathbf{f}_u \cdot \mathbf{u})| \geq 1$ holds the required perturbations δp_i, will, in general, grow until they exceed the maximum allowed value δp_{max}, and the range of control will be left. To avoid this instability (i.e., the growing of δp_i) we propose an alternative approach. We try to find a control law for δp_i such that δp_{i+1} automatically will become zero. This is done by demanding that the system stabilizes only the next but one step, $i + 2$, and that δp_{i+1} equals zero, i.e., by the requirements $\mathbf{f}_u \cdot \delta\xi_{i+2} = 0$ and $\delta p_{i+1} = 0$.

Using the linearization twice, these requirements yield the second modification of the control formula,

$$\delta p_i = -\frac{\lambda_u^2}{\lambda_u \mathbf{f}_u \cdot \mathbf{u} + \mathbf{f}_u \cdot \mathbf{v}} \mathbf{f}_u \cdot \delta\xi_i - \frac{\lambda_u \mathbf{f}_u \cdot \mathbf{v}}{\lambda_u \mathbf{f}_u \cdot \mathbf{u} + \mathbf{f}_u \cdot \mathbf{v}} \delta p_{i-1}. \qquad (3)$$

The control formulas introduced above have been applied to simulations of a Duffing oscillator [12] given by $\ddot{x} + d\dot{x} + x + x^3 = f \cos \omega t$. This system has been numerically integrated. As a measurement function the displacement of the oscillator $z(t) \equiv x(t)$ is chosen. We use a three-dimensional embedding with delay time $\tau = T/4$ with $T = 2\pi/\omega$. The experimental surface of section was obtained by taking $[\mathbf{X}(t_i)]_1 = z(t_i) = \text{const}$. For the localization of fixed points, the standard method described in Refs. [9,10] is used (for details see also [13]). To obtain the vectors $\mathbf{u}$ and $\mathbf{v}$ the perturbations δp_i are alternately switched on and off at every piercing of the surface of section such that $\delta p_i = 0$ for i odd and $\delta p_i = \bar{p}$ for i even, $\bar{p}$ small, respectively. Regarding all pairs (ξ_i, ξ_{i+1}) with even i as one group and the pairs with odd i as another, it is now possible to fit affine mappings in the neighborhood of ξ_F to $P(\cdot, p_0 + \bar{p}, p_0)$ using only pairs (ξ_{r_i}, ξ_{r_i+1}), r_i odd, and to $P(\cdot, p_0, p_0 + \bar{p})$ using only pairs (ξ_{r_i}, ξ_{r_i+1}), r_i even, respectively. These fits then determine $\mathbf{u}$ and $\mathbf{v}$ by the relations $P(\xi_F, p_0 + \bar{p}, p_0) \cong \xi_F + \mathbf{v}\bar{p}$ and $P(\xi_F, p_0, p_0 + \bar{p}) \cong \xi_F + \mathbf{u}\bar{p}$.

To compare the performances of the three different control formulas we tried to stabilize the three fixed points ξ_{F1}, ξ_{F2}, and ξ_{F3} which were determined embedded in a chaotic attractor of the Duffing oscillator (see Figure 1). To stabilize these orbits we choose as accessible parameter p the amplitude of the driving f and a maximal allowed perturbation $\delta p_{max} = 0.5$. In Figure 2, the three different control formulas are successively applied to stabilize ξ_{F1}. Only the second modification (3) was able to stabilize ξ_{F1}. The coefficients of the control formulas (see Table 1) explain why our first modification (2) of the OGY algorithm did not work. The criterion for a stable control algorithm $|b_2| < 1$ was hurt. The large absolute value of $b_2 = (\mathbf{f}_u \cdot \mathbf{v})/(\mathbf{f}_u \cdot \mathbf{u})$, indicates further that the influence of the change of the preceding parameter p_{i-1} is relatively larger than that of the actual one p_i. But this is exactly what is neglected if one applies the original approach of OGY without considering the meaning of the time delay coordinates.

The stabilization of the second fixed point shows different features. Here the generic condition $(\mathbf{f}_u \cdot \mathbf{w} \neq 0)$ of the OGY formula is almost violated. Because of the resulting large value of the coefficient a there were only rare cases where the control requirement $\delta p_i < \delta p_{max}$ was met. But even then the control range was soon left without succeeding in

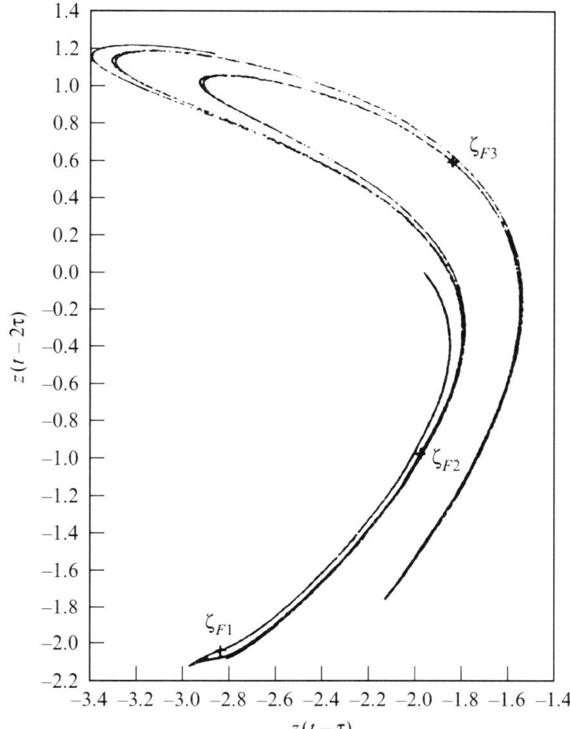

Figure 1 A chaotic attractor of the Duffing oscillator ($d = 0.2$, $f = p = 36$, $\omega = 0.661$) in the surface of section. The surface of section was obtained by the conditions $z(t_i) = 1$, $\dot{z}(t_i) > 0$, and $z(t_i - \tau) < 0$. The three unstable fixed points observed are indicated by the crosses. For further reference they are called ξ_{F1}, ξ_{F2}, and ξ_{F3}.

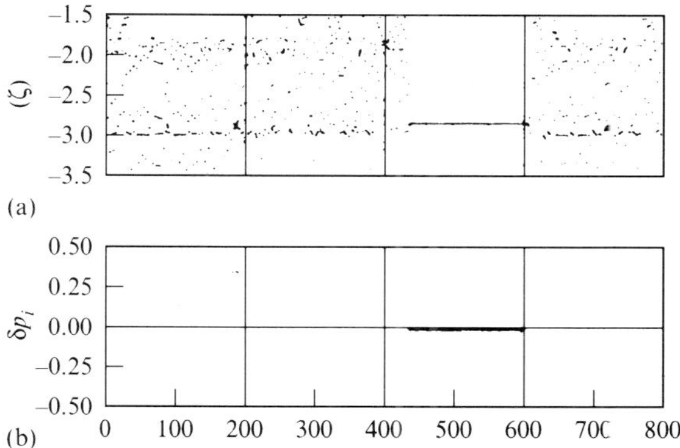

Figure 2 (a) The first component $(\xi_i)_1$ or the points in the surface of section vs i. In order to stabilize the fixed points ξ_{F1} the OGY control formula (1) was switched on from $i = 1$ to 200, the first modification (2) from $i = 201$ to 400, the second (3) from $i = 401$ to 600, and again OGY's control formula from 601 to 800. As can be seen only procedure (3) was able to stabilize ξ_{F1}. (b) The parameter perturbations δp_i vs i used for control. The maximal allowed disturbance was $\delta p_{max} = 0.5$ and $p_0 = 36$.

control. The coefficient b_2 just violates the stability criterion. Indeed, the used perturbations δp_i increased at the beginning. But finally, probably due to nonlinear effects, the control procedure stabilized and the algorithm was able to achieve control. The second modification Eq. (3) was again able to achieve control but with pertubations drastically smaller than the one used for Eq. (2).

The third fixed point ξ_{F3} could be stabilized by any of the three versions of the control formula. For ξ_{F3} the coefficients of the control formulas are very similar (see Table 1). The coefficient b_2 is relatively small which indicates the small influence of δp_{i-1} compared to δp_i. So one can expect that all three algorithms will work.

The algorithms were also tested using further surfaces of section. Among others we also investigated the stroboscopic surface of section map which was used by Ditto, Rauseo, and Spano in [5], i.e., as time series we took a stroboscopic measurement $x(t_i)$, $t_i - t_{i-1} = T$, and obtained a surface of section with points $\xi_i = (x(t_i), x(t_{i-1}))$. In this surface of section the periodic motion corresponding to ξ_{F1} could be stabilized by all three algorithms. They were almost equivalent because b_2 and c_2 were nearly zero (of the order of 10^{-4}), so the other coefficients were practically the same ($a \approx b_1 \approx c_1 \approx 2.7$). The periodic motion corresponding

to ξ_{F2} could not be stabilized because the embedding in the neighborhood of the fixed point was bad (not injective). The third fixed point finally could only be stabilized using second modification (3).

Altogether the numerical investigations show that the possibility of stabilizing a fixed point is not an intrinsic property of a fixed point, as the eigenvalues λ_u and λ_s are, for example. The coefficients of the control formulas differ for different surfaces of section and so do their performances. We always observed that the performance of the first modification (2) is superior to the one of the original OGY formula and the second modification outperforms the latter two. However, their performances are similar whenever the influence of the preceding parameter perturbation δp_{i-1} is small which results in a small value of $\mathbf{f}_u \cdot \mathbf{v}$. But we did observe that the OGY formula failed and the applications of one of the modifications could stabilize the desired fixed point. As a rule this happened when the influence of the changes of the preceding parameter was noticeable, which resulted in a non-negligible value of $\mathbf{f}_u \cdot \mathbf{v}$.

In conclusion, we introduced two modifications of the

Table 1 The numerically obtained values of the coefficients in the control formulas for the three fixed points considered. The coefficients are introduced implicitly by writing the OGY control formula as $\delta p_i = a\hat{\mathbf{f}}_u \cdot \delta\xi_i$, the first modification Eq. (2) as $\delta p_i = b_1\hat{\mathbf{f}}_u \cdot \delta\xi_i + b_2\delta p_{i-1}$, and the second modification Eq. (3) as $\delta p_i = c_1\hat{\mathbf{f}}_u \cdot \delta\xi_i + c_2\delta p_{i-1}$ with $\hat{\mathbf{f}}_u = \mathbf{f}_u/\|\mathbf{f}_u\|$

	a	b_1	b_2	c_1	c_1	λ_u
ξ_{F1}	-16.43	-228.4	-12.9	38.9	2.2	-1.87
ξ_{F2}	164.6	-7.35	1.04	-9.4	1.33	4.82
ξ_{F3}	-2.46	-1.97	0.20	-1.79	0.18	-1.85

control formula of OGY which can lead to a better performance of the control in the case that the dynamical system is reconstructed using time delay coordinates. Therefore these modifications extend the range of applicability of the OGY control method. With these modifications all remarkable advantages of the OGY control method are preserved; e.g., the dynamics equation is not required, the perturbations of the accessible parameter can be very small, different periodic points can be stabilized in the same parameter range for the same system, and after having determined the control coefficients the computational effort at every iteration is negligible which opens the possibility of real time applications. We expect that the OGY control method will yield important applications in the future for technical systems also.

We acknowledge fruitful discussions with C. Mohrdieck and U. Parlitz. One of us (G.N.) was supported by the Studienstiftung des Deutschen Volkes.

Notes

[1] E. Ott. C. Grebogi, and J.A. Yorke, *Phys. Rev. Lett.* **64**, 1196–1199 (1990).

[2] Here we want to mention the resonant control method (using no feedback) introduced by Hübler and Lücher [A. Hübler and E. Lücher. *Naturwissenschaften* **76**, 67 (1989); A. Hübler, *Helv. Phys. Acta* **62**, 291 (1989)]. Hübler suggests a modification of the underlying dynamics such that a goal dynamics becomes a stable solution. To do this one must have or construct a model equation for the dynamics and one must be able to modify experimentally the driving force of these equations with a possibly quite large modification. There are approaches to deal with these difficulties [J. Breeden and A. Hübler, *Phys. Rev. A* **42**, 5817 (1990)].

[3] N.H. Packard, J.P. Crutchfield, J.D. Farmer, and R.S. Shaw, *Phys. Rev. Lett.* **45**, 712–716 (1980).

[4] F. Takens, *Dynamical Systems and Turbulence* (Springer-Verlag, Berlin, 1981), p. 230.

[5] W.L. Ditto, S.N. Rauseo, and M.L. Spano, *Phys. Rev. Lett.* **65**, 3211–3214 (1990).

[6] F.J. Romeiras, E. Ott, C. Grebogi, and W.P. Dayawansa, in *Proceedings of the American Control Conference* (IEEE, New York, to be published).

[7] E. Ott, C. Grebogi, and J.A. Yorke, in *Chaos/XAOC Soviet–American Perspectives on Nonlinear Science*, edited by D.K. Campbell (American Institute of Physics, New York, 1990), pp. 153–172.

[8] We note that the OGY control formula looks different from the one in Ref. [1]. The reason for this is that we use the linearization around ξ_F and p_0 (as is used in [6]) and do not estimate the new position of the fixed point $\xi_F(p)$ as one changes the parameter p, $\xi_F(p) = \xi_F(p_0) + \mathbf{g}\delta p$, as OGY do. But it is easy to show that these approaches are equivalent and $\mathbf{g}$ and $\mathbf{w}$ are related through $\mathbf{g} = [1 - D_\xi P(\xi F, p_0)]^{-1}\mathbf{w}$.

[9] D.P. Lathrop and E.J. Kostelich, *Phys. Rev. A* **40**, 4028–4031 (1989).

[10] K. Pawelzik and H.G. Schuster, *Phys. Rev. A* **43** 1808–1812 (1991).

[11] D. Auerbach, P. Cvitanovic, G. Gunaratne, J.-P. Eckmann, and I. Procaccia, *Phys. Rev. Lett.* **58**, 2387–2389 (1987).

[12] U. Parlitz and W. Lauterborn, *Phys. Lett.* **107A**, 351–335 (1985).

[13] G. Nitsche and U. Dressler (to be published).

Paper 4
Experimental control of chaos

W.L. Ditto, S.N. Rauseo and M.L. Spano
Naval Surface Warfare Center, Silver Spring, Maryland 20903-5000 USA

Received 27 August 1990

We have achieved control of chaos in a physical system using the method of Ott, Grebogi, and Yorke [*Phys. Rev. Lett.* **64**, 1196 (1990)]. The method requires only small time-dependent perturbations of a single-system parameter and does not require that one have model equations for the dynamics. We demonstrate the power of the method by controlling a *chaotic* system around unstable periodic orbits of order 1 and 2, switching between them at will.

In a recent Letter, Ott, Grebogi, and Yorke (OGY) [1] demonstrated that one can convert the motion of a chaotic dynamical system to periodic motion by controlling the system about one of the many unstable periodic orbits embedded in the chaotic attractor, through only small time-dependent perturbations in an accessible system parameter. They demonstrated their method numerically by controlling the Hénon map.

Far from being a numerical curiosity that requires experimentally unattainable precision, we believe this method can be widely implemented in a variety of systems including chemical, biological, optical, electronic, and mechanical systems. In this Letter we report the control of chaos in a physical system, a parametrically driven magnetoelastic ribbon, using the method of OGY.

Theoretical background

The method is based on the observation that unstable periodic orbits are dense in a typical chaotic attractor. Their method assumes only the following four points. First, the dynamics of the system can be represented as arising from an n-dimensional non-linear map (e.g., by a surface of section or time one return map), the iterates given by $\xi_{n+1} = \mathbf{f}(\xi_n, p)$, where p is some accessible system parameter. Second, there is a specific periodic orbit of the map which lies in the attractor and around which one wishes to stabilize the dynamics. Third, there is maximum perturbation δp_* in the parameter p by which it is acceptable to vary p from the nominal value p_0. Finally, one assumes that the position of the periodic orbit is a function of p, but that the local dynamics about it do not vary much with the allowed small changes in p. Note that while the dynamics is assumed to arise from a map, one needs no model for the global dynamics. These assumptions would seem to allow for the control of any chaotic system for which a faithful Poincaré section can be constructed. The construction of a map from and the location of periodic orbits in [2] experimental data are straightforward processes.

To control chaotic dynamics one only needs to learn the *local* dynamics around the desired periodic orbit by observing iterates of the map near the desired orbit and fitting them to a local linear approximation of the map $\mathbf{f}$ [3]. From this, one can find the stable and unstable eigenvalues as well as the local stable and unstable manifolds (given by the eigenvectors). Next, by changing p slightly and observing how the desired orbit changes position, one can estimate the partial derivatives of the orbit location with respect to p.

To control the chaos, one attempts to confine the iterates of the map to a small neighborhood of the desired orbit. When an iterate falls near the desired orbit, we change p from its nominal value p_0 by δp, thereby changing the location of the orbit and its stable manifold, such that the *next* iterate will be forced back toward the stable manifold of the *original* orbit for $p = p_0$. [Figure 1 illustrates this method for the case of a saddle fixed point located at $\xi_F(p_0)$.] That the method of OGY rests on attempting to force the dynamics to stay in the neighborhood of an unstable periodic orbit in the attractor makes it quite different from other previously published methods from removing chaos [4].

Experimental setup and results

The experimental system consisted of a gravitationally buckled, amorphous magnetoelastic ribbon. The ribbon material belongs to a new class of amorphous magnetostrictive materials [5] that have been found to exhibit very

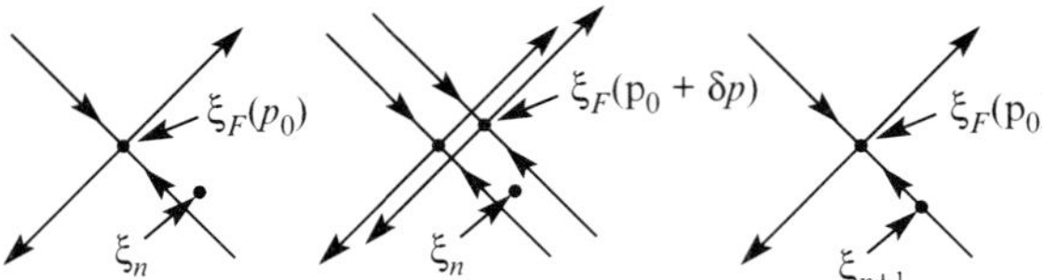

Figure 1 Schematic of the OGY control algorithm for a saddle fixed point: (a) The nth iterate ξ_n falls near the fixed point $\xi_F(p_0)$. (b) Turn on the perturbation of p to move the fixed point. (c) The next iterate is forced onto the stable manifold of $\xi_F(p_0)$. Turn off the perturbation.

large reversible changes of Young's modulus $E(H)$ with the application of small magnetic fields [6,7]. The ribbon was clamped at the base to yield a free vertical length greater than the Euler buckling length, thus giving an initially buckled configuration. The ribbon was placed within three mutually orthogonal pairs of Helmholtz coils, which allowed us to compensate for the Earth's magnetic field and to apply an approximately uniform vertical magnetic field along the ribbon. The Young's modulus of the ribbon was varied by applying a vertical magnetic field having the form $H = H_{dc} + H_{ac} \cos(2\pi f t)$. To lowest order, the ribbon was not driven by magnetic forces, but was forced by gravity as $E(H)$ was varied. The magnetic-field amplitudes were typically set in the range 0.1–2.5 Oe. A sensor measured the curvature of the ribbon near its base. Other details of the experiment can be found in Refs. 6 and 7.

The data were time-series voltages $V(t)$ acquired from the output of the sensor. Voltages were sampled at the drive period of the ac field (at times $t_n = n/f$) by triggering a voltmeter off the ac signal.

By considering the sampled voltages as arising from iterates of a map, $\mathbf{X}_n = V(t_n)$, we are able to directly apply the control theory outlined above. We selected H_{dc} to be the parameter to be varied to achieve control (i.e., $p = H_{dc}$). First, we chose a parameter region (H_{ac}, H_{dc}, and f) such that the ribbon was oscillating chaotically. In order to simplify the comparison with the theory, the parameter region chosen was one in which the dynamics of the iterates near the orbits of interest clearly appears to be two dimensional (i.e., the two-dimensional return map, $\mathbf{X}_{n+1}$ vs $\mathbf{X}_n$, is always single valued in the neighborhood of the orbits of interest). The first 2350 iterations (in gray) in Figure 2(a) are of the uncontrolled time-series data for $H_{ac} = 2.050$ Oe, $H_{dc} = 0.112$ Oe ($= p_0$), and $f = 0.85$ Hz (from 1 to 2350 iterations). In Figure 2(b), the return map for the uncontrolled system is shown in gray. We estimate the dynamical noise in our system, i.e., the deviation of the motion of the ribbon away from deterministic chaos, to be ±0.005 V, since any structure on the attractor below this scale is blurred out.

We found the approximate location X_F of an unstable period-1 orbit of the map (i.e., a fixed point) by noting that any fixed point of the dynamics must lie along the $\mathbf{X}_{n+1} = \mathbf{X}_n$ line in the plot of the return map. To stabilize this fixed point we next examined the data series and found all pairs of iterates *both* of which fell within 0.05 V of the

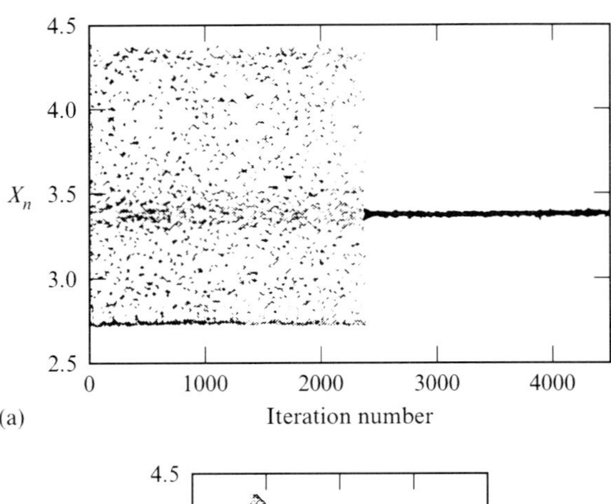

(a)

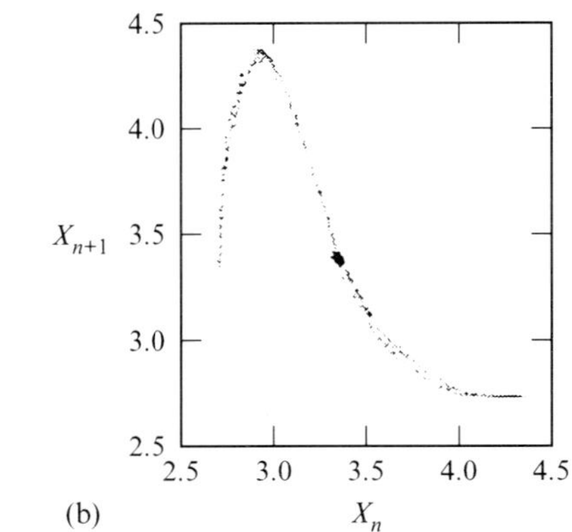

(b)

Figure 2 (a) Time series of $\mathbf{X}_n = V(t_n)$ for H_{dc}(nominal) = 0.112 Oe, $H_{ac} = 2.050$ Oe, and $f = 0.85$ Hz. Control was initiated after iteration 2350. (b) The first return map ($\mathbf{X}_{n+1}$ vs $\mathbf{X}_n$) for the controlled system (in black) is superimposed on the map for the uncontrolled system (in gray). The large density of points of low values of $\mathbf{X}_n$ is due to the saturation of the sensor for large excursions of the ribbon away from the sensor.

approximate fixed point. To these pairs of iterates we fit the approximate local *linear* map $\mathbf{M}$, where

$$\xi_{n+1} - \xi_F = \mathbf{M}(\xi_n - \xi_F), \quad \xi_{n+1} = \begin{pmatrix} X_{n+2} \\ X_{n+1} \end{pmatrix},$$

$$\xi_n = \begin{pmatrix} X_{n+1} \\ X_n \end{pmatrix}, \quad \xi_F = \begin{pmatrix} X_F \\ X_F \end{pmatrix}.$$

Knowing $\mathbf{M}$, we could extract the stable and unstable eigenvalues (λ_s, λ_u) and eigenvectors ($\mathbf{e}_s$, $\mathbf{e}_u$). We actually only needed λ_u and the unstable *contravariant* eigenvector[1] $\mathbf{f}_u$, given by $\mathbf{f}_u \cdot \mathbf{e}_u = 1$ and $\mathbf{f}_u \cdot \mathbf{e}_s = 0$.

Next, we changed H_{dc} slightly ($H_{dc} = 0.120$ Oe) and collected another set of data. We again found the precise location of the fixed point and calculated $\mathbf{g} = \partial \xi_F / \partial p \approx \delta \xi_F / \delta H_{dc}$.

To control the oscillations of the ribbon, we set $p = p_0$; when $|\xi_n - \xi_F| < \delta \xi_*$, we attempted control. Here, $\delta \xi_* \approx [\lambda_u - 1)/\lambda_u]\delta p \cdot (\mathbf{g} \cdot \mathbf{f}_u)$ is the maximum distance from the stable manifold of ξ_F for which one can achieve control for a given δp. As long as the iterate was within $\delta \xi$, of ξ_F, we perturbed p from p_0 by $\delta p = C(\xi_n - \xi_F) \cdot \mathbf{f}_u$, where Ref. 1 gives $C = [\lambda_u/(\lambda_u - 1)]/\mathbf{g} \cdot \mathbf{f}_u$. Since noise and errors in determining ξ_F, $\mathbf{f}_u$, $\mathbf{g}$, and λ_u, as well as any inaccuracies due to the linear approximation, prevented us from getting the next iterate exactly on the stable manifold, a new δp was calculated for each iterate. Note that both $\delta \xi$, and C can be computed at the start of the run, and that the calculations at each iterate are very simple. We could apply the changes to the applied magnetic field and change the Young's modulus of the ribbon in under 1 ms. Thus, our change in p was effectively instantaneous in relation to the 1.2 s period of the ac drive.

At the values of H_{ac}, H_{dc}, and f mentioned above, we calculated $X_F = 3.398 \pm 0.002$, $\delta X_F/\delta H_{dc} = -337 \pm 50$, $\mathbf{f}_u = \binom{-0.2}{1.2} \pm \binom{0.2}{0.1}$, and $\lambda_u = -1.2 \pm 0.2$. These numbers are typical of our data in that the fixed point can be determined with a great deal of accuracy, but the computed values of the eigenvalues and eigenvectors are sensitive to the noise on the attractor. Fortunately, the control is quite insensitive to variations in λ_u and $\mathbf{f}_u$ (e.g., using $\lambda_u = -1.4$ yielded results similar to those using $\lambda_u = -1.2$).

We have been able to control the oscillations of our ribbon for over 200 000 iterates (> 64 h), with a maximum allowed perturbation of 0.01 Oe. Figure 2(a) (after 2350 iterations) shows the controlled time series (in black) and Figure 2(b) the return map (superimposed, in black, on the attractor for the uncontrolled system). The control was to ± 0.015 V of the desired fixed point, about triple the dynamic noise present on the uncontrolled attractor.

We have also controlled the motion about a period-2 oscillation (again for over 50 000 iterates and with δp. = 0.01 Oe). The same procedure outlined above is followed except using $\xi_n = \binom{X_{2n+2}}{X_{2n}}$. The control was adjusted only at every other data iterate, about the periodic point at $X_F = 3.926 \pm 0.004$. As a demonstration of the versatility of the method, Figure 3 shows time-series data while the system was switched between no control and control about the fixed point or the period-2 orbit, again with the same values of H_{dc}(nominal), H_{ac}, and f as for Figure 2.

In conclusion, we have demonstrated the first control of chaos in a physical system, using the method of Ott, Grebogi, and Yorke. Some advantages of this method are the following: (1) No model for the dynamics is required; (2) the computations required at each iterate are minimal; (3) the required changes in the parameter can be quite small; (4) different periodic orbits can be stabilized for the *same* system in the *same* parameter range; (5) control can be achieved even with imprecise measurements of the eigenvalues and eigenvectors; and (6) this method is *not* restricted to periodically driven mechanical systems, but extends to any system whose dynamics can be characterized by a nonlinear map.

This work was supported by the Naval Surface Warfare Center Independent Research Program and by ONR through the Navy Dynamics Institute Program. We wish to thank Ed Ott for helpful comments and Heather D Lynn and Bryan D. Lee for their assistance.

Notes

[1] E. Ott, C. Grebogi, and J.A. Yorke, *Phys. Rev. Lett.* **64**, 1196 (1990); in *Chaos*, edited by D.K. Campbell (American Institute of Physics, New York, 1990), pp. 153–172.

[2] D.P. Lathrop and E.J. Kostelich, *Phys. Rev. A* **40**, 4028 (1989).

[3] J.-P. Eckman and D. Ruelle, *Rev. Mod. Phys.* **57**, 617 (1985).

[4] These include T.B. Fowler, *IEEE Trans. Autom. Control* **34**, 201 (1989); B.A. Huberman and E. Lumer, *IEEE Trans. Circuits Syst.* **37**, 547 (1990); A. Hübler and E. Lüscher, *Naturwissenschaften* **76**, 67 (1989); R. Georgii, W. Eberl, E. Lüscher, and A. Hübler, *Helv. Phys. Acta* **62**, 291 (1989); A. Hübler, *Helv. Phys. Acta* **62**, 343 (1989). In particular, we contrast this method with the resonant control method of Hübler and co-workers that has been applied to nonlinear pendula and other oscillators with nonlinear potentials. Their method differs from the method described above in at least three important respects. (1) One must have or construct model equations for the dynamics. (2) One must be able to modify the *driving force* of these equations, and these modifications can be rather large. The method of OGY requires no model equations and the perturbations could be to *any* accessible system parameter. (3) Rather than apply corrections as the dynamics wanders from a given unstable orbit, the resonant control method seeks to modify the underlying dynamical system such that the goal dynamics become *stable* solutions of the system (and thus uses no feedback).

[5] C. Modzelewski, H.T. Savage, L.T. Kabacoff, and A.E. Clark, *IEEE Trans. Magn.* **17**, 2837 (1981).

[6] W.L. Ditto, S. Rauseo, R. Cawley, C. Grebogi, G.-H. Hsu, E. Kostelich, E. Ott, H.T. Savage, R. Segnan, M.L. Spano, and J.A. Yorke, *Phys. Rev. Lett.* **63**, 923 (1989).

[7] H.T. Savage and C. Adler, *J. Magn. Magn. Mater.* **58**, 320 (1986); H.T. Savage and M.L. Spano, *J. Appl. Phys.* **53**, 8002 (1982); H.T. Savage, W.L. Ditto, P.A. Braza, M.L. Spano, S.N. Rauseo, and W.C. Spring, III, *J. Appl. Phys.* **67**, 5619 (1990).

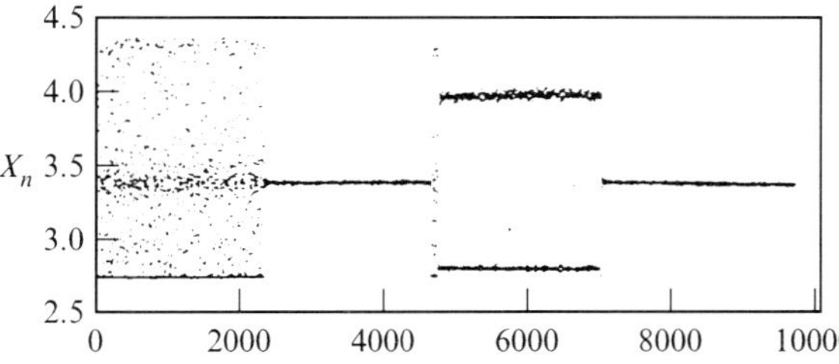

Figure 3 Time-series data as the system is switched from no control to control about the fixed point (at $n = 2360$), to control about the period-2 orbit (at $n = 4800$), and back to control around the fixed point (at $n = 7100$).

Paper 5
Controlling transient chaos

Tamás Tél

Institute for Theoretical Physics, RWTH, D-5100 Aachen, and Institut für Festkörperforschung, Research Center Jülich, D-5170 Jülich, Federal Republic of Germany

Received 27 August 1991

The method of Ott, Grebogi and Yorke is extended to control transient chaos. The controlled signal then exhibits a periodic behavior which is qualitatively different from that of the actual attractor. The time needed to achieve control is shown to be constant and to lie in the order of magnitude of the transient lifetime. The number of controlled trajectories, however, decreases the maximum perturbation according to a power law. Applicability to experimental situations and comparison with permanent chaotic cases are discussed.

The problem of controlling chaos has attracted recent interest [1–11]. The method of Ott, Grebogi and Yorke (OGY) [3] has the unique feature that it enables one to select a *predetermined* time-periodic behavior by making only *small* time-dependent perturbations. They show that permanent chaos can always be depressed by stabilizing one of the many unstable periodic orbits embedded in the chaotic attractor. The idea is to start with any initial condition, wait until the trajectory falls into a target region around the desired periodic orbit and then apply feedback control.

In this letter we demonstrate that transient chaos can also be controlled in the same spirit. This has the striking consequence that in such cases a behavior completely *different* from that of the actual attractor(s) can be selected. Furthermore, the controlling process exhibits *novel* features: the time needed to achieve control turns out to be constant, independent of the maximum perturbation, as a consequence of the finite chaotic lifetime in the unperturbed system; the number of trajectories controlled does, however, depend on the maximum perturbation and follows a power law.

In systems exhibiting transient chaos there exists in phase space an invariant set called a *chaotic saddle* or *repeller* [12–15], together with an attractor which is often simple, i.e. periodic. Trajectories start from randomly chosen initial points then approach the attractor with probability one. Before reaching it, however, they might come close to the strange repeller and stay in its vicinity for a shorter or longer time. This results in the appearance of chaotic motion with an average lifetime of $1/\kappa$ where κ is the *escape rate*, a basic characteristic of the chaotic repeller. Furthermore, the strange set has a fractal structure along both stable and unstable directions and, just like a chaotic attractor, appears to be the closure of an infinity of hyperbolic periodic orbits [16–18]. As an example, Figure 1

shows the invariant sets (attractor and chaotic repeller) for the Hénon map $x_{n+1} = a - x_n^2 + by_n$, $y_{n+1} = x_n$ at a parameter setting where the attractor is a period-5 orbit. Methods for constructing chaotic repellers are available [13–15]. Long-lived chaotic transients – the best candidates for experimental observability – are present around crisis configurations [12], at parameter values just beyond the disappearance of the chaotic attractor. It is worth

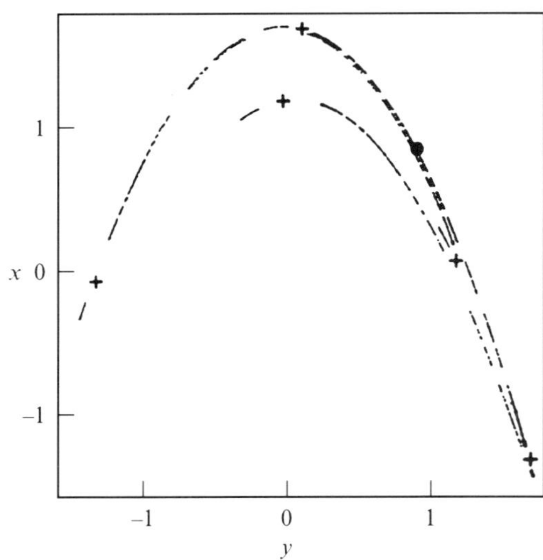

Figure 1 Invariant sets of the Hénon map $x_{n+1} = a - x_n^2 + by_n$, $y_{n+1} = x_n$ at the parameter setting $a = 1.45$, $b = 0.2$ where the attractor is a 5-cycle (black crosses). The fractal set is the chaotic repeller. Any periodic orbit on it can be stabilized by applying the method of OGY. We select the fixed point denoted by a dot.

mentioning that systems with fractal basin boundaries [19] are also accompanied by transient chaos since such boundaries are, in general, the stable manifolds of chaotic saddles.

The astonishing feature in controlling transient chaos is that one stabilizes an *atypical* behavior associated with a chaotic repeller, a set with a measure zero basin of attraction. In order to achieve this, one has to use a large *ensemble* of points starting from some region of phase space including the repeller and concentrate on long-lived chaotic transients. We assume that the dynamics can be represented by a k-dimensional ($k \geq 1$) nonlinear map $\xi_{n+1} = f(\xi_n, p)$ where p is some accessible system parameter. We choose to stabilize a periodic orbit *on* the chaotic repeller and specify a target region around it. For simplicity, we take here always a fixed point but note that any of the hyperbolic periodic orbits of the strange set can be chosen, providing the method with a high degree of flexibility. Without loss of generality we set $\xi = 0$ and $p = 0$ at the desired fixed point.

Next, take a ball around the repeller (or some part of it), choose randomly a large number of points in it, and iterate them forward. Some will stay around the repeller over many time steps and might fall near the desired fixed point at $\xi = 0$. Therefore, wait until ξ_n of any trajectory enters the target region around the origin and then change the actual value p_n of the perturbation parameter p to be different from zero. Pick p_n so that the next iterate $\xi_{n+1} = f(\xi_n, p_n)$ falls on the stable manifold of the origin of the uncontrolled map. If this is the case, the parameter perturbation can again be set to zero ($p_{n+1} = 0$) and the orbit will approach the fixed point according to a geometric progression. This mechanism is exactly the same as for chaotic attractors; therefore, the result for the appropriate choice of p_n can be taken over from OGY. The computation based on the *linearized* dynamics around a fixed point of a two-dimensional map says [3] that

$$p_n = \frac{\lambda_u}{\lambda_u - 1} \frac{\xi_n f_u}{g f_u} \qquad (1)$$

Here λ_u and f_u are the unstable eigenvalue of the fixed point in the uncontrolled map ($p = 0$) and the corresponding left eigenvector, respectively. The quantity gp yields the shift of the position of the fixed point when changing the perturbation parameter by a small amount of p. It is supposed that the parameter p can be varied in a small range $|p| < p_*$ only. Thus, if $|p_n|$ happens to be greater than the maximum perturbation p_* we set $p_n = 0$. This last condition also specifies the size of the target where control is activated.

Using this algorithm the control of chaotic transients can be carried out. Figure 2 exhibits an uncontrolled transient chaotic signal of the Hénon map with a period-5 attractor, and its controlled version obtained by applying rule (1) with $p_* = 0.1$. This illustrates, by modifying a statement of OGY, that improvement is *possible* via small control even in systems with periodic attractors, provided they coexist with chaotic repellers in phase space. If the only invariant set is a periodic attractor, small perturbation can change the orbit only slightly [3]. If, however, weak chaos is present, i.e. the

topological entropy of the system is positive, one can choose to stabilize any of the periodic orbits of the chaotic repeller. This leads to a behavior, selected according to some criterion [3], which is completely different from that of the attractor.

The average time τ needed to achieve control of *permanent* chaos was found [3] to depend on the maximum perturbation p_* according to a power law: $\tau \sim p_*^{-\gamma}$ with an exponent $\gamma \geq 1$ for small p_*. Here we show that this rule is no longer valid for *transient* chaos. The reason is that *not all trajectories* will now be controlled since the majority escapes the repeller before reaching the desired fixed point.

This can be best demonstrated by the example of one-dimensional maps. We consider a single humped map $f(x)$ defined on some support interval and having a maximum outside this interval (see Figure 3). Such maps generate transient chaotic signals and possess Cantor-like sets as their invariant repellers. The position of the actual attractor depends on the form of $f(x)$ outside the support interval. Since it does not play any role in what follows, we do not specify this form. Let us start with a large number N_0 of initial points distributed uniformly on the support. Control sets in if, after any number of interations, a trajectory falls into the target region, an interval I of length Δ around the fixed point $x_F(\neq 0)$. In general, Δ is proportional to the maximum perturbation p_*. A single humped map $f(x)$ can be embedded in two dimensions by considering the recursions $x_{n+1} = f(x_n)$, $y_{n+1} = x_n$. The vectors f_u and g then point along the x-axis and the diagonal, respectively, and (1) can be applied. (In the example of the controlled parabola map $x_{n+1} = a + p_n - x_n^2$, (1) yields $p_n = 2x_F(x_n - x_F)$ if $|p_n| < p_*$, and thus $\Delta = p_*/x_F$.) The number of trajectories controlled in the first step is proportional to the lengths Δ_1 and Δ_0 of the two pre-images I_1 and I_0 of the interval I, respectively, as shown in Figure 3. One of the pre-images of the interval I_0, which contains the fixed point, falls into itself, the other one into I_1. Therefore, when counting the number of controlled trajectories in later steps, it is sufficient to follow the pre-images of I_1 without those of I_0. The number of trajectories controlled in the nth step is thus proportional to the sum of the lengths $l_i^{(n-1)}$, $i = 1, 2, \ldots, 2^{n-1}$ of the $(n-1)$th pre-images of I_1. Note that in transient chaotic cases these pre-images do *not* overlap for sufficiently small Δ-values, i.e. for maximum perturbation p_* much less than unity. The independence of the controlling time on p_* for $p_* \to 0$ relies essentially on this property.

The total number of trajectories controlled at any time step can thus be expressed as

$$N = N_0 \left(\Delta_0 + \Delta_1 + \sum_{n>1} \sum_{i=1}^{2^{n-1}} l_i^{(n-1)} \right). \qquad (2)$$

The total number of steps before control is then

$$T = N_0 \left(\Delta_1 + \sum_{n>1} \sum_{i=1}^{2^{n-1}} n l_i^{(n-1)} \right). \qquad (3)$$

from which the average time to achieve control is computed as $\tau = T/N$. Next, let us observe that Δ_0 and Δ_1 can be obtained by dividing Δ with the slopes c and c_1 of the map taken at the fixed point x_F and its pre-image x_1 (Figure 3),

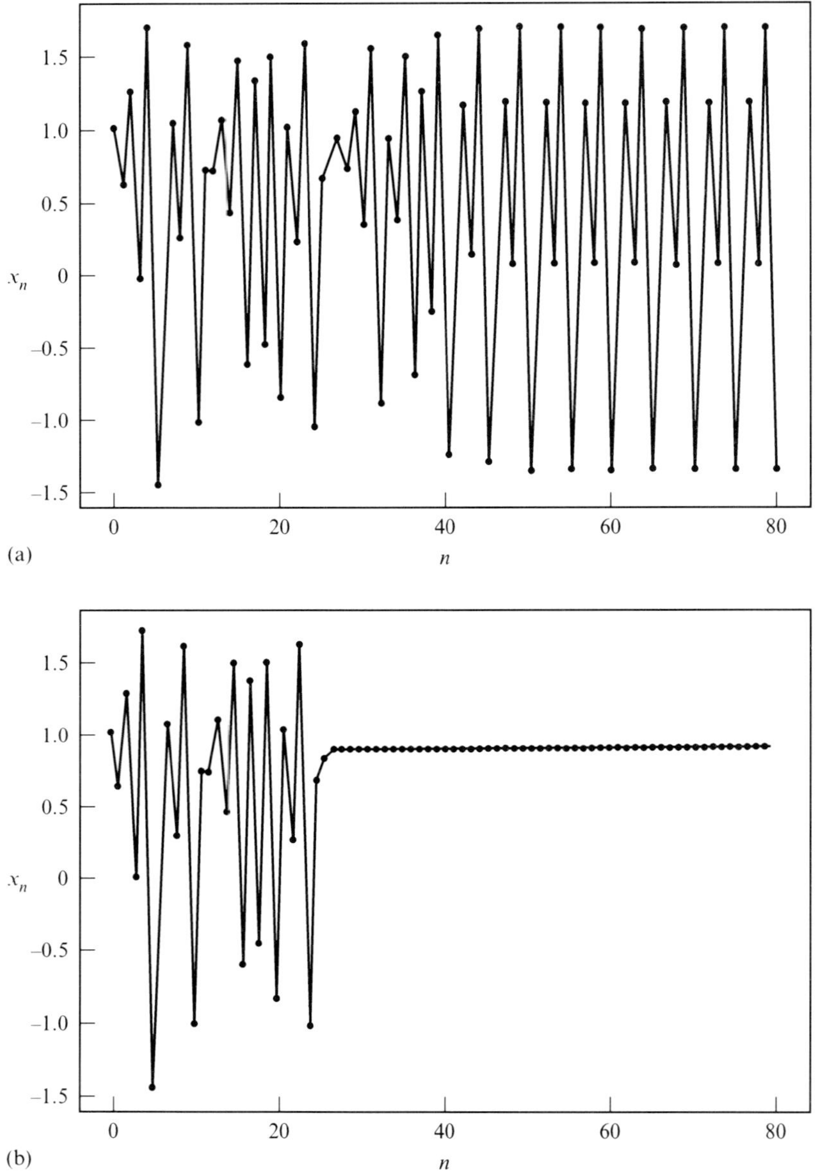

Figure 2 (a) Transient chaotic signal x_n versus n starting from the point $x_0 = y_0 = 1.014\,782$ in the Hénon map of Figure 1. The trajectory ceases to be chaotic at about the 38th time step where it comes to the neighborhood of the attractor. (The average lifetime of chaotic transients is $1/\kappa = 22$ at these parameters [13].) (b) Controlled signal started from the same initial point. The Hénon map was taken in the form given in the text with $a = 1.45 + p_n$ where the perturbation parameter p_n is specified by (1) with $p_* = 0.1$. The fixed point is at $x_F = y_F = 0.868\,858$, and $\xi_n \equiv (x_n - x_F, y_n - x_F)$. Control sets in at the 26th step and the fixed point becomes stabilized.

respectively, for Δ sufficiently small. In general, the length scales $\{l_i^{(n)}, i = 1, \ldots, 2^n\}$ can similarly be expressed by means of the derivative of the n-fold iterated map f^n taken at the nth pre-images of x_1. It has been shown [20] that the sum $\Sigma_1^{2^n} |f^{n'}(f^{-n}(x))|^{-1}$ scales for large n at *any* value of x as $\exp(-\kappa n)$ where κ is the escape rate. Therefore, the number of controlled trajectories can be rewritten for $\Delta \to 0$ as

$$N = N_0 \Delta \left(A + \frac{b}{c_1} \sum_{n=1}^{\infty} \exp[-\kappa(n-1)] \right)$$

$$= N_0 \Delta \left(A + \frac{B}{c_1} \frac{1}{1 - \exp(-\kappa)} \right) \tag{4}$$

with A and B as constants, because the sum converges for $\kappa > 0$. Similarly, we obtain

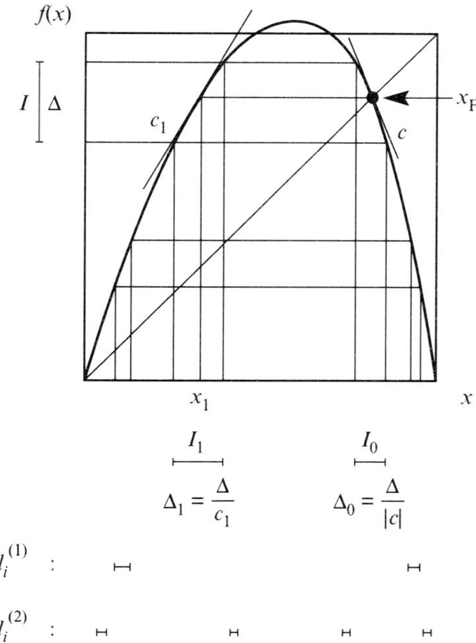

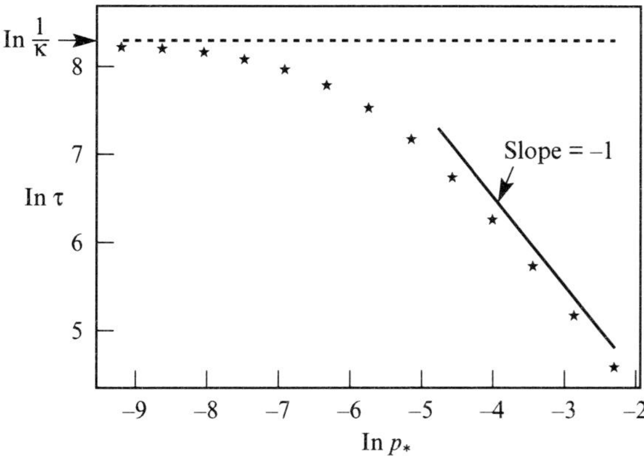

Figure 3 One-dimensional map generating transient chaos. Control is achieved if a trajectory falls into interval I having length Δ. Some pre-images of I and their sizes relevant for computing the average time of control are also shown.

Figure 4 Average time τ to achieve control versus the maximum perturbation in the parabola map $x_{n+1} = a - x_n^2$ with $a = 2 + 10^{-6}$. 10^5 initial points uniformly distributed on the support were iterated either up to 10^8 steps or until they did not reach the interval I of length p_*/x_F around the fixed point $x_F = 1 + 10^{-6}/3$. The logarithm of τ is plotted against $\ln p_*$ in the range $10^{-4} \le p_* \le 10^{-1}$. The straight line shown has a slope -1 and corresponds to the scaling in permanent chaos. Note the crossover into saturation for $p_* < 10^{-3}$. The broken line represents the average lifetime of transient chaos $1/\kappa = 3920$ in the uncontrolled system. The crossover into saturation cannot be seen any longer if $a > 2 + 10^{-4}$.

$$T = N_0\Delta\left(C + \frac{B}{c_1}\frac{1}{[1 - \exp(-\kappa)]^2}\right) \tag{5}$$

where C is another constant. Since both N and T are now proportional to Δ, the average time τ to achieve control turns out to be *independent* of Δ and, therefore, of the maximum perturbation p_*. For long-lived chaotic transients, i.e. for $\kappa \ll 1$, we find from (4) and (5) that $\tau \approx 1/\kappa$, which says that the time of control and the chaotic lifetime then coincide for $p_* \to 0$.

The argument breaks down at $\kappa = 0$ since the geometric series does not converge and the pre-image intervals overlap, so that all trajectories will be controlled. Therefore, one expects for $\kappa \to 0$ a crossover to the permanent chaos rule [3] $\tau(\kappa = 0) \sim 1/p_*$ of one-dimensional maps if p_* is not infinitesimally small. Interestingly, the crossover can be seen in an extremely close neighborhood of the crisis configuration only. To illustrate this, Figure 4 shows τ as a function of p_* for the parabola map at $a = 2 + 10^{-6}$. The behavior can be understood by applying the following argument for κ and p_* small. Both the decay from the repeller and the decay into the target region are essentially random processes. In cases when the pre-images of the target region overlap, the rate constant of the control is essentially the same as that of permanent chaos, i.e. $1/\tau(\kappa = 0)$. These decay processes are, in a first approximation, independent, therefore, the combined process is described by the sum of the rates, and thus $1/\tau = \kappa + 1/\tau(\kappa = 0)$. For one-dimensional maps $1/\tau(\kappa = 0) = cp_*$ where c is a constant, and we obtain $\tau = 1/(\kappa + cp_*)$, which gives a very good fit to the data of Figure 4 with $c = 0.0973$. This rule also implies that the saturation range where $\tau = 1/\kappa$ holds is reached for p_* smaller than some crossover value which scales with parameter a in the same way as the escape rate, i.e. as $(a - 2)^{1/2}$.

The essential features in determining τ were the following: (i) the pre-images of the fixed point's surrounding did not overlap, (ii) their characteristic lengths along the unstable direction could be summed up, yielding an exponential dependence on the escape rate, and (iii) the infinite sum of such factors was converging. Since these properties also hold for repellers of any higher-dimensional map, the time of control $\tau(p_*)$ at a maximum perturbation p_* is expected to be found in general as

$$\tau(p_*) = \text{constant} \le \frac{1}{\kappa} \tag{6}$$

for p_* sufficiently small. Results for τ of the Hénon map at the period-5 attractor are shown in the lower part of Figure 5. The independence on the maximum perturbation is clear.

The number $N(p_*)$ of the trajectories controlled is influenced by the shape and size of the region where control is activated. For two-dimensional maps, this is typically a parallelogram situated around the fixed point having some lengths $l_1(p_*)$ and $l_2(p_*)$ along the unstable and stable manifolds of this orbit, respectively. $N(p_*)$ is obviously proportional to the probability of falling into the control parallelogram. Since we use an ensemble of trajectories distributed uniformly on a neighborhood of the repeller, subsequent images of this neighborhood will tend towards

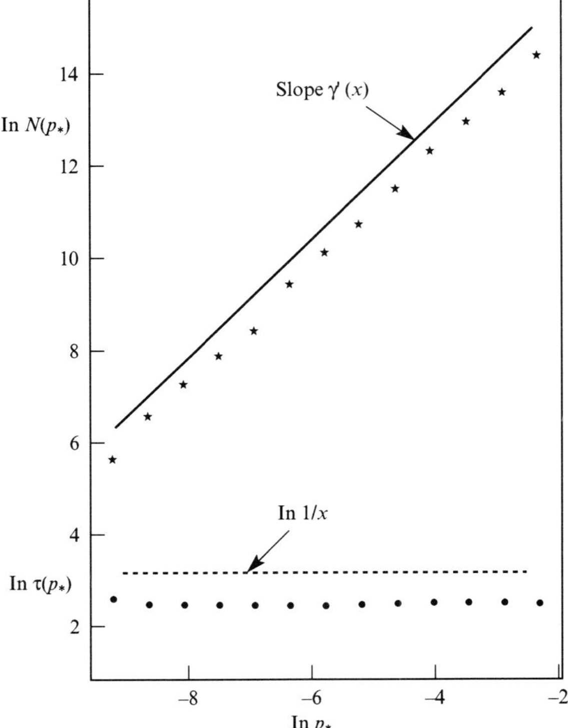

Figure 5 Average time to achieve control and number of controlled trajectories in the Hénon map with a period-5 attractor. Parameters as in Figures 1 and 2. 10^8 initial points were distributed on the square $(0 < x < 1, -1 < y < 0)$. Control was not activated during the first 10 steps. The number of further iterates of the non-escaping trajectories was computed until they did not fall within a circle around the fixed point with radius $2.9p_*$, in the range $10^{-4} \le p_* \le 10^{-1}$. Lower part: the logarithm of the average of these times as a function of $\ln p_*$. The broken line corresponds to the chaotic lifetime $1/\kappa = 22$. Upper part: the logarithm of $N(p_*)$ against $\ln p_*$. The straight line has the slope $\gamma'(\kappa) = 1.256$ given by (12). The results do not depend on the shape of the region where trajectories start from.

the unstable manifold of the chaotic set and the limiting distribution will be *smooth* on the manifold. The measure generated in this way is proportional to the so-called conditionally invariant measure [18, 21] (c-measure for short). Another relevant measure is the natural measure [13, 15, 18] obtained by taking the normalized restriction of the c-measure to the repeller, which is therefore not smooth along the unstable direction. The fractal properties of these measures along the stable manifold are, however, identical.

The c-measure of the control parallelogram can thus be written as

$$\mu(p_*) \sim l_1(p_*)l_2(p_*)^{\alpha_2} \tag{7}$$

with α_2 being the crowding index (pointwise dimension) along the stable direction at the hyperbolic fixed point of the

repeller. Because of smoothness, $\alpha_1 = 1$. The non-trivial crowding index has been determined by means of the periodic orbit theory of strange sets and reads [18]

$$\alpha_2 = \frac{\ln|\lambda_u| - \kappa}{\ln 1/|\lambda_s|} \tag{8}$$

where $\lambda_u(\lambda_s)$ denotes the larger (smaller) eigenvalue at the fixed point. Since for $p_* \to 0$ the dimensions of the control parallelogram should be proportional to some powers of p_*, we find the number of controlled trajectories to follow a power law

$$N(p_*) \sim p_*^{\gamma(\kappa)} \tag{9}$$

with an exponent also depending on, besides local properties, the escape rate of the repeller. Although derived for maps of the plane, this law is conjectured to hold for higher dimensional systems, too.

The control parallelogram mainly used by OGY has sizes proportional to p_* and $p_*^{1/2}$ along the unstable and stable directions, respectively. With the same choice of the control parallelogram one thus finds the exponent for transient chaos to be

$$\gamma(\kappa) = 1 + \frac{1}{2}\frac{\ln|\lambda_u| - \kappa}{\ln 1/|\lambda_s|} \tag{10}$$

Another way of specifying the condition for turning on the control is to take a ball of radius proportional to p_*. The corresponding exponent is then

$$\gamma'(\kappa) = 1 + \frac{\ln|\lambda_u| - \kappa}{\ln 1/|\lambda_s|} \tag{11}$$

For one-dimensional maps $\lambda_s = 0$ and $\gamma(\kappa) = 1$ follows in agreement with (4).

In order to check the prediction, we initiated an ensemble of trajectories around the repeller of Figure 1. They were first iterated up to 10 steps in order to have time to approach the distribution corresponding to the c-measure. It was then determined how many trajectories fall in later steps into a circle of radius proportional to p_* centred at the fixed point, where iteration was stopped. (For the analogous permanent chaos case see [3].) The upper part of Figure 5 shows $\ln N(p_*)$ obtained this way against $\ln p_*$, along with the predicted slope.

At this point a simple argument can be given from which the difference between the control of permanent and transient chaos clearly follows. The number of trajectories controlled per time steps is proportional to the c-measure (natural measure in the permanent case) $\mu(p_*)$ of the control parallelogram. The number $N(p_*)$ of *all* controlled trajectories can then be estimated – up to a constant factor – as the measure multiplied by the number of iterates $\tau(p_*)$ needed to achieve control in average. Therefore,

$$N(p_*) \sim \mu(p_*)\tau(p_*) \tag{12}$$

is expected to be valid for both permanent and transient chaos. For chaotic attractors all trajectories are controlled, $N(p_*) = $ constant; therefore, $\tau(p_*) \sim 1/\mu(p_*)$ and the rule $\tau(p_*) \sim p_*^{-\gamma(0)}$ derived by OGY is recovered. For transient chaos not extremely close to crisis we saw, on the contrary, that $\tau(p_*) = $ constant, from which $N(p_*) \sim \mu(p_*)$ follows.

Note that (12) seems to hold even if neither N nor τ is constant. One can easily check that in the parabola map at $a = 2 + 10^{-6}$ N first grows linearly with the maximum perturbation but then (for $p_* > 10^{-3}$) goes into saturation. Nevertheless, $N(p_*)/\tau(p_*)$ is proportional to p_* in the *entire* range investigated.

In conclusion, we can see that controlling transient chaos is more difficult than permanent chaos as one has to use ensembles of trajectories, but it is also simpler since the time needed does not grow with decreasing perturbation and remains bounded by the average transient lifetime.

The assumption concerning the existence of a nonlinear map allows the investigation of any system, including experiments, for which a faithful Poincaré section can be constructed. Ensembles of trajectories can be generated by repeating the experiment several times with different initial conditions. Both the construction and the analysis of periodic orbits is nowadays straightforward [25] from measured time series. Unfortunately, the experimental investigation of transient chaos has received disproportionately little attention. Experimental control of permanent chaos has, however, been carried out [6–8] exactly in those systems (a driven magnetoelastic ribbon [22], a convection loop [23] and a spin wave [24] experiment) which had earlier been studied (at other parameter values, of course) in detail from the transient chaotic point of view. All these dynamics can be approximated by low-dimensional maps. Thus, we can hope that the possibility of stabilizing a non-trivial state in the presence of simple periodic attractors can be verified experimentally in these systems.

Finally, we mention that the results also hold in the Hamiltonian limit of transient chaos, which corresponds to chaotic scattering [26]. The method described here, together with the construction of the chaotic repeller, then provides us with the ability to stabilize intermediate complexes of driven classical scattering systems (e.g. chemical reactions) in time-periodic states.

The author is indebted to M. Eisele, C. Grebogi, Z. Kovács and K.G. Szabó for useful comments and a critical reading of the manuscript. He thanks the referee for pointing out that the crossover process can be described by the sum of two decay rates. The kind hospitality at the IFF of the KFA Jülich and at the Institute of Theoretical Physics of the RWTH Aachen is acknowledged.

References

[1] Fowler T.B. 1989 *IEEE Trans. Autom. Control* **34** 201

[2] Hübler A. and Lüscher E. 1989 *Naturwissenschaften* **76** 67
Hübler A. 1989 *Helv. Phys. Acta* **62** 343
Plapp B.B. and Hüber A. 1990 *Phys. Rev. Lett.* **65** 2302
Jackson E.A. 1991 *Physica* **50D** 341

[3] Ott E, Grebogi C. and Yorke J.A. 1990 *Phys. Rev. Lett.* **64** 1196; *Chaos* ed D.K. Campbell (New York: American Institute of Physics) pp 153–72

[4] Huberman B.A. and Lumer E. 1990 *IEEE Trans. Circuits Systems* **37** 547
Hogg T and Huberman B.A. 1991 *Controlling Chaos in Distributed Systems IEEE Trans. Systems, Man. Cyber.* special issue on distributed artificial intelligence, in press

[5] Sinha S., Ramaswamy R. and Rao J.S. 1990 *Physica* **D43** 118
Sinha S. 1991 *Phys. Lett.* **156A** 475

[6] Ditto W.L., Rauseo S.N. and Spano M.L. 1990 *Phys. Rev. Lett.* **65** 3211

[7] Singer J., Wang Y-Z and Bau H.H. 1991 *Phys. Rev. Lett.* **66** 1123
Vincent T.L. and Yu J. 1991 *Dynamics Control* **1** 35

[8] Azevedo A. and Rezende S.M. 1991 *Phys. Rev. Lett.* **66** 1342

[9] Braiman Y. and Goldhirsch I. 1991 *Phys. Rev. Lett.* **66** 2545

[10] Romeiras F.J., Ott E., Grebogi C. and Dayawansa W.P. 1991 *Controlling Chaotic Dynamics Systems Proc of the 1991 American Control Conference* (IEEE) in press

[11] Hentschel H.G.E. and Jiang Z. 1991 *Learning to control period-doubling bifurcations Preprint*

[12] Grebogi C., Ott E. and Yorke J.A. 1982 *Phys. Rev. Lett.* **48** 1507; 1983 *Physica* **7D** 181

[13] Kantz H. and Grassberger P. 1985 *Physica* **17D** 75

[14] Nusse H.E. and Yorke J.A. 1989 *Physica* **36D** 137; 1991 *Ergod. Theor. Dyn. Syst.* **11** 189

[15] Tél T. 1990 *Directions in Chaos* vol 3, ed Bai-lin Hao (Singapore: World Scientific) pp 149–221

[16] Grebogi C., Ott E. and Yorke J.A. 1987 *Phys. Rev. A* **36** 3522

[17] Cvitanović P. 1988 *Phys. Rev. Lett.* **61** 2729

[18] Tél T. 1989 *J. Phys. A: Math. Gen.* **22** L691

[19] Grebogi C., Ott E. and Yorke J.A. 1983 *Phys. Rev. Lett.* **50** 935

[20] Tél T. 1987 *Phys. Rev. A* **36** 1502, 2507

[21] Pianigiani G. and Yorke J.A. 1979 *Trans. Am. Math. Soc.* **252** 351
Ott E., Sauer T. and Yorke J.A. 1989 *Phys. Rev. A* **39** 4212

[22] Ditto W.L. *et al* 1989 *Phys. Rev. Lett.* **63** 923

[23] Gorman M., Widmann P.J. and Robbins K.A. 1984 *Phys. Rev. Lett.* **52** 2241; 1986 *Physica* **D19** 255
Widmann P.J., Gorman M. and Robbins K.A. 1989 *Physica* **D36** 157

[24] Carroll T.L., Pecora L.M. and Rachford F.J. 1987 *Phys. Rev. Lett.* **59** 28; 1989 *Phys. Rev. A* **40** 377, 4149; 1988 *J. Appl. Phys.* **64** 5396
Carroll T.L., Rachford F.J. and Pecora L.M. 1988 *Phys. Rev. B* **38** 2938
Rödelsperger F., Weyrauch T. and Benner H. 1991 *Preprint*

[25] Auerbach D *et al* 1987 *Phys. Rev. Lett.* **58** 2387
Lathrop D.P. and Kostelich E.J. 1989 *Phys. Rev. A* **40** 4028

[26] Smilansky U. 1990 *Chaos and Quantum Physics* ed M.-J. Giannoni *et al* (New York: Elsevier) in press

Paper 6
Using chaos to direct trajectories to targets

Troy Shinbrot,[a] Edward Ott,[b],[c] Celso Grebogi,[b],[d] and James A. Yorke[d]
University of Maryland, College Park, Maryland 20742, USA

[a]*Department of Physics.*
[b]*Laboratory for Plasma Research.*
[c]*Departments of Electrical Engineering and of Physics.*
[d]*Institute for Physical Science and Technology and Department of Mathematics.*

Received 1 June 1990

A method is developed which uses the exponential sensitivity of a chaotic system to tiny perturbations to direct the system to a desired accessible state in a short time. This is done by applying a small, judiciously chosen, perturbation to an available system parameter. An expression for the time required to reach an accessible state by applying such a perturbation is derived and confirmed by numerical experiment. The method introduced is shown to be effective even in the presence of small-amplitude noise or small modeling errors.

Chaotic systems exhibit extreme sensitivity to initial conditions. This characteristic is often regarded as an annoyance, yet it provides us with an extremely useful capability without a counterpart in nonchaotic systems. In particular, the future state of a chaotic system can be substantially altered by a tiny perturbation. If we can accurately sense the state of the system and intelligently perturb it, this presents us with the possibility of rapidly directing the system to a desired state. As we will show below for a particular example, an initial condition for a chaotic system which, if left on its own, would require more than 6000 time steps to reach a small target region can be directed with small perturbations to the desired region in only 12 time steps. In this paper, we report a method for achieving such targeting, and we show that the method can be effective even in the presence of small-amplitude noise and small modeling errors.

For simplicity, we consider a three-dimensional continuous-time dynamical system, $dX/dt = F(X)$. The extension to higher dimensions is given at the end of the paper. We assume that model equations describing the system are known, although they need not be exact. We now employ a surface of section, and denote the coordinates in the surface of section by ξ, and the Poincaré map of ξ by

$$\xi_{n+1} = f(\xi_n, \alpha), \tag{1}$$

where the map f is necessarily invertible and α is a system parameter [1]. Suppose that we wish to go from a source point X_s to a small region about a target point, X_t. Following the trajectory from X_s forward in time, we find its first intersection with the surface of section and denote this point ξ_s. Following the trajectory through X_t backward in time, we similarly determine its first intersection with the surface of section and denote this point ξ_t. Thus we have reduced our problem to that of a two-dimensional map in which we desire to go from ξ_s to the vicinity of ξ_t. We assume that the system parameter α is available for adjustment at each iterate. Thus we can replace α in the map (1) by α_n. However, we also suppose that only small adjustments of α are allowed. That is, $\alpha_n = \bar{\alpha} + \delta_n$, where $\bar{\alpha}$ is a nominal value of α and the deviation from $\bar{\alpha}$, denoted δ_n, is restricted to be small.

If the process under study is ergodic, then in the absence of perturbations (i.e., for $\alpha_n = \bar{\alpha}$) the time required to travel from a source point ξ_s, to a small neighborhood $\bar{\varepsilon}_t$ of linear size ε_t about a target point ξ_t in the ergodic set is typically $\tau_0 \sim 1/\mu(\bar{\varepsilon}_t)$, where μ denotes the natural probability measure of the chaotic set. The measure $\mu(\bar{\varepsilon}_t)$ typically scales with the information dimension [2], D, so the time required obeys

$$\tau_0 \sim 1/\mu(\bar{\varepsilon}_t)^D, \tag{2}$$

for small ε_t. Thus in the absence of perturbations, the amount of time required to reach a desired target increases according to a power law as the size of the target region decreases. We will see that this power law can be converted to a much weaker logarithmic increase by applying a carefully chosen small perturbation.

Now suppose that perturbations of α can be applied. After one iteration of the return map, the change of the state, $\delta\xi$, relative to the point $f(\xi_s, \bar{\alpha})$, due to a small perturbation, δ_1, is given by the Taylor expansion

$$\delta\xi \cong \left. \frac{\partial f(\xi_s, \alpha)}{\partial \alpha} \right|_{\bar{\alpha}} \delta_1. \tag{3}$$

Letting δ_1 vary through a small interval, $|\delta_1| < \delta_*$, Eq. (3) defines a small line segment through the point $f(\xi_s, \bar{\alpha})$. We denote this line segment $\overline{\delta\xi}$, and we denote its length $\delta\xi$.

Since our system is chaotic, the length of the image of this line segment will grow roughly geometrically with each successive iteration of the map $\mathbf{f}(\xi, \alpha)$. Let n_1 denote the number of iterates required for the small line segment to stretch to a length of order 1. This typically happens when $\delta\xi \exp(n_1\lambda_1) \sim 1$ if $\delta\xi$ is small, where λ_1 is the positive Lyapunov exponent obtained for typical initial conditions on the attractor. Defining $\tau_1 = \lambda_1^{-1}\ln(1/\delta\xi)$, the length of the line segment becomes of order 1 after about τ_1 iterates (i.e., $n_1 \sim \tau_1$) if $\delta\xi$ is small. Without loss of generality, we take the size of the relevant ergodic region to be of the order of 1 so that τ_1 is approximately the number of iterates required for the line segment to span the ergodic region. Likewise, if we map the region $\bar{\varepsilon}_t$ backward in time, we find that its preimage spans the ergodic region after a number of preiterates which is typically of the order of $\tau_2 = |\lambda_2|^{-1}\ln(1/\varepsilon_t)$ if ε_t is small, where λ_2 is the negative Lyapunov exponent for the map $\mathbf{f}$ for typical initial conditions on the attractor.

Thus we adopt the following procedure. We iterate the segment $\overline{\delta\xi}$ forward using $\alpha_n = \bar{\alpha}$ for n_1 iterates until its length becomes of order unity. We then iterate the region $\bar{\varepsilon}_t$ backward for n_2 iterates until it first intersects the n_1 forward iteration of the line segment $\overline{\delta\xi}$. Typically for small $\delta\xi$ and ε_t we have $n_1 \cong \tau_1$ and $n_2 \cong \tau_2$. Iterating a point in the middle of this intersection backward n_1 times, we find a point on the line segment $\overline{\delta\xi}$ which is mapped to the target region $\bar{\varepsilon}_t$ in $n_1 + n_2$ iterates. Knowing this point we can then determine the required perturbation δ_1 to be applied on the first iterate from ξ_s by using Eq. (3). Note that, for the situation considered so far, we assume no noise and no modeling error, and consequently we can achieve targeting with $\delta_n = 0$ for $n \geq 2$. For small $\delta\xi$ and ε_t the total time to go from ξ_s to $\bar{\varepsilon}_t$ by this method scales as

$$\tau = \tau_1 + \tau_2 = \lambda_1^{-1}\ln(1/\delta\xi) + |\lambda_2|^{-1}\ln(1/\varepsilon_t). \tag{4}$$

For example, for $\delta\xi \sim \varepsilon_t$, we have $\tau \sim \ln(1/\varepsilon_t)$ in contrast to the power law (2).

In practice, we cannot actually iterate either the line $\overline{\delta\xi}$ or the region $\bar{\varepsilon}_t$. Rather, we iterate discrete approximations to them and make successive refinements until a sufficiently accurate intersection is obtained. We do this by putting a fixed number $\hat{N}_s \gg 1$ of equally spaced points on $\overline{\delta\xi}$, iterating these points, and joining their images with straight-line segments. Similarly, we iterate $\hat{N}_t \gg 1$ points on the perimeter of $\bar{\varepsilon}_t$ backward in time and join their images with straight-line segments. Once an intersection is detected, we refine its accuracy by repeatedly halving the intersecting forward and backward line segments and determining which of the halves actually contain the intersection. We must achieve accuracy sufficient to strike the n_2 backward iterate of the target, which is a long, thin region with width of order $\varepsilon_t \exp(-\lambda_1 n_2)$. Since we have normalized the size of the attractor to be of order 1, the curvature of the line $\overline{\delta\xi}(n_1)$ and the long sides of the n_2 backward iterate of $\bar{\varepsilon}_t$ are also of order 1. Thus, taking account of the curvature, to resolve the intersection within a distance of $\varepsilon_t \exp(-\lambda_1 n_2)$, we require that the distance between points on $\overline{\delta\xi}(n_1)$ and on the n_2 backward iterate of $\bar{\varepsilon}_t$ be of order $[\varepsilon_t \exp(-\lambda_1 n_2)]^{1/2}$. The square root results

because the maximum distance between the curve and its discrete straight-line approximation is quadratic in the length of the straight-line segment. Each time we halve our line segments, we increase the resolution by a factor of 2 at the expense of including three additional points [one on $\overline{\delta\xi}(n_1)$ and one each on the two segments bounding the backward iterate of $\bar{\varepsilon}_t$ near the intersection]. Thus to resolve the intersection, we require a number of points N' additional to the original $\hat{N} = \hat{N}_s + \hat{N}_t$ points, where N' obeys

$$[\varepsilon_t \exp(-\lambda_1 n_2)]^{1/2} > 2^{-N'/3}. \tag{5}$$

Using the relation $\varepsilon_t \exp(|\lambda_2|n_2) \sim 1$, this can be rewritten as

$$N' \gtrsim \tfrac{3}{2} D \ln(1/\varepsilon_t), \tag{6}$$

where $D = 1 + \lambda_1/|\lambda_2|$ is the Lyapunov dimension of the attractor. We stress that $\hat{N}$ is fixed (typically we took $\hat{N} \sim 100$) and does not depend on ε_t or $\delta\xi$. Consequently, as ε_t is reduced, the required computational effort increases logarithmically in ε_t, as shown in Eq. (6).

In order to show why our method of using forward and backward iterations was employed, we now contrast it with another conceivable procedure. If one iterated the line segment $\overline{\delta\xi}$ forward until it first intersected the region $\bar{\varepsilon}_t$, it would do so on iterate $n_1 + n_2$. One could then choose a point in this intersection, iterate the point backward $n_1 + n_2$ steps to find the corresponding point on the original line segment $\overline{\delta\xi}$, and then determine δ_1 from Eq. (3). While this works in principle, the numerical requirements of this pure forward scheme are needlessly more severe than when we determine an intersection by iterating $\overline{\delta\xi}$ forward n steps and $\bar{\varepsilon}_t$ backward n_2 steps. In the pure forward method, to detect an intersection between the target and the $n_1 + n_2$ iterate of the source, we require that the approximation of $\overline{\delta\xi}(n_1 + n_2)$ obtained by joining the N_s points with straight-line segments intersect the region $\bar{\varepsilon}_t$. Since the curvature of $\overline{\delta\xi}(n_1 + n_2)$ is typically of order 1, we thus require

$$\delta\xi(n_1 + n_2)/N_s < \varepsilon_t^{1/2}. \tag{7}$$

The source line will have length unity after n_1 iterates, and will then expand by roughly $\exp(n_2\lambda)$ during the next n_2 iterates, where λ is the topological entropy [3]. So we require

$$N_s \gtrsim \varepsilon_t^{-1/2} \exp(n_2\lambda) \tag{8}$$

or

$$N_s \gtrsim (1/\varepsilon_t)^{\lambda/|\lambda_2|+1/2}. \tag{9}$$

Thus the number of points required by the pure forward method increases exponentially with $1/\varepsilon_t$, but only increases logarithmically with $1/\varepsilon_t$ in the forward–backward method [4].

We now illustrate the method with a specific chaotic system. In particular, we deal with the Hénon map [5] in the form $x_{n+1} = a + 0.3y_n - x_n^2$ and $y_{n+1} = x_n$, with $\bar{a} = 1.4$. As an initial example, we choose the target region to be a small square centered on ξ_t with edge length $\varepsilon_t = 0.0038$. We find that for a representative pair of source and target points, say, $\xi_s = (0.4772, -1.188)$ and $\xi_t = (0.1371, -1.328)$, without applying a perturbation, 6062 iterations are required before the orbit from ξ_s strikes within the target neighborhood, $\bar{\varepsilon}_t$. However, if we are permitted to vary α by up to 1 part in 1000 about its nominal value, we find that our targeting method directs the trajectory to the target neighborhood in only twelve iterations.

To confirm the predicted logarithmic behavior in Eq. (4), the following numerical experiment was performed. Source and target locations were chosen at random with respect to the natural measure on the Hénon attractor. Then we fix a target size ε_t, and for each pair of source and target points, our targeting algorithm, described above, was applied. The total number of iterations required to go from ξ_s to $\bar{\varepsilon}_t$ was determined for each pair, and the results were then averaged over many source–target pairs. This process was repeated for several values of ε_t. The neighborhood $\bar{\varepsilon}_t$ was chosen to be a small square of edge length ε_t centered on ξ_t. The result of this experiment is shown in Figure 1. The solid line of slope $\lambda_1^{-1} + |\lambda_2|^{-1}$, predicted by Eq. (4), is consistent with the data. Also shown as a dashed line is the power-law dependence expected without control from Eq. (2) with $D \cong 1.26$ (the information dimension for the Hénon attractor). The logarithmic dependence of the time to reach the target on ε_t with control shows dramatic improvement over the power-law dependence without control.

The preceding discussion demonstrates that targeting can be achieved for chaotic systems using only small controls. It remains to be shown, however, that the method discussed can be effective in the presence of noise or modeling errors. Thus we suppose that the real system obeys $\xi_{n+1} = \mathbf{g}(\xi_n, \alpha) + \Delta_n$. Here we imagine that our model map $\mathbf{f}(\xi_n, \alpha)$ is slightly in error, and that, unknown to us, the correct form is $\mathbf{g}(\xi_n, \alpha)$. We, furthermore, allow small-amplitude random noise to disturb the system at each iteration as indicated by the term Δ_n.

To investigate the effect of noise alone, we take $\mathbf{f} = \mathbf{g}$. The following test was performed. Source and target locations were chosen at random on the Hénon attractor, and a trajectory between the source and the target

neighborhood was found for the case without noise as previously described. As an example, the neighborhood size ε_t was chosen to be 0.01 and the time required to hit the target in the absence of noise and with only $\delta_1 \neq 0$ was ten iterations. Then for each of the ten iterations, a random amount of noise was applied with Δ_n distributed uniformly in the interval $|\Delta_n| < \Delta_*$. As shown in Figure 2 for the case $\Delta_* = 0.01$, the noise displaced the tenth iteration to a point (denoted $\hat{\xi}_{10}$ in the figure) far away from ξ_t. Since the noise was applied at every iteration, we next compensated by recomputing the trajectory at every iteration and adjusting the applied perturbation correspondingly. That is, at *each* iterate we used the map $\mathbf{f}$ to determine δ_n by calculating the intersection of the forward iteration of the line determined from $\delta\xi_{n+1} = \delta_{n+1}\,\partial\mathbf{f}/\partial\alpha$ with the backward iteration of the region $\bar{\varepsilon}_t$. The result of this procedure is shown in the inset of Figure 2. The tenth iteration (denoted $\hat{\xi}'_{10}$ in the figure) now lies within the target region. Thus we have shown that our method can be effective in the presence of small-amplitude noise provided that we apply a correction δ_n at each iterate.

It can similarly be shown that targeting can also be achieved even when the system is imperfectly modeled, i.e., when $\mathbf{f}$ differs slightly from the true map, $\mathbf{g}$. After each iteration, we have to compensate for the difference $\mathbf{g} - \mathbf{f}$. For example, let us consider the Hénon map for the case without noise, where $\mathbf{f}$ is the Hénon map with $\bar{\alpha} = 1.4$, and $\mathbf{g} - \mathbf{f} = 0.014$; ε_t is still 0.01, and we use the same source and target as in our noise example (cf. Figure 2). If we apply our procedure only on the first iterate from ξ_s (as we would if $\mathbf{f} = \mathbf{g}$), then the trajectory again ends at a point (denoted ξ'_{10} in Figure 2) far from the target. As before, however, if our targeting algorithm is applied at every iteration, the tenth iterate (denoted ξ'_{10} in Figure 2) arrives in the target neighborhood despite the modeling error.

To generalize our method to higher dimensions, consider that the map $\mathbf{f}$ is N dimensional and the attractor has k

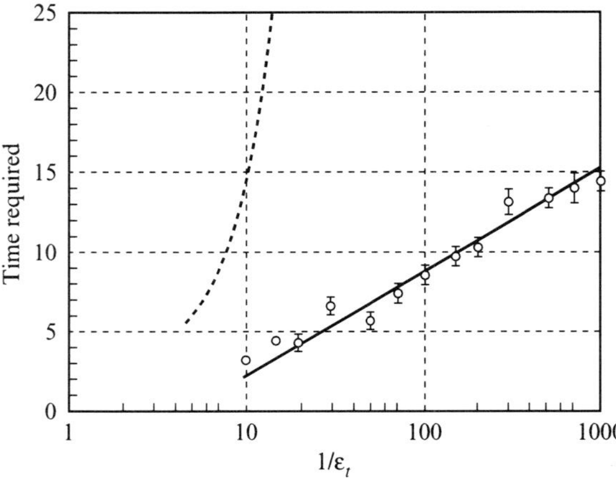

Figure 1 Average time required to reach typical target neighborhood from typical source with control vs neighborhood size, ε_t ($\delta\xi = \varepsilon_t$). Solid line has slope predicted from Eq. (4); dashed line indicates expected behavior without control. Error bars are standard error for 25-point means.

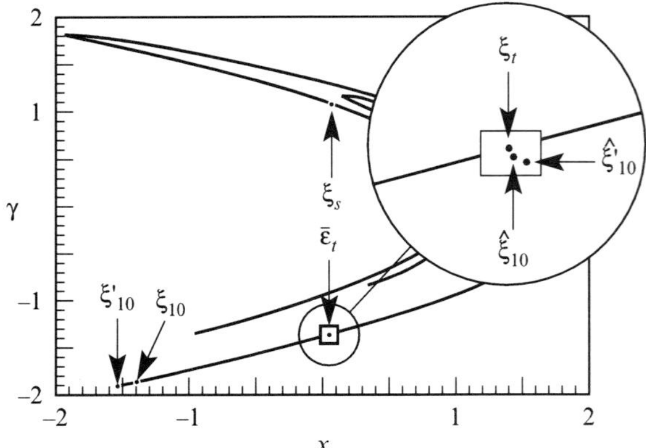

Figure 2 Source ξ_s and target region $\bar{\varepsilon}_t$ on the Hénon attractor. Inset: When the targeting procedure is applied at every iteration, the noise or modeling error can be compensated for, and ξ_{10} and ξ'_{10} both lie within the target region.

expanding and $N - k$ contracting directions at typical points. We note that when a k-dimensional surface and an $(N - k)$-dimensional surface intersect, generically they do so at isolated points, and small smooth perturbations will not destroy these intersections or create new ones. Thus, for a typical point and a typical small k-dimensional disk D^k centered at this point, the nth iterate of the disk $\mathbf{f}^n(D^k)$ will be a k-dimensional surface and its k-dimensional area will increase with n. Similarly, if we take an $(N - k)$-dimensional disk, D^{N-k}, centered at a typical point, then $\mathbf{f}^{-n}(D^{N-k})$ will be an $(N - k)$-dimensional surface whose area will increase with n. As these areas increase, typically they will intersect. We emphasize that targeting can typically be achieved with any dimensionality N even if we only have one available adjustable scalar parameter α. To see this we note the following. Consider a trajectory $\xi_i = \mathbf{f}^i(\xi_0, \alpha)$. If we perturb α from $\bar{\alpha}$ by an infinitesimal amount δ_i at time i, then at time $m > i$, a perturbation of ξ_m given by $\mathbf{v}_{i,m}\delta_i$ results, where $\mathbf{v}_{i,m}$ is an N-dimensional vector which is determined by the partial derivatives of the map along the trajectory. For typical ξ_0 and $\mathbf{f}$, the vectors $\mathbf{v}_{0,k}, \mathbf{v}_{1,k}, \ldots, \mathbf{v}_{k-1,k}$ are linearly independent and thus can be used to create the k disk D^k.

In conclusion, we have demonstrated that it is possible to rapidly reach a small, accessible target region in a chaotic system by applying small perturbations to an available parameter. The method used is robust against small-amplitude noise and small modeling errors, making it especially suited to practical applications. We emphasize that the problem addressed in this Letter is a very general one and can be expected to arise often [6].

This research was supported by the U.S. Department of Energy (Scientific Computing Staff Office of Energy Research). The computation was done at the National Energy Research Supercomputer Center.

Notes

[1] Alternatively, if model equations are not available, the map $\mathbf{f}$ might be derived from experimental data by use of the delay coordinate embedding technique [F. Takens, in *Dynamical Systems and Turbulence*, edited by D. Rand and L.S. Young (Springer-Verlag, Berlin, 1981), p. 230; N.H. Packard *et al.*, *Phys. Rev. Lett.* **45**, 712 (1980); J.-P. Eckmann and D. Ruelle, *Rev. Mod. Phys.* **57**, 617 (1985)].

[2] See, for example, J.D. Farmer, E. Ott, and J.A. Yorke, *Physica (Amsterdam)* **7D**, 153 (1983).

[3] S. Newhouse, in *Physics of Phase Space*, edited by Y.S. Kim and W.W. Zachary (Springer-Verlag, Berlin, 1987), p. 2; Y. Yomdin, *Isr. J. Math.* **57**, 285 (1987). The length of a *small* line segment typically grows at the exponential rate λ_1. After the length becomes of order 1, however, it grows at the exponential rate λ.

[4] We note that some improvement can be obtained by using higher-order fitting (e.g., parabolic rather than linear) of the curves to the iterated points. The exponential and logarithmic dependences of the pure forward and the forward-backward methods remain, however.

[5] Similar results have been obtained for the area-preserving 'standard map.'

[6] For example, see E. Ott, C. Grebogi, and J.A. Yorke, *Phys. Rev. Lett.* **64**, 1196 (1990); in *Chaos*, edited by D.K. Campbell (American Institute of Physics, New York, 1990), pp. 153–172.

Paper 7
Continuous control of chaos by self-controlling feedback

K. Pyragas[1,2]
Institute for Theoretical Physics, Technical University of Berlin, W-1000 Berlin, Germany

[1]*Alexander von Humboldt Fellow.*
[2]*On leave of absence from Institute of Semiconductor Physics, 2600 Vilnius, Lithuania.*

Received 17 August 1992; accepted for publication 16 September 1992
Communicated by J. Flouquet

Two methods of chaos control with a *small time continuous* perturbation are proposed. The stabilization of unstable periodic orbits of a chaotic system is achieved either by combined feedback with the use of a specially designed external oscillator, or by delayed self-controlling feedback without using of any external force. Both methods do not require an a priori analytical knowledge of the system dynamics and are applicable to experiment. The delayed feedback control does not require any computer analyses of the system and can be particularly convenient for an experimental application.

1. Introduction

Dynamic chaos is a very interesting nonlinear effect which has been intensively studied during the last two decades. The effect is very common; it has been detected in a large number of dynamic systems of various physical nature. In practice, however, this effect is usually undesirable. It restricts the operating range of many electronic and mechanic devices. Ott, Grebogi and Yorke [1] (OGY) have suggested an efficient method of chaos control that can eliminate chaos. The method is based on the idea of the stabilization of unstable periodic orbits (UPOs) embedded within a strange attractor. This is achieved by making a small time-dependent perturbation in the form of feedback to an accessible system parameter. The method turns the presence of chaos into an advantage. Due to the infinite number of different UPOs embedded in a strange attractor, a chaotic system can be tuned to a large number of distinct periodic regimes by switching the temporal programming of *small* parameter perturbation to stabilize different periodic orbits. Recently the OGY method has been successfully applied to some experimental systems [2–4].

An experimental application of the OGY method requires, as a rule, a permanent computer analysis of the state of the system. The changes of the parameter, however, are discrete in time since the method deals with the Poincaré map. This leads to some limitations. The method can stabilize only those periodic orbits whose maximal Lyapunov exponent is small compared to the reciprocal of the time interval between parameter changes. Since the corrections of the parameter are rare and small, the fluctuation noise leads to

occasional bursts of the system into the region far from the desired periodic orbit, and these bursts are more frequent for large noise [1]. Therefore, the idea of a *time-continuous* control seems attractive in this context.

The response of chaotic systems to continuous periodic and aperiodic perturbations have been considered in many investigations [5–9] to suppress chaos in the system [5, 6], to achieve some desired behavior [7], to synchronize some subsystems in a complex chaotic system [8, 9]. But none of these investigations considered the perturbation in the form of the feedback. The methods developed cannot be applied to the UPO stabilization. They can eliminate the chaos in the system, but the resulting periodic orbits obtained by the methods differ from the UPOs of the initial system and, therefore, they require a comparatively large perturbation.

In the following two methods of permanent control in the form of feedback are suggested. Both methods are based on the construction of a special form of a time-continuous perturbation, which does not change the form of the desired UPO, but under certain conditions can stabilize it. A combined feedback with a periodic external force of a special form is used in the first method. The second method does not require any external force; it is based on a self-controlling delayed feedback. The block diagrams of these methods are shown in Figure 1.

2. External force control

Let us consider a dynamic system which can be simulated by ordinary differential equations. We imagine that the

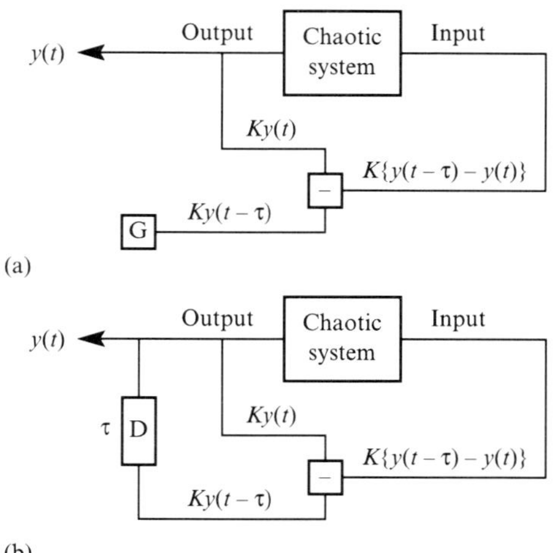

(a)

(b)

Figure 1 Block diagram of (a) external force control, and (b) delayed feedback control. G is a special external periodic oscillator, D is a delay line.

equations are unknown, but some scalar variable can be measured as a system output. We also suppose that the system has an input available for external force. These assumptions can be met by the following model,

$$\frac{dy}{dt} = P(y,x) + F(t), \quad \frac{dx}{dt} = Q(y,x). \tag{1}$$

Here y is the output variable and the vector x describes the remaining variables of the dynamic system which are not available or not of interest for observation. It is assumed for simplicity that the input signal $F(t)$ disturbs only the first equation, corresponding to the output variable. We suppose that the considered system without an input signal ($F = 0$) has a strange attractor.

It has been demonstrated using a standard method of delay coordinated that a large number of distinct UPOs on a chaotic attractor can be obtained from one scalar signal [10–12]. Applying this method to our system, we can determine from the experimentally measured output signal $y(t)$ various periodic signals of different form $y = y_i(t)$, $y_i(t + T_i) = y_i(t)$ corresponding to different UPOs. Here T_i is the period of the ith UPO. Then we examine these periodic signals and select the one which we intend to stabilize. To achieve this goal we have to design a special external oscillator, which generates the signal proportional to $y_i(t)$. The difference $D(t)$ between the signal $y_i(t)$ and the output signal $y(t)$ is used as a control signal:

$$F(t) = K[y_i(t) - y(t)] = KD(t). \tag{2}$$

Here K is an experimentally adjustable weight of the perturbation. The perturbation has to be introduced into the system input as a negative feedback ($K > 0$). An experimental realization of such a feedback presents no difficulties for many physical systems. The important feature of perturbation (2) is that it does not change the solution of eq. (1) corresponding to the UPO $y(t) = y_i(t)$. By selecting the weight K, one can achieve the stabilization. When this stabilization is achieved the output signal is very close to $y_i(t)$

and the perturbation $F(t)$ becomes extremely small. Therefore here, as well as in the OGY method, only a *small* external force is used to stabilize the UPOs. We do not intend to prove the validity of this method for the general case, but we have verified it for many chaotic systems such as the Rossler [13], Lorenz [14], Rabinovich and Fabrikant [15], Duffing oscillator [5, 16] systems and others.

The main results presented here are illustrated for the Rossler system:

$$\frac{dy}{dt} = -y - z, \quad \frac{dy}{dt} = x + 0.2y + F(t),$$

$$\frac{dz}{dt} = 0.2 + z(x - 5.7). \tag{3}$$

Here $F(t)$ is the perturbation defined in Eq. (2). For definiteness y is chosen as an output signal. The results do not depend on the choice of output variable. Figure 2(a) shows the results of stabilization of the period-five UPO of the Rossler attractor. The origin of the curve F corresponds to the time when the perturbation is switched on. As it was expected, the perturbation becomes small after a transient process and the system comes into the periodic regime corresponding to an initially unstable orbit. To illustrate the

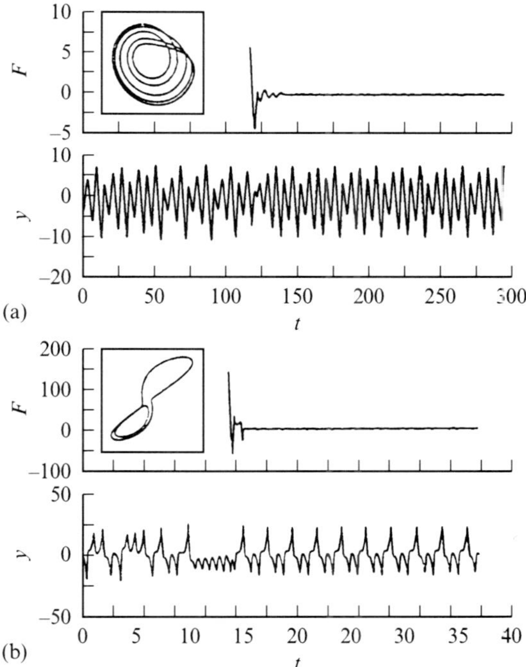

(a)

(b)

Figure 2 Dynamics of the output signal $y(t)$ and perturbation $F(t)$ (a) for the Rossler system (Eq. (3)), $K = 0.4$, $y_1(t)$ corresponds to the period-five cycle, and (b) for the Lorentz system: $dx/dt = 10(x - y)$, $dy/dt = -xz + 28x - y + F(t)$, $dz/dt = xy - \frac{8}{3}z$, the perturbation $F(t)$ is determined by Eq. (2), $y_i(t)$ corresponds to the period-two cycle. The origin of curve F corresponds to the moment of switching on the perturbation. The implement shows the x–y phase portrait of the system in the post-transient regime.

validity of the method for other chaotic systems Figure 2(b) shows the results of stabilization of the period-two UPO of the Lorenz system.

The amplitude of perturbation in a post-transient regime depends on two factors, on the accuracy of the UPO $y_i(t)$ reconstruction, and on the fluctuation noise. In an ideal case the perturbation has to be vanishingly small when the system moves along its periodic orbit, and the stabilization can be achieved with a very small signal of the external oscillator. To investigate the influence of noise, we add terms $\varepsilon\xi_x(t)$, $\varepsilon\xi_y(t)$, and $\varepsilon\xi_z(t)$ to the right-hand sides of Eq. (3). The random functions ξ_x, ξ_y, and ξ_z are independent of each other, having mean value 0 and mean-squared value 1. Figure 3 shows the results of the stabilization of the period-one cycle of the Rossler attractor for two different levels of noise. Since the control is permanent, the system does not experience any bursts into the region far from the UPO even for sufficiently large noise. The increase in noise leads to the increase of the amplitude of perturbation and to the smearing-out of the periodic orbit.

Note one difference between the OGY and the above method. The perturbation in the OGY method is applied only when the state of the system is close to the fixed point, since it uses a linear approximation for the deviations from the fixed point. Here we do not need to wait until the state of the system comes close to the desired periodic orbit. The perturbation can be switched on at any moment. The Rossler system synchronizes with the external oscillator even when the initial conditions are far from the periodic

orbit. Then the initial perturbation can be rather large. However, we do not expect that this will be the case for all dynamic systems. More complicated periodically driven dynamic systems along with the stabilized UPO can have alternative stable solutions belonging to different basins of initial conditions. Such multistability can be an undesired feature for the purpose considered here. Large initial values of the perturbation can be also undesired for some experiments. In many cases both these problems can be solved by restriction of the perturbation. Introducing some nonlinear element into the feedback circuit it is possible to achieve the saturation of the perturbation $F(t)$ for large values of the deviation $D(t)$:

$$
\begin{aligned}
F(t) &= -F_0, & KD(t) &\le -F_0, \\
&= KD(t), & -F_0 &< KD(t) < F_0, \\
&= F_0, & KD(t) &\ge F_0.
\end{aligned}
\tag{4}
$$

Here $F_0 > 0$ is the saturating value of the perturbation. Although in proximity to the UPO both perturbations (2) and (4) work identically, they lead to different transient processes. Figure 4 illustrates the influence of restriction (4) to the system dynamics. The perturbation in this case is always small including the transient process. However, the transient process on average is now much longer. The system 'waits' until the trajectory comes close to the periodic orbit and only then synchronizes with an external oscillator. As in the OGY method the mean duration of the transient process increases rapidly with the decrease of F_0. The efficiency of restriction (4) to eliminate the multistability will be illustrated in Section 3.

To analyze the local stability of the system we have calculated the maximal Lyapunov exponent of the UPOs using the linearization of system (3) with respect to small deviations from the corresponding UPOs. The dependence of the leading Lyapunov exponent λ of the period-one and period-two orbits on the parameter K is shown in Figure 5. The negative values of $\lambda(K)$ determine the interval of K corresponding to the stabilized UPO. The period-one UPO is stable in the finite interval $K = [K_{\min}, K_{\max}]$, but the period-two UPO has an infinite interval $K = [K_{\min}, \infty]$ of stabilization. Here the values $K_{\min}$ and $K_{\max}$ define the

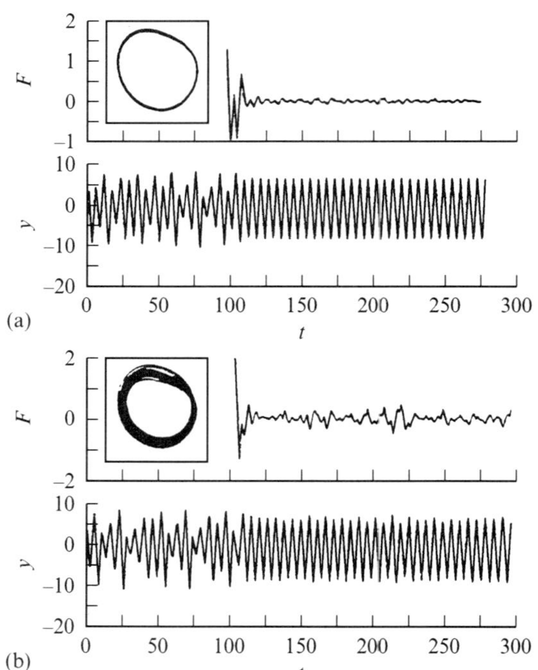

Figure 3 Results of stabilization of the period-one circle of the Rossler system at two different levels of noise. $K = 0.4$; (a) $\varepsilon = 0.1$; (b) $\varepsilon = 0.5$.

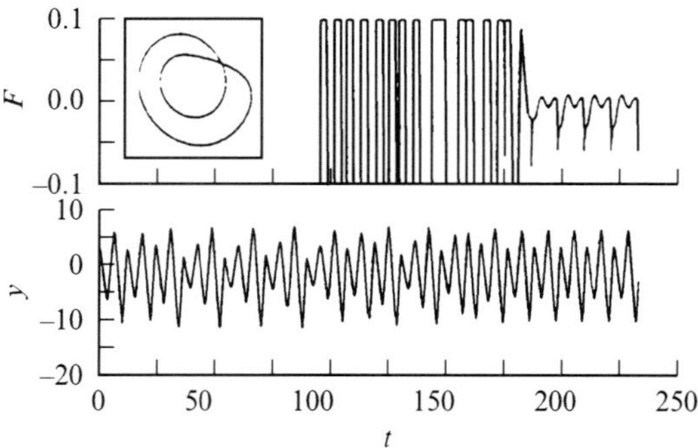

Figure 4 Results of stabilization of the period-two circle of the Rossler system with the restricted perturbation (4). $K = 0.4$, $F_0 = 0.1$.

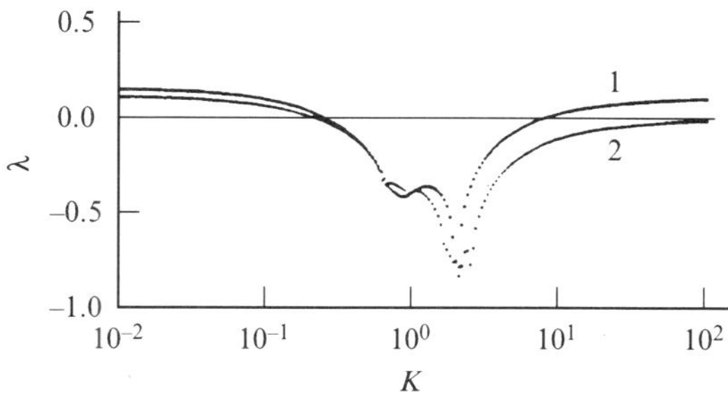

Figure 5 Dependence of the Lyapunov exponents of the period-one and period-two UPOs of the Rossler system on the weight K of the perturbation.

threshold of the stabilization: $\lambda(K_{min}) = \lambda(K_{max}) = 0$. The Lyapunov exponent $\lambda(K)$ of both orbits have minima at some value of $K = K_{op}$ providing an optimal control. Note that for all values of $K > 0$ the perturbation decreases the Lyapunov exponent of the initial system, $\lambda(K) < \lambda(0)$, but not for all values of K this perturbation is sufficiently efficient to invert the sign of λ. The presence of the minimal threshold of the stabilization is well understood. The weight K of the perturbation has to be sufficiently large to compensate the divergence of the trajectories close to the UPO. A rather large value of K deteriorates the control. This is related to the fact that the perturbation disturbs only one variable of the system. For large K the changes of this variable are very fast and the remaining variables have no time to follow these changes. To support this assumption we have considered multivariable control. A perturbation in the form of Eq. (2) with corresponding variables have been added to each equation of the system (3). As a result the monotonously decreasing characteristics $\lambda(K)$ for both orbits have been obtained.

The experimental application of this method can be divided into two stages. In the first, preparatory, stage the output signal should be investigated and the oscillator generating a periodical signal proportional to $y_i(t)$ should be designed. In the second stage the control is achieved simply by combining the scheme shown in Figure 1(a). A combined feedback using the difference between an output signal and the signal of the external oscillator performs here a self-controlling function.

3. Delayed feedback control

The complexity of the experimental realization of the above method is mainly in the design of a special periodic oscillator. The second method which we have considered has no such shortcoming. The idea of this method consists in substituting the external signal $y_i(t)$ in Eq. (2) for the delayed output signal $y(t - \tau)$. In other words, we use a perturbation of the form

$$F(t) = K[\,y(t - \tau) - y(t)] = KD(t). \qquad (5)$$

Here τ is a delay time. If this time coincides with the period of the ith UPO $\tau = T_i$ then the perturbation becomes zero for

the solution of system (1) corresponding to this UPO $y(t) = y_i(t)$. This means that the perturbation in the form (5) as well as in the form (2) does not change the solution of system (1) corresponding to the ith UPO. Choosing an appropriate weight K of the feedback one can achieve the stabilization. The results of such a stabilization for the Rossler system and for the Duffing oscillator are shown in Figure 6. These results are very similar to those in the case of an external force control. However, an experimental realization is simpler in this case. No external perturbation or computer is needed for this control. This control is achieved by the use of the output signal, which is fed in a special form into the system input. The difference between the delayed output signal and the output signal itself is used as a control signal. This feedback performs the function of self-control. Only a simple delay line is required for this feedback. To achieve the stabilization of the desired UPO, two parameters, namely, the time of delay τ and the weight K of the feedback, should be adjusted in experiment. The amplitude of the feedback signal can be considered as a criterion of UPO stabilization. When the system moves along its UPO this amplitude is extremely small. The dependence of this amplitude on the delay time for the Rossler system is illustrated in Figure 7(a). Excluding the transient process, the dispersion of the perturbation $\langle D^2(t) \rangle$ has been calculated for each value of τ with 20 different initial

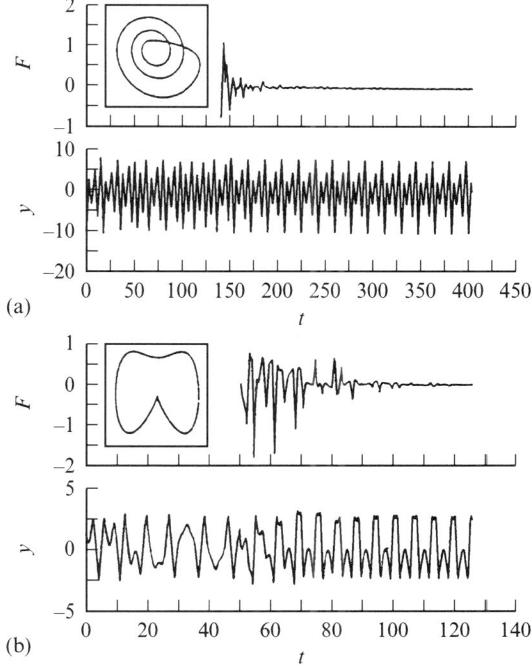

Figure 6 Results of the stabilization of (a) the period-three cycle of the Rossler system, $K = 0.2$, $\tau = 17.5$, and (b) the period-one cycle of the nonautonomous Duffing oscillator: $dx/dt = y$, $dy/dt = x - x^3 - dy + f\cos(\omega t) + F(t)$, $f = 2.5$, $\omega = 1$, $d = 0.2$, $K = 0.4$, $\tau = 2\pi/\omega$, in the case of delayed feedback control with the use of a perturbation in the form of Eq. (5).

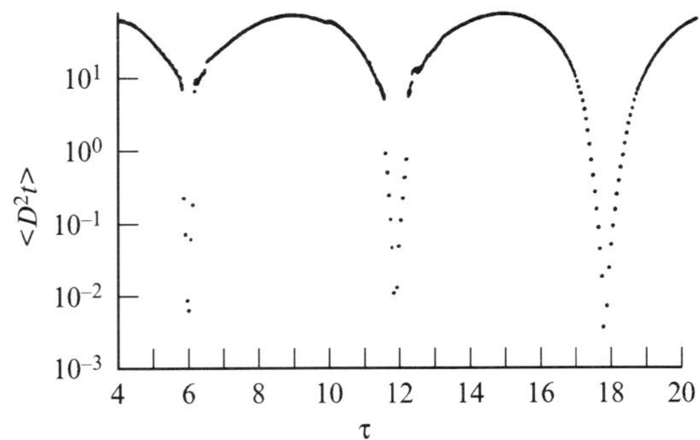

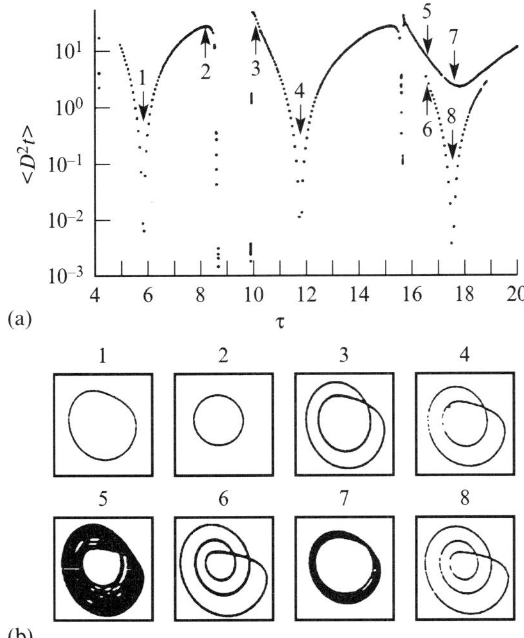

Figure 7 (a) Dependence of the dispersion of perturbation on delay time and (b) the x–y phase portraits of the Rossler system in the post-transient regime for some values of the delay time. $K = 0.2$.

conditions, and the corresponding 20 values of this dispersion for each τ have been depicted. The resulting figure represents the sequence of resonance curves with very deep minima. These minima are located at the points of delay time coinciding with the periods of the UPO $\tau = T_i$. The phase portraits for these values of delay time are shown in Figure 7(b1, 4, 8). They correspond to initially unstable period-one, -two and -three cycles. The resonance curves are separated by additional minima intervals, corresponding to the steady-state solution of the Rossler system, i.e. an unstable fixed point. Therefore, the method can stabilize unstable fixed points as well as UPO. When the delay time differs considerably from the period of the UPO, the output oscillations of the system can be chaotic (Figure 7(b3, 5–7) of periodic (Figure 7(b2)). The periodic orbits obtained far from resonance (Figure 7(b2)) differ considerably from the UPOs. They correspond to new periodical solutions of the system caused by a large perturbation. The periods of these orbits differ from the delay time τ.

The problem of multistability arises for the Rossler system with delayed perturbation. As can be seen in Figure 7(b), the Rossler system for large values of the delay time has two stable solutions depending on initial conditions. The phase portraits 7(b5) and 7(b6) as well as 7(b7) and 7(b8) have been obtained for the same values of the delay time but with different initial conditions. As has been mentioned in the previous section this problem can be avoided by restriction of the perturbation. The influence of restriction (4) on the results presented in Figure 7(a) can be seen from Figure 8. Due to the restriction the upper branch

Figure 8 The same as in the Figure 7(a), but for the case of restricted delayed perturbation. $F_0 = 0.1$.

of points in proximity to the period-three resonance disappears. An asymptotical behavior of the system becomes unambiguous for all values of K. The windows of K corresponding to the stabilization of the fixed point also disappear. This is because the trajectories of an unperturbed Rossler attractor do not reach the fixed point and to reach it a large perturbation is needed.

The dependences of the dispersion $\langle D^2(t)\rangle$ and the Lyapunov exponents λ on K for the two first periodic orbits are shown in Figures 9 and 10. In the case of delayed feedback each of the two orbits can be stabilized in a finite interval of K. These intervals are much narrower than those obtained with an external force control. This means that the delayed feedback control is more sensitive to the fitting of the parameters. The external force control is more efficient since the perturbation always tends to attract the current trajectory to the desired periodic orbit, determined

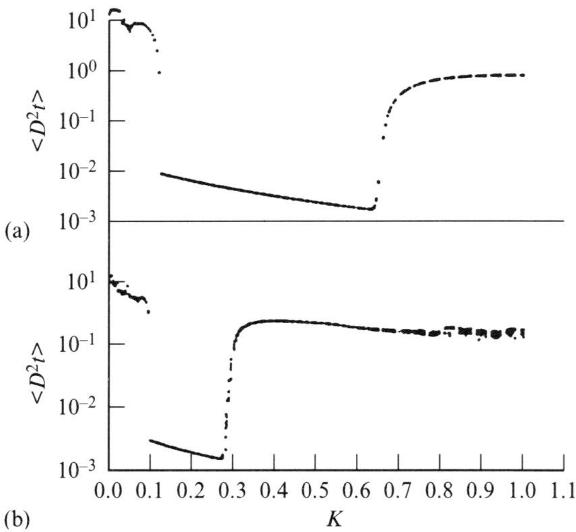

Figure 9 Dependence of the dispersion of the perturbation on K for two values of the delay time τ, coinciding with the periods of the first two periodic orbits: (a) $\tau = 5.9$, (b) $\tau = 11.75$.

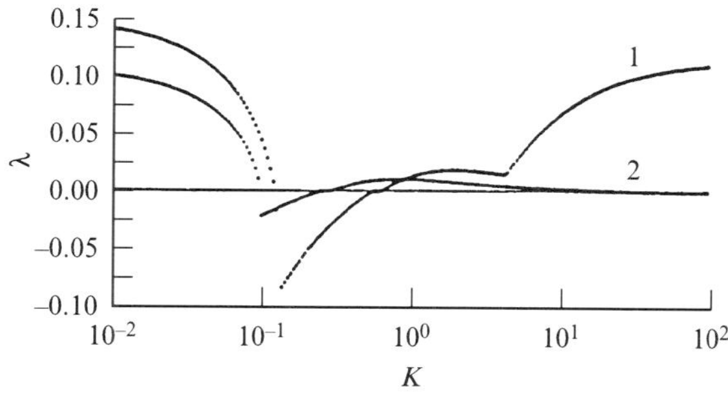

Figure 10 Dependence of the Lyapunov exponents of the two first periodic orbits of the Rossler system on K in the case of delayed feedback: (1) $\tau = 5.9$, (2) $\tau = 11.75$.

beforehand. The delayed feedback perturbation tends to decrease the distance between the current trajectory and the delayed trajectory which in the process of stabilization does not coincide exactly with the UPO.

4. Discussion and conclusions

Note that the perturbation in both forms (2) and (5) expands the dimension of the originally low-dimensional system. In the case of an external force control the perturbation increases the dimension by one, as any external periodical signal $y_i(t)$ can be presented by one additional ordinary differential equation. The delayed feedback perturbation increases the dimension to infinity. Therefore, one can conclude that the stabilization in both methods is achieved through additional degrees of freedom introduced in the system with the perturbation. The perturbation does not change the projections of the UPOs on an original low-dimensional phase space. The additional degrees of freedom change only the Lyapunov exponent of the UPOs, so that they become stable.

This can be illustrated with a simple analytical example. An unperturbed ($F_n = 0$) one-dimensional logistic map

$$x_{n+1} = 4x_n(1 - x_n) + F_n \qquad (6)$$

has the unstable fixed point $x_n = \frac{3}{4}$ with the eigenvalue $\lambda = -2$. The perturbation in the form of a delay $F_n = K(x_{n-1} - x_n)$ does not change the x-coordinate of this fixed point, but increases the dimension of the map to two. The analysis of this two-dimensional map shows that the absolute values of both eigenvalues of the fixed point are less than 1 in the interval of the parameter $K = [-1, -0.5]$. Therefore, for these values of K a 'one-dimensional' unstable fixed point turns into a 'two-dimensional' stable fixed point. A more

detailed theory of this stabilization is in progress and will be reported elsewhere.

In conclusion, we have shown that the UPO of a chaotic system can be stabilized by a *small time-continuous* perturbation. The permanent control is noise resistant. The stabilization can be achieved by the use of a specially designed external periodic oscillator, or by the use of delayed self-controlling feedsback without use of any external force. The multistability of the system under control can be avoided by restriction of the perturbation. An experimental realization of the second method is very simple and this method should be applicable to a wide variety of systems.

Acknowledgement

The author acknowledges E. Schöll for useful discussions and the Alexander von Humboldt Foundation for financial support.

References

[1] E. Ott, C. Grebogi and A. Yorke, *Phys. Rev. Lett.* **64** (1990) 1196.

[2] W.L. Ditto, S.N. Rauseo and M.L. Spano, *Phys. Rev. Lett.* **65** (1990) 3211.

[3] E.R. Hunt, *Phys. Rev. Lett.* **67** (1991) 1953.

[4] E.R. Hunt, *Phys. Rev. Lett.* **68** (1992) 1259.

[5] R. Lima and M. Pettini, *Phys. Rev. A* **41** (1990) 726.

[6] Y Braiman and I. Goldhirsch, *Phys. Rev. Lett.* **66** (1991) 2545.

[7] A. Hubler and E. Luscher, *Naturwissenschaften* **76** (1989) 67.

[8] L.M. Pecora and T.L. Carrol, *Phys. Rev. Lett.* **64** (1990) 821.

[9] L.M. Pecora and T.L. Carrol, *Phys. Rev. A* **44** (1991) 2374.

[10] C. Modzelewski, H.T. Savage, L.T. Kabacoff and A.E. Clark, *IEEE Trans. Magn.* **17** (1981) 2837;

[11] W.L. Ditto, S. Rauseo, R. Cawley, C. Grebogi, G.-H. Hsu, E. Kostelich, E. Ott, H.T. Savage, R. Segnan, M.L. Spano and Y.A. Yorke, *Phys. Rev. Lett.* **63** (1989) 923.

[12] D.P. Lathrop and E.J. Kostelich, *Phys. Rev. A* **40** (1989) 4028.

[13] O.E. Rossler, *Phys. Lett. A* **57** (1976) 397.

[14] E.N. Lorenz, *J. Atmos. Sci.* **20** (1963) 130.

[15] M.I. Rabinovich and A.L. Fabrikant, *Zh. Eksp. Teor. Fiz.* **77** (1979) 617 [*Sov. Phys. JETP* **50** (1979) 311].

[16] P. Holmes and D. Whitley, *Physica D* **7** (1983) 111.

Paper 8
On the control of complex dynamic systems

E. Atlee Jackson

Department of Physics, University of Illinois at Urbana-Champaign, 1110 West Green St., Urbana, IL 61801, USA and Center for Complex Systems Research, Beckman Institute, University of Illinois at Urbana-Champaign, 405 North Mathews, Urbana, IL 61801, USA

Received 16 July 1990
Revised manuscript received 12 December 1990
Accepted 26 December 1990
Communicated by H. Flaschka

A method is described for the limited control of the dynamics of systems which generally have several dynamic attractors, associated either with maps or first order ordinary differential equations (ODE) in $\mathbb{R}^n$. The control is based on the existence of 'convegent' regions, C_k ($k = 1, 2, \ldots$), in the phase space of such systems, where there is 'local convergence' of all nearby orbits. The character of the control involves the 'entrainment' and subsequent possible 'migration' of the experimental system from one attractor to another. Entrainment means that $\lim_{t \to \infty} |x(t) - g(t)| = 0$, where $x(t) \in \mathbb{R}^n$ is the system's controlled dynamics, and the goal dynamics, $g(t) \in G_k$, has any topological form but is limited dynamically and to regions of phase space, G_k, contained in some C_k, $G_k \subset C_k$. The control process is initiated only when the system enters a 'basin of entrainment', $BE_k \supset G_k$, associated with the goal region G_k. Aside from this 'macroscopic' initial-state information about the system, no further feedback of dynamic information concerning the response of the system is required. The experimental reliability of the control requires that the regions, BE_k, be convex regions in the phase space, which can apparently be assured if $G_k \subset C_k$. Simple illustrations of these concepts are given, using a general linear and a piecewise-linear ODE in $\mathbb{R}^n$. In addition to these entrainment-goals, 'migration-goal' dynamics is introduced, which intersects two convergent regions $G \cap C_i \neq \varnothing$, $G \cap C_j \neq \varnothing$ ($i \neq j$), and permits transferring the dynamics of a system from one attractor to another, or from one convergent region to another. In the present study these concepts are illustrated with various one-dimensional maps involving one or more attractors and convergent regions. Several theorems concerning entrainment are derived for very general, continuous one-dimensional maps. Sufficient conditions are also established which ensure 'near-entrainment' for a system, when the dynamic model of the system is not exactly known. The applications of these concepts to higher dimensional maps and flows will be presented in subsequent studies.

1. Introduction

While the terminology 'complex dynamic system' is frequently associated with any system which is capable of exhibiting some form of chaotic motion, many of the more important complex systems in nature undoubtedly require wide variability in their response to their environment (e.g., refs. [16, 26, 30]). Dynamic systems which are capable of such varied as well as complex dynamics generally must have a number of dynamic attractors, A_k ($k = 1, 2, \ldots$) (i.e. ω-limit sets), possibly of different topological types (e.g., stable fixed points, stable limit cycles, intermittent attractors, semi-periodic attractors, or strange (chaotic) attractors). Such systems will be referred to as multiple attractor systems (MAS). They are known to exist in fluid dynamics [3, 4], the heart [8, 9, 11, 34], optics (reviewed by Lugiato [24] and Hammel, Jones, and Moloney [12] and are undoubtedly widespread in neural networks and biological systems. The present study concerns a method which allows one to impose a variety of dynamic motions on such systems, provided that their dynamics can be accurately described either by maps or a system of first order ordinary differential equations, involving differentiable functions in $\mathbb{R}^n$.

This control method was first introduced by Hübler [14] and used by Hübler and Lüscher [15], and Lüscher and Hübler [23] in studies involving the logistic map and nonlinear damped oscillators. Based on Hübler's method, and the results obtained from the study of the logistic system by Jackson and Hübler [19], the present generalized formulation addresses the basic issues: (1) For what types of systems can this control method be used? (2) What types of controls can be established for these systems? (3) What are the limitations concerning when (in time) these controls can reliably be initiated? (4) How can this control dynamics be used to transfer the dynamics of a MAS from one attractor to another, possibly 'more beneficial', attractor? This type of (finite time) control of truly complex MAS clearly could be of great importance in the future.

To answer such questions, the present formulation focuses on the existence of convergent regions, C_k ($k = 1, 2, \ldots$), in the phase space of attractor systems. In each of these regions all nearby orbits 'locally converge' toward each other. ('Local convergence' need not imply uniform convergence, as will be discussed in Section 2.) Thus, while many of the above attractors have positive Lyapunov exponents, they nonetheless also have these limited regions in their basins of attraction where nearby orbits generally converge; the Lyapunov exponents are only an average aspect of the dynamics. Based on all examples studied to date, it is conjectured that every system which has attractors also has at least one convergent region within each basin of attraction; however, a proof of this conjecture is not presently known.

These convergent regions make it possible to employ the system's natural (albeit, localized) dynamics to cause the system to tend to a limited set of desired 'goal' dynamics $g(t) \in \mathbb{R}^n$. If the system's dynamics is $x(t)$, and

$$\lim_{t \to \infty} |x(t) - g(t)| = 0 \qquad (1.1)$$

we say that the system is 'entrained' to this goal dynamics, $g(t)$. The goal dynamics may be selected to have any 'topological' character (fixed point, periodic, knotted, chaotic, etc.), but it is restricted quantitatively to be in some goal region, G_k, which intersects one of the convergent regions $g(t) \in G_k$, $G_k \cap C_{l_k} \neq \varnothing$ and C_l is system-dictated (i.e. a property of the autonomous system). While it is not generally necessary, we will initially simplify matters by requiring that $G \subset C$,

$$g(t) \in G_k \subset C_{l_k}. \qquad (1.2)$$

The set of contiguous initial conditions (when the control is initiated) for which the system is entrained to the goal (1.2),

$$\mathrm{BE}_k = \left\{ x(0) \big| \lim_{t \to \infty} |x(t) - g(t)| = 0 \right\} \qquad (1.3)$$

will be called the basin of entrainment for that goal. Thus, each region G_k has some $\mathrm{BE}_k \supset G_k$.

The fact that the system has this convergent property, which gives rise to entrainment (1.1), means that, once the control is initiated there is no need to further monitor the system's dynamics, nor to feedback this information in order to sustain the control. This is obviously very important in systems which have chaotic dynamics, since their sensitivity to small errors makes them very difficult, and probably impossible, to control using conventional feedback methods over all of their phase space. However, by restricting the goal regions, we can use the system's natural tendency to locally converge in some phase-space regions, to eliminate the need for this feedback of dynamic information.

In any such complex system, it is important to have a control method which is both reliable, and also does not require a precise knowledge of the state of the system to achieve this control. The reliability aspect means that when the control is initiated, we want to be able to ensure the future behavior of the system. We do not, for example, want the system to be 'scattered' into some unexpected attractor and acquire dynamics which differs from the goals we set. In short, we want to be able to ensure that the system will be entrained to $g(t)$, and we want to accomplish this without a precise knowledge of the system's initial state, $x(t = 0)$. On the other hand, if we want reliability, we clearly must know something about the system's initial state, $x(t = 0)$, before we initiate the control. Specifically we must know that $x(t = 0)$ is in the desired basin of entrainment, $x(t = 0) \in \mathrm{BE}_k$, before the control is turned on; therefore, it is important to determine these basins of entrainment. Moreover, not all basins of entrainment are acceptable [19]. In keeping with the spirit of reliability and our lack of detailed knowledge of the state of the system, we do not want this basin of entrainment to be 'full of holes'. More precisely, we will require that all basins of entrainment be convex regions. This means that if (x, y) are two points in BE then all the points $z = x(1 - \theta) + \theta y$ ($0 \leq \theta \leq 1$, $x,y,z \in \mathbb{R}^n$) are also in BE. The purpose of this is to be able to reliably predict entrainment when we only require the initial state to satisfy some inequality (e.g., $l < x(t = 0) < u$). This type of 'macro' initial-state information is commonly the only experimental information available, and we want to be able to make $|u - l|$ as large as possible.

This method of control also affords the opportunity to cause a permanent change in the dynamics of multiple attractor systems (MAS), using a control for only a finite time. This can be accomplished by causing the system to transfer its dynamic activity from one attractor to another. To do this, one needs to introduce migration-goal dynamics which moves from one convergent region to another. Once the system has been entrained in one convergent region, it may be possible (with sufficient care) to cause the system to follow the migration goal dynamics outside of the convergent region to another convergent region. If these two regions are associated with different attractors, the control can then be terminated, leaving the system in the dynamic state of the second attractor. There are many situations in which it would be very desirable to be able to reliably transfer a system from one behavior to another (one thinks of such classic cases as heart fibrillations, or 'epileptic-like' states). The ability to accomplish this transfer in systems which can be modeled by maps and ODE, appears to be likely in many cases, if the migration-goal dynamics is selected intelligently and noise effects are not too large. Examples of this will be given in Sections 5 and 6.

To make these points particularly clear, we list both the restrictions and objectives of the present method:

Restrictions
(R1) This analysis applies to dynamic systems which can be described by maps or first-order differential equations, involving differentiable functions in $\mathbb{R}^n$.
(R2) These systems must have one or more attractors (ω-limit sets). Thus this control method does not apply to 'conservative' systems such as area-preserving maps, or any Hamiltonian flow.
(R3) There are system-dictated quantitative (but not topological) limitations on the goal dynamics which can be permanently imposed on the system.

Subject to these restrictions, the objectives of this method are:

Objectives
(O1) Entrainment of the system dynamics, $x(t)$, to a prescribed goal dynamics, $x(t) \to g(t)$, provided that

$g(t) \in G \subset C$, where C is a convergent region associated with an attractor.

(O2) The goal and system dynamics may have any topological character (e.g., any periodicity, chaotic character, etc.).

(O3) The entrainment is reliable even if the initial state of the system (when the control is initiated) is only known to lie in a prescribed 'macroscopic' basin of entrainment, BE(G) associated with G.

(O4) Aside from this initial information, no further information concerning the state of the system needs to be used as a feedback to the control.

(O5) Once near-entrainment is obtained ($|x(t) - g(t)| \leq \varepsilon_e$), another form of control can be initiated, which uses a migration-goal dynamics between different convergent regions, $G \cap C_k \neq \varnothing$ ($k = 1,2$). In particular, this will permit the system to be transferred from one attractor to another attractor in multiple attractor systems (MAS).

(O6) The method should be stable to finite errors in the modeling of the dynamics of the system, and also to the influence of limited amounts of external noise on the system.

The program outlined above represents a unification and extension of a number of pioneering studies which have used this control method for particular systems [1, 6, 15, 19, 21, 23, 31], which exhibit examples of entrainment. The present study examines general systems and studies the limitations of the goal dynamics, the role of convergent regions, the extent of basins of entrainment, and the possible application of migration-goals. The study of the control of the logistic-map chaos by Jackson and Hübler [19] first explored the nature of basins of entrainment, and many other responses of this particular system, but not for general goals, nor the other issues noted above.

The application of the above objectives to many systems is too extensive to be described in the present study. These applications to maps in two-dimensions, and to flows in one to three dimensions have been established and will be presented in subsequent studies [18, 20]. In Section 2 the general definitions of convergent regions and basins of entrainment which are applicable to all of these studies will be presented. The distinction between the present emphasis on regions of phase space, and the numerous discussions of the variability of contracting features of autonomous solutions of equations of motion will also be discussed. Following these general considerations, the analysis and illustrations in the remaining sections will be limited to one-dimensional maps. This will allow for the presentation of a variety of control-features in some detail. Thus in Section 3, several theorems concerning the basins of entrainment are proved for very general (differentiable) maps. The distinction is also made between entrainment and uniform entrainment. In Section 4, a simple example, involving the logistic map, is used to illustrate the topological freedom of the character of the entrainment dynamics *vis-à-vis* the system's dynamics. A Gaussian map is introduced in Section 5, to illustrate global basins of entrainment, which are particularly nice for controls. Finally Section 6 illustrates the

use of migration goal, for MAS, by transferring a system between a number of convergent regions.

2. Convergent regions, goal dynamics, and basins of entrainment

The control method presented in this study involves the interrelationship of several dynamic regions in the phase space of an 'experimental' (E) system whose dynamics can be represented either by a map

$$x_{k+1} = E(x_k), \quad x \in \mathbb{R}^n, \quad k = -1, -2, -3, \ldots, \quad (2.1)$$

or a system of first-order ordinary differential equations (ODE)

$$dx/dt = E(x), \quad x \in \mathbb{R}^n, \quad t < 0. \quad (2.2)$$

The dynamic equations for positive 'times' will involve controls, to be discussed below.

The first regions which are required to exist for these systems are various attracting sets (ω-limit sets), which we designate by A_k ($k = 1, 2, 3, \ldots$), together with their basins of attraction, BA_k. Common examples of such attractors are stable fixed points, stable limit cycles, semi-periodic attractors, intermittent attractors, or strange (chaotic) attractors. In the case of flows, (2.2), the latter attractors only occur if their phase space dimension is $n \geq 3$. If the systems (2.1) and (2.2) do not have such attractors then the following method will not apply. In particular, 'conservative' (area-preserving) dynamics do not have attractors. Classic examples of such systems are Chirikov's standard map ($x \in \mathbb{R}^2$) or Arnold's cat map ($x \in \mathbb{R}^2$) or any Hamiltonian system (a special case of system (2.2)).

Next we introduce another type of region of phase space which appears to exist for all systems which have attractors. Whether this is true or not, we will define a convergent region, $C_k(E)$, which is dependent on $E(x)$, to be related to the 'local convergence' of all solutions of (2.1) or (2.2) in $C_k(E)$. This is a much stronger condition than the familiar concept of contraction of volume elements in the phase space, as will become clear later. We will use the adjective 'convergence' to describe the approach of two orbits, and 'contraction' to describe the decrease in volume of a region of phase space. The character of 'local convergence' will be clarified by an example following Theorem $\emptyset$ below.

With these rough ideas, we now give the formal definitions. First for maps, (2.1), a convergent region is a connected set

$$C(E) = \left\{ x \,\middle|\, x \in \mathbb{R}^n, \right.$$
$$\left\| \partial E_i / \partial x_j - \delta_{ij} \mu(x) \right\| = 0,$$
$$\left. |\mu(x)| < 1 \text{ for all } \mu(x) \right\}. \quad (2.3)$$

In the case of the ODE system, (2.2), a convergent region is defined by

$$C(E) = \left\{ x \,\middle|\, x \in \mathbb{R}^n, \right.$$
$$\left\| \partial E_i / \partial x_j - \delta_{ij} \lambda(x) \right\| = 0,$$
$$\left. \text{Re } \lambda(x) < 0 \text{ for all } \lambda(x) \right\}. \quad (2.4)$$

Thus these regions are determined by the fact that all of the local characteristic multipliers, $\mu(x)$, or exponents, $\lambda(x)$, of the variational equations of the maps (2.1) or ODE (2.2) have a converging character. It should be emphasized that the phase functions $\mu(x)$ and $\lambda(x)$ define connected regions of phase space, and do not imply that the system cannot concurrently have positive Lyapunov exponents. Hence, even systems with strange attractors have limited convergent regions in their phase space. It, of course, has been appreciated for some time that the dynamics near strange attractors can have fluctuating, or non-uniform divergence rates around the attractor (e.g. refs. [10, 25, 27]). It has also been recently recognized [22] that even regular attractors, such as the Bonhoeffer–van der Pol limit cycle can have vector fields with converging and diverging regions in their phase space. In contrast to these observations, the present focus (in order to introduce controls) is on regions of phase space, $C(E)$, rather than properties of the autonomous dynamics. As will be made clear in this and future articles, the above fluctuations can often be related to the fact that these autonomous dynamics continually enter and leave the regions $C(E)$.

The relationship between convergent regions and attractors is interesting and apparently non-trivial. From an examination of many maps and flows in $\mathbb{R}^n$, the following conjectures are made:

Conjecture 1. Every system which has an attractor, A_k, also has at least one convergent region, C_l, in its basin of attraction, $BA_k \supset C_l$.

Conjecture 2. Every attractor of a system, A_k, has associated with it at least one convergent region, C_{l_k}, which intersects A_k, $A_k \cap C_{l_k} \neq \varnothing$.

At present no proofs of these conjectures are known. They would be nice to know, but the present method of control only rests on the presence of convergent regions.

Finally, it will be shown in subsequent studies that the convergent regions (2.4) can be determined without determining the eigenvalues, $\lambda(x)$. This can be done, with the help of the Routh–Hurwitz theorem (e.g., Ref. [5]), which represents an important simplification in the application of this concept.

The 'goal' dynamics is what we wish the systems (2.1) or (2.2) to do. We define a family of such goal dynamics by the fact that they occur in some goal regions in phase space,

$$\{g_k | k = 0, 1, 2 \ldots\} \subset G$$

$$\text{or } \{g(t) | t \geq 0\} \subset G. \tag{2.5}$$

By introducing goal regions all specific characteristics of any particular goal set becomes irrelevant; in other words, the topological character of some goal set $\{g_k\}$ is not a significant factor for the following control.

We will also distinguish between two classes of goal dynamics, the entrainment-goals (e-goals) and the migration-goals (m-goals).

E-goals

The first class of goal dynamics are those which we want the system to be entrained to in the future;

$$\lim_{k \to 0} |x_k - g_k| = 0$$

or

$$\lim_{t \to \infty} |x(t) - g(t)| = 0. \tag{2.6}$$

To accomplish this entrainment, we will consider control equations of the form

$$x_{k+1} = E(x_k) + F(g_{k+1}, g_k)S(k) \tag{2.7a}$$

or (using the notation $\dot{x} \equiv dx/dt$)

$$\dot{x} = E(x) + F(\dot{g}, g)S(t), \tag{2.7b}$$

where the system is autonomous for negative 'times'. $S(k) = 0$ $(k < 0)$ and $S(t) = 0$ $(t < 0)$, and $S = 1$ or 0 otherwise (so S is an on/off Heaviside switch). The advantages of using 'softer' switching functions, will be discussed in subsequent studies. The function $F(u, v)$ only depends on the goal set, as indicated, and will be discussed below. The control equations (2.7) are very different from the usual control equations

$$\dot{x} = E(x) + G(x, y(t)) \tag{2.8}$$

in which the control function, $G(x, y(t))$ depends both upon a prescribed function $y(t)$ and the instantaneous state of the system, $x(t)$ (e.g. see Ref. [28]). The control (2.7) does not make use of this instantaneous information about the state of this system. It is sometimes referred to as a 'no-feedback' or 'open-loop' control.

In order for (2.6) to be possible, Eqs. (2.7) (with $S = 1$) must have as particular solutions, $x_k = g_k$ or $x(t) = g(t)$. If these are substituted into (2.7a) and (2.7b) respectively, one finds that the functions $F(u, v)$ must be given by

$$F(g_{k+1}, g_k) = g_{k+1} - E(g_k), \tag{2.9a}$$

$$F(\dot{g}, g) = \dot{g} - E(g). \tag{2.9b}$$

Controls of the form (2.9) were first introduced by Hübler [14] and used by Hübler and Lüscher [15] and Lüscher and Hübler [23] in the study of controls and stimulations of the logistic dynamics and damped nonlinear oscillators. The controls (2.9) are necessary for (2.7) to yield entrainment, (2.6). Also the entrainment (2.6) is only possible if the Lyapunov exponents associated with these goals are negative. For example, for one-dimensional maps it is necessary that

$$\log \left| \prod_{k=0}^{\infty} \partial E(g_k)/\partial g_k \right| < 0. \tag{2.10}$$

This condition ensures the stability of the goal dynamics $\{g_k\}$ to infinitesimal perturbations. However, it does not ensure that entrainment, (2.6), will occur for finite initial differences

$$|x_0 - g_0| = \varepsilon_i \quad \text{or} \quad |x(0) - g(0)| = \varepsilon_i, \tag{2.11}$$

which is of basic interest in the present control method. This issue was first considered by Jackson and Hübler [19], in the case of the logistic map.

To discuss this issue more generally, we introduce the concept of the basin of entrainment associated with the e-goal dynamics (2.5) and the control dynamics (2.7). The basin of entrainment is the set of initial states of (2.7) which satisfy (2.6),

$$\mathrm{BE(G)}$$
$$= \left\{ x_0 \Big| \lim_{k \to \infty} |x_k - g_k| = 0; \right.$$
$$x_{k+1} = E(x_k) + F(g_{k+1}, g_k),$$
$$\left. g_k \in \mathrm{G},\, k = 0, 1, 2 \dots \right\} \tag{2.12}$$

or for the ODE

$$\mathrm{BE(G)}$$
$$= \left\{ x(0) \Big| \lim_{t \to \infty} |x(t) - g(t)| = 0; \right.$$
$$\dot{x} = E(x) + F(\dot{g}, g);$$
$$\left. g(t) \in \mathrm{G},\, t \ge 0 \right\} \tag{2.13}$$

and where the functions F are given by (2.9). In order to ensure that the following control method is reliable, we will also require that any BE of interest be a convex region:

$$\text{if } (x, y) \in \mathrm{BE} \text{ then } x\theta + (1 - \theta)\, y \in \mathrm{BE},$$
$$0 \le \theta \le 1. \tag{2.14}$$

Therefore, one of the basic conditions for entrainment is that the initial state of the system is in the BE. When the system is in this region we can initiate the control (set $k = 0$ or $t = 0$, and turn on the switch S in (2.7)). We therefore require that

$$x_0 \subset \mathrm{BE}, \quad x(0) \subset \mathrm{BE} \tag{2.15}$$

so that the determination of the basin of entrainment is of basic importance. In order to be able to use a 'macroscopic' condition for the initiation of the control which is in the form of inequalities

$$l < x_0 < u \quad \text{or} \quad l < x(0) < u,$$
$$x, u, l \in \mathbb{R}^n \tag{2.16}$$

it is necessary for the basin of entrainment to have the convex property (2.14). We will require that the initiation of the control can be determined by conditions (2.16) and (2.15), hence the BE must satisfy (2.14). This is what is meant by a 'reliable control', based on experimental data which can only be determined to the accuracy (2.16).

To illustrate the above concepts in their simplest context, the following entrainment theorem will be proved for any linear ODE:

Theorem $\emptyset$. Let $\dot{x} = E(x) \equiv A \cdot x$ $(x \in \mathbb{R}^n)$, where $A = (a_{ij})$ is a constant matrix, and all roots λ_k of $\|a_{ij} - \lambda\delta_{ij}\| = 0$ have negative real parts (so $C(E) = \mathbb{R}^n$). Let $g(t) \in G \equiv \mathbb{R}^n$ be any differentiable function. Then all solutions of

$$\dot{x} = A \cdot x + \dot{g} - A \cdot g \quad (t \ge 0)$$

are entrained to $g(t)$, $\lim_{t \to \infty} |x(t) - g(t)| = 0$; that is, the basin of entrainment of G is global ($\mathrm{BE(G)} = \mathbb{R}^n$).

The proof is trivial. Let $u = x - g$, so that $\dot{u} = A \cdot u$. Because the roots λ_k all have negative real parts, any

solution of this equation satisfies $\lim_{t \to \infty} |u(t)| = 0$, proving entrainment of $x(t)$ to $g(t)$. It should be noted, however, that even in this simple case $|u(t)|$ need not decrease uniformly (i.e., $d|u(t)|/dt > 0$ can occur for an initial period of time), if $n \ge 2$. In the case of maps this nonuniformity can even occur if $n = 1$ (Section 3). Thus the expression 'local convergence' does not necessarily imply uniform convergence, but rather an asymptotic convergence if the two solutions remain in the convergent region. The generalization to cases where the $a_{ij} \equiv (\partial E_i/\partial x_j)$ are not constant in $\mathbb{R}^n$, cannot be treated with such ease, and quite different theorems need to be developed. Illustrations of these will be given below and in subsequent studies [18].

Some of the issues which arise in the control of nonlinear systems can be illustrated by considering the 'artificial' nonlinear (piecewise-linear) system

$$\dot{x} = E(x), \quad x \in \mathbb{R}^n:$$
$$E(x) = A \cdot x, \quad |x| \le 1,$$
$$E(x) = B \cdot x, \quad |x| \le 1, \tag{2.17}$$

where A is given in Theorem $\emptyset$, and $\|b_{ij} - \lambda\delta_{ij}\| = 0$ has some roots with positive real parts. In this case $C(E) = \{x \mid |x| \le 1\}$. Now it is not generally true (if $n \ge 2$) that, if $g(t \ge 0) \in C(E)$, the solutions of $\dot{x} = E(x) + \dot{g} - g(t)$ $(t \ge 0)$ will be entrained to $g(t)$, (2.6), even if $x(t = 0) \in C(E)$. This problem arises from the fact that $|x(t) - g(t)|$ may initially increase, as noted above. If $x(t)$ moves into the region $|x| > 1$, the controlled dynamics may (or may not) have unbounded solutions. To ensure that $x(t) \in C(E)$ for all t (so Theorem $\emptyset$ again applies) requires both that $g(t) \in G = \{x \mid |x| < r_g < 1\}$, and that the control only be initiated when $x \in \mathrm{BE(G)} = \{x \mid |x| < r_e\}$. Where the values (r_g, r_e) might be, for example, $(\tfrac{1}{2}, \tfrac{3}{4})$. They are dictated by the matrices A and B. Note that BE(G) is a convex region, (2.14), and moreover it is possible for $r_e > 1$ (that is, for the basin of entrainment to contain $C(E)$), because all solutions of $\dot{x} = B \cdot x + \dot{g} - A \cdot g$ might be entrained. If, indeed, they are entrained for $r_e > 1$, then all solutions in $\mathbb{R}^n$ are entrained, because of the piecewise-linear nature of this nonlinearity. In that case the basin of entrainment would again be global. In any case, for sufficiently small (r_g, r_e) this nonlinear system can certainly be controlled. More realistic examples will be given in a subsequent study [17].

M-goals

Finally, we introduce the concept of a migration-goal dynamics. This is a finite set of goal states $(g_0, g_1, \ldots, g_n)$, with $g_0 \in C_i$ contained in one convergent region and $g_n \in C_j$ contained in another convergent region $(i \ne j)$. Thus, the migratory set is defined for maps or ODE:

$$\mathrm{M}_{ij} = \left\{ g_k \mid k = 0, 1, \ldots, n; \right.$$
$$\left. g_0 \in \mathrm{C}_i,\, g_n \in \mathrm{C}_j\,(i \ne j) \right\},$$
$$\mathrm{M}_{ij} = \left\{ g(t) \mid 0 \le t \le T, g(0) \in \mathrm{C}_i, \right.$$
$$\left. g(T) \in \mathrm{C}_j\,(i \ne j) \right\}. \tag{2.18}$$

The idea is that, once a system has been nearly entrained, say

$$|x_k - g_k| < \varepsilon_c, \qquad |x(t) - g(t)|\lambda\varepsilon_c, \tag{2.19}$$

it may be possible to introduce a migratory control, which allows us to 'migrate' the system from one convergent region, C_i, to another C_j. To do this, the system generally needs to exist in regions where the states g are unstable for some time. These ideas will be discussed further, and illustrated in Section 6.

We now will consider very general one-dimensional maps, and begin by considering first the basic problem of ensuring entrainment of the system to a suitable goal-dynamic region, using (2.7a) and (2.9a).

3. Entrainment theorems for one-dimensional maps

In this section two types of entrainment theorems will be presented for the case of one-dimensional map dynamics. The first theorem generally gives a very conservative estimate for the basin of entrainment, but it has the advantage that it can be extended to higher-dimensional maps, as will be shown in subsequent studies. The second type of theorems gives a much more precise determination of the basins of entrainment associated with a more precise determination of the basins of entrainment associated with a goal region, but their extension to higher dimensions is not presently obvious.

The first type of theorem is simply based upon the use of the mean value theorem within a convergent region, coupled with a suitably restricted goal region.

Theorem 1. Let $x_{n+1} = E(x_n)$ $(x \in \mathbb{R})$, where $E(x)$ is differentiable, and assume that $E(x)$ has a restricted converging region, defined by

$$\overline{C}(E)$$

$$= \{I = (l < x < u)\}$$

$$m \equiv \max_I |\partial E/\partial x| < 1,$$

$$M \equiv \max_{(x,y)\in I}[E(x) - E(y)] < \tfrac{1}{2}(u - l)\}. \tag{3.1}$$

Moreover, consider a governing set

$$\overline{G} = \{g_n | g_n \in \mathbb{R}, l + M < g_n < u - M, n \geq 0\}. \tag{3.2}$$

If $x_0 \in \overline{C}$, and

$$x_{n+1} = E(x_n) + g_{n+1} - E(g_n), \quad n = 0, 1, \ldots, \tag{3.3}$$

then the dynamics is entrained,

$$\lim_{n\to\infty}|x_n - g_n| = 0. \tag{3.4}$$

We note that the set $\overline{C}(E)$, (3.1), is generally more restrictive than the converging set $C(E)$, (2.3); that is, $\overline{C}(E) \subseteq C(E)$. Also note that $\overline{C}(E)$ is a connected region (i.e., all $l < x < u$ satisfy the conditions).

The proof of this theorem is quite simple. We first show that if $x_0 \in \overline{C}$, the set (3.2) guarantees that $x_n \in \overline{C}$ for all

$n \geq 0$, if (3.3) holds. We note first, that the condition $M < \tfrac{1}{2}(u - l)$ in (3.1) ensures that $\overline{G}$ is not an empty set (i.e., $u - M > l + M$). If $x_n \in \overline{C}$, the definitions (3.1) and (3.2) yield

$$x_{n+1} = E(x_n) - E(g_n) + g_{n+1} < M + g_{n+1} < u \tag{3.5}$$

and $x_{n+1} > g_{n+1} - M > l$; so $l < x_{n+1} < u$. Since $x_0 \in \overline{C}$, $x_n \in \overline{C}$ for all $n \geq 0$.

Next we note that

$$x_{n+1} - g_{n+1} = E(x_n) - E(g_n). \tag{3.6}$$

Since $E(x)$ is differentiable, we can use the mean value theorem to conclude that

$$x_{n+1} - g_{n+1} = (x_n - g_n)(\partial E/\partial x)_m, \tag{3.7}$$

where the derivative is evaluated at some point x_m which lies between x_n and g_n (and hence $x_m \in \overline{C}$, because $\overline{C}$ is a connected region). It is important to note that (3.7) does not require that $x_n - g_n$ be small; it is globally valid within $\overline{C}$. We now make use of the converging condition in (3.1), to conclude that

$$|x_{n+1} - g_{n+1}| \leq |x_n - g_n|m < m^n(u - l) \tag{3.8}$$

and since $m < 1$, we have the desired result (3.4).

It is not difficult to see that, without additional restrictions on the function $E(x)$ or on the governing set $\{g_k\}$, the region $\overline{C}(E)$ and $\overline{G}$ are not only sufficient for entrainment, but also necessary. To see this, assume that $E(g_0) - E(x_0) = M$ and $g_k = l + M - \varepsilon$ $(k = 1, 2 \ldots)$ where $\varepsilon > 0$ is arbitrarily small. Thus we violate the condition $\{g_k\} \in \overline{G}$. Then $g_1 - x_1 = M$, so $x_1 = l - \varepsilon$ lies below $\overline{C}$. If $\partial E/\partial x > 1$ for $x < l$, then since $g_2 - x_2$ $(g_1 - x_1)(\partial E/\partial x)_m$ it is possible for $x_2 < x_1$, depending on how rapidly E decreases when $x < l$. In such cases, $x_{k+1} < x_k$ can occur, so entrainment certainly does not occur.

This type of difficulty is clearly due to a terrible choice of the governing set, coupled with an exceptionally uncommon function E (it must decrease precipitously when $x < l$, to ensure that $(\partial E/\partial x)_m > 1$). More commonly both the converging region $\overline{C}(E)$, (3.1), and the governing set $\overline{G}$, (3.2), are often over-conservative. This will soon be made clear.

A second class of theorems improves significantly on the mean-value approach of Theorem 1, at least for one-dimensional maps, and it introduces a clear distinction between a basin of entrainment and a basin of uniform entrainment. The latter concept is very useful for several reasons, as will be demonstrated. These theorems, however, require a more global knowledge of $E(x)$ than required for theorem 1.

We begin with a very special case, which is nonetheless instructive for what follows. For future reference, this will be stated as a lemma:

Lemma 1. Let $E(x) = \Sigma_{k=0}^{\infty} A_k x^{2k}$, so there is a convergent region $C = \{x | |\partial E/\partial x| < 1, |x| < a; |\partial E/\partial x|_{x=a} = 1\}$ for some $a > 0$. The goal dynamics is taken to be the fixed point at the origin, $g_k = g_0 = 0$ $(k \geq 0)$. Let x^* be the root of $E(x^*) - E(0) = x^*$, such that $|E(x) - E(0)| < |x|$ for all $|x| < |x^*|$. If $x_{n+1} = E(x_n) - E(0)$, a basin of entrainment to $g = 0$ is $BE = \{x | |x| < |x^*|\}$, and moreover $BE \supset C$ (i.e., $|x^*| > a$).

The proof is straightforward. If $|x| \leq \delta |x^*|$ for some $1 > \delta > 0$, then $|E(x) - E(0)| \leq \varepsilon(\delta) |x|$ for some $1 > \varepsilon(\delta) > 0$. Therefore, if $x_n \leq \delta |x^*|$ for some n, $|x_{n+1}| = |E(x_n) - E(0)| < \varepsilon |x_n| < \varepsilon \delta |x^*| < \delta |x^*|$, so $|x_{n+1}| \leq \delta |x^*|$ also holds. It follows that, if $|x_0| \leq \delta |x^*|$, $|x_n| \leq \varepsilon^n |x_0|$ for all $n \geq 0$, so $\lim_{n \to \infty} |x_n| = 0$, proving entrainment. Finally $|E(x) - E(0)| = \left| \int_0^x (\partial E/\partial x) \, \mathrm{d}x \right| < |x|$ for all $|x| \leq a$. Hence $a < |x^*|$.

While Lemma 1 applies only to symmetric functions $E(x)$ and a fixed-point goal at $x = 0$, several points should be noted which will apply generally. First of all, values of x_0 which are not in BE may also be entrained to $x = 0$, but they will lie in regions which are generally (but not necessarily) disjoint from BE, and hence (by definition) are not part of BE. It should be emphasized that what is always obtained is only 'a' basin of entrainment, not 'the' basin of entrainment. Secondly, the fact that $BE \supset C$ in this case means that one could use the criterion $x_0 \in C$ for initiating the control, if this is more convenient (e.g., more easily determined). Thirdly, it should be noted that global comparison of distant points of $E(x)$ is used (i.e., $|E(x) - E(0)| < |x|$, $\forall |x| < |x^*|$), which is different from the local properties of $E(x)$ only in $\overline{C}(E)$, used in Theorem 1.

Next it should be noted that the bound $|x_n| \leq \varepsilon^n |x_0|$ could be established for all n, so that not only is entrainment proved, but it has a uniform bound (good for all n). By establishing the relation between $\varepsilon(\delta)$ and δ we could also obtain families of basins of uniform entrainment, each with a 'guaranteed' rate of convergence to the fixed point goal dynamics. This clearly could be of great practical interest, and will be returned to later in the consideration of migration goals (Section 6).

To see the distinction between a basin of entrainment (BE) and a basin of uniform entrainment (BUE) in its simplest context, we now shift the fixed-point goal to $|g_0| < a$ (required for stability).

Consider the situation illustrated in Figure 1. A coordinate system has been introduced where $E(x) - E(g_0)$ has been plotted versus $x - g_0$. Moreover two lines $x - g_0 = E(x) - E(g_0)$ and $x - g_0 = E(g_0) - E(x)$ have been drawn in this reference system. If x_n is in the region such that

$$|x_n - g_0| > |E(x_n) - E(g_0)|, \tag{3.9}$$

so that $E(x_n) - E(g_0)$ falls between these two lines, then $|x_{n+1} - g_0| = |E(x_n) - E(g_0)| < |x_n - g_0|$, and therefore x_{n+1} will be closer to g_0, than is x_n. This is a necessary condition for uniform entrainment, but it is not sufficient, nor is it necessary for entrainment.

Figure 1 shows why x_0 may be in the basin of entrainment even if (3.9) does not hold for $n = 0$, using a simple symmetric function $E(x)$. This function has the two roots $x = l$ and $x = u$, which satisfy $l - g_0 = E(l) - E(g_0)$ and $u - g_0 = E(g_0) - E(u)$ respectively. If $|l| > x_0 > u$, then because of the symmetry of $E(x)$, $x_1 > l$, but $|x_0 - g_0| < |x_1 - g_0|$, so x_1 is further from g_0 than x_0. However, since $x_1 > l$, and because x_n never gets larger than u for this particular function $E(x)$ (see below), all subsequent x_n will satisfy $|x_{n+1} - g_0| < |x_n - g_0|$ ($n \geq 1$). Hence, after a delay of one step, the entrainment becomes uniform. Therefore, even if (3.9) is not satisfied for $n = 0$, x_0 may fall within the basin of entrainment.

However, this analysis assumed that $x_n < u$ for all $n \geq 1$. What would happen in this process if $x_n > u$ for some n? If it still satisfies $x_n < |l|$, the previous analysis shows that x_{n+2} is again closer to g_0 than x_{n+1} so it might seem that again there has only been a one-step break in the uniform entrainment (i.e., $|x_n - g_0| < |x_{n+1} - g_0|$). However, since $|x_2 - g_0| < |x_1 - g_0|$, and $|x_0 - g_0| < |x_1 - g_0|$, it is clear that for some x_0, $x_0 = x_2$ is a possible situation (see Figure 2). In other words, a period-two dynamics might exist in this spatial region. Moreover, if $|(\partial E/ \partial x_0)(\partial E/ \partial x_1)| < 1$ this period-two dynamics is stable, and hence it could be an attractor with a finite basin of attraction. This means that the basin of entrainment would be disjoint, at best. Hence we must generally require that $E(x) - E(g_0) < u - g_0$ for all $l < x < g_0$ (see Figure 2). In this case $l < x < u$ is the basin of uniform entrainment and $|x| < |l|$ is a basin of entrainment, for the symmetric function illustrated in Figure 1.

There is clearly nothing special about the symmetry of $E(x)$; it was only introduced for simplicity. We next generalize these results, by removing the symmetry feature of $E(x)$. What follows is still specialized, but is useful as a heuristic example before a general theorem is given.

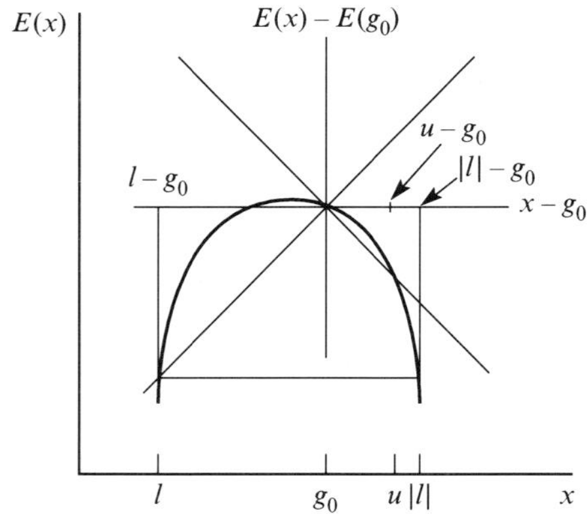

Figure 1 The graphical determination of the basin of entrainment to the fixed-point goal g_0, for symmetric $E(x)$, when $E(x) - E(g_0) < u - g_0$.

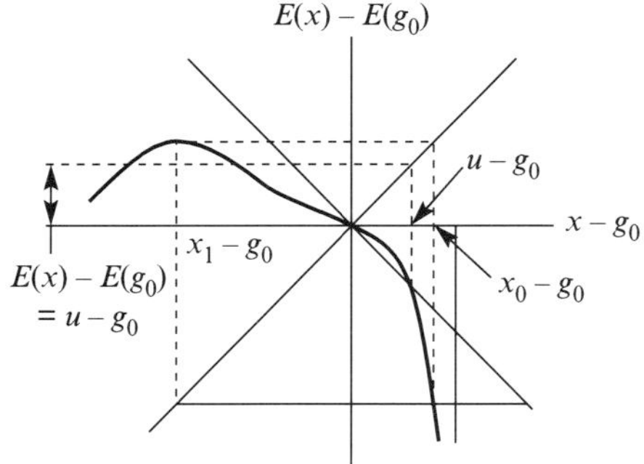

Figure 2 This figure illustrates the possibility of a stable period-two attractor, for a fixed-point goal, g_0, if the condition $E(x) - E(g_0) < u - g_0$ is violated for $l < x < g_0$. For simplicity, the primary coordinates $(E(x), x)$ are no longer shown.

Lemma 2. Let $E(x)$ be a piecewise differentiable function of x and assume that for some g_0

$$u - g_0 = E(g_0) - E(u) > 0, \tag{L.1}$$

$$g_0 - l = E(g_0) - E(l) > 0, \tag{L.2}$$

where

$$\left| E(x) - E(g_0) \right| < \left| x - g_0 \right|, \quad l < x < u, \quad x \ne g_0. \tag{L.3}$$

Also assume that $\left| u - g_0 \right| < \left| l - g_0 \right|$, and

$$E(x) - E(g_0) < u - g_0, \quad l < x < g_0. \tag{L.4}$$

Then the (maximum) basin of uniform entrainment to g_0 is

$$\mathrm{BUE}(g_0) = \left[x \middle| l(g_0) < x < u(g_0) \right].$$

Also, if $\hat{x}$ is the smallest root of

$$E(\hat{x}) = l \quad (\text{so } E(x) > E(l), \, \forall x < \hat{x}) \tag{L.5}$$

and

$$E(x) = E(g_0) < \hat{x} - g_0, \quad \forall u < x < \hat{x} \tag{L.6}$$

then a (minimal) basin of entrainment to g_0 is

$$\mathrm{BE}(g_0) = \left[x \middle| l(g_0) < x < \hat{x}(g_0) \right].$$

The condition (L.6) ensures that not only is $x_{n+1} < \hat{x}$ if $x_n < \hat{x}$, but moreover no periodic orbits are possible, as discussed above. Note that the $\mathrm{BUE}(g_0)$ is as large as possible, whereas $\mathrm{BE}(g_0)$ may be larger.

The content of Lemma 2 is illustrated by an example in Figure 3. If $E(x)$ is piecewise differentiable (useful in the cases of tent-like maps) the definition of a convergent region can be easily generalized in terms of one-sided limits for the associated derivatives at each point. Such a convergent region will be denoted by $\hat{C}$. The mean-value theorem cannot be generally used in such a region, hence $\hat{C}$ cannot be used in Theorem 1. The conditions (L.3) and (L.6) apply in the context indicated in Figure 3.

Note that if $l = -\infty$ and $u = +\infty$, then $\mathrm{BUE}(g_0) = \mathbb{R}$ (all of the space). Even if u is finite, if $l = -\infty$ and $E(x)$ is bounded, then $\hat{x} = +\infty$ and $\mathrm{BE}(g_0) = \mathbb{R}$. Such global basins of

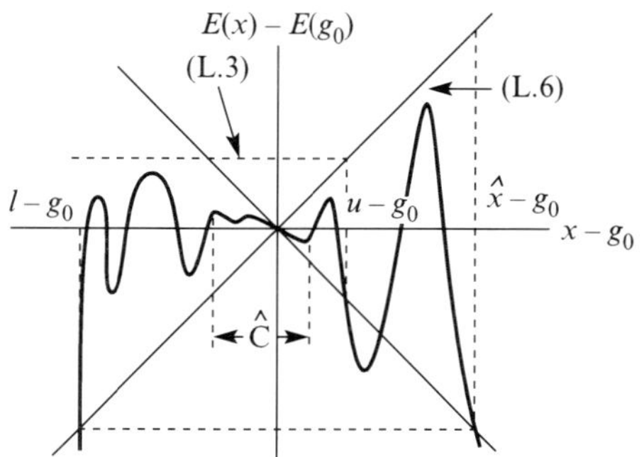

Figure 3 A more general $E(x)$, satisfying $E(x) - E(g_0) < u - g_0$ for all $l < x < g_0$ is illustrated. The basin of uniform entrainment extends over the range $l < x < u$, whereas the basin of entrainment extends at least over the range $l < x < \hat{x}$. The conditions (L.3) and (L.6) are related to the indicated features. $E(x)$ needs only to be piecewise differentiable, as illustrated, leading to a generalized convergent region, $\hat{C}$.

entrainment are clearly of great importance, and will be discussed in Section 5.

The cases covered by Lemma 2 are limited in several respects, which we now remove. The condition (L.3) is essential for uniform convergence to g_0, but (L.1) and (L.2) represent only one of the four ways that (L.3) can first be broken namely

$$u - g_0 = \left| E(g_0) - E(u) \right|,$$

$$g_0 - l = \left| E(g_0) - E(l) \right|,$$

Moreover the example in Lemma 2 (and Figure 3) was for the case $\left| u - g_0 \right| < \left| l - g_0 \right|$. Finally, what happens if a condition such as (L.4) is not valid for all $l < x < g_0$? The following theorem removes all these restrictions, and generalizes conditions such as (L.4). However, for simplicity it will be limited to the maximum basin of uniform entrainment.

Theorem 2. Let $E(x)$ be a piecewise differentiable function, and assume that g_0 is in a convergent region; then there exists $(+\infty \ge u > l \ge -\infty)$ such that

$$\left| E(x) - E(g_0) \right| < \left| x - g_0 \right|,$$

$$\forall l < x < u; \, x \ne g_0,$$

and if (u, l) are finite,

$$u - g_0 = \left| E(u) - E(g_0) \right|,$$

$$g_0 - l = \left| E(l) - E(g_0) \right|. \tag{I}$$

Let $\{l^*, u^*\}$ be the set of roots of

$$E(l^*) - E(g_0) = u - g_0, \quad l^* < g_0,$$

$$E(u^*) - E(g_0) = l - g_0, \quad g_0 < u^*. \tag{II}$$

The maximum basin of uniform entrainment to g_0 is

$$\mathrm{BUE}(g_0)$$

$$= \left[x \middle| \max(l, \{l^*\}) < x < \min(u, \{u^*\}) \right] \tag{3.10}$$

That is, if $x_{n+1} - g_0 = E(x_n) - E(g_0)$, then for any $1 > \delta > 0$, if $\delta[\max(l, \{l^*\}) - g_0] \le x_0 - g_0 \le \delta[\min(u, \{u^*\}) - g_0]$ there exists an $1 > \varepsilon(\delta) > 0$ such that $\left| x_n - g_0 \right| \le \varepsilon^n \left| x_0 - g_0 \right|, \, \forall n \le 0$.

The first conditions, (I), ensure that the range of x is the largest possible for which $\left| x_{n+1} - g_0 \right| < \left| x_n - g_0 \right|$, if x_n is in this range. The second conditions, (II), help ensure that, if x_n is in this range of convergence, then x_{n+1} will also be in this range. It is not hard to see, for example, that if

$$E(x) - E(g_0) < u - g_0, \quad l < x < g_0, \tag{3.11}$$

then, if $l < x_n < u$, one also has $l < x_{n+1} < u$. Moreover (L.1) is ensured if $\left| u - g_0 \right| > \left| g_0 - l \right|$. In the example of Lemma 2, this was not the case, so (L.4) = (3.11) was assumed to hold. If one does not want this restriction, then it is necessary to introduce the first of conditions (II) – actually, as $\mathrm{BUE}(g_0)$ indicates, one only is interested in the largest $l^* < g_0$, and only if the largest l^* is larger than l. If it is not larger than l, then (L.4) does indeed hold, and we recover a result like in Lemma 2. The second of the conditions in (II) likewise concerns the smallest u^* such that, if $x_n < u^*$, then $x_{n+1} < l$. Moreover $\min\{u^*\} > u$ if $\left| g_0 - l \right| > \left| u - g_0 \right|$, which was the case in the example of Lemma 2. Hence this second condition of (II) did not arise in Lemma 2. Perhaps it should be emphasized that

$$E(x) - E(g_0) < u - g_0, \quad l < x < g_0$$

$$\text{if } |u - g_0| > |g_0 - l|$$

and

$$E(x) - E(g_0) < l - g_0, \quad g_0 < x < u$$

$$\text{if } |g_0 - l| > |u - g_0|$$

so that, at most, only one of the conditions (II) will apply in any particular case. Finally, these results show that it is not generally true that the entire convergent region is contained in $\mathrm{BUE}(g_0)$, so the initiation of the control cannot always be based on $x_0 \in \mathrm{C}(g_0)$, but must specifically involve $\mathrm{BE}(g_0)$.

The rate of converge of the entrainment can be established by modifying (I) and (II) of Theorem 2. We note this useful fact as a corollary:

Corollary. If $(u - g_0)$, $(g_0 - l)$ and $|x - g_0|$ are each multiplied by γ $(1 > \gamma > 0)$ in (I) and (II) of Theorem 2, so the basin of uniform entrainment, (3.10), depends both on g_0 and γ, $\mathrm{BUE}(g_0, \gamma)$, then one has the explicit bounds on the rate of convergence,

$$|x_n - g_0| \leq \gamma^n |x_0 - g_0|. \tag{3.12}$$

The corresponding modification of the graphical method, illustrated in Figure 3, can easily be seen to correspond to decreasing the angle between the two diagonal lines.

While Theorem 2 is much more precise concerning the size of the basin of entrainment than what is given in Theorem 1, the latter describes any goal dynamics in a region G, whereas Theorem 2 only treats the case of a fixed-point goal, g_0. While it is possible to generalize Theorem 2 to include more general goal sets $\{g_k\}$, by generalizing the graphical method shown in Figure 3, these details will have to be deferred to future studies. Moreover, the fixed-point goal is the most important goal for many purposes, as will be shown in Sections 5 and 6.

4. Logistic-entrainment: topological freedom

A detailed study of the response of the chaotic logistic map to periodic goals, using the present control method, was made by Jackson and Hübler [19]. Here we compare some of these results with the entrainment theorems of Section 3. This illustrates some of the conservative aspects of these theorems, which on the other hand treat goals and dynamic systems which are topologically much more general than the logistic map, as will be illustrated.

Figure 4 is based on the analysis in Ref. [19]. Here the goal dynamics is a simple fixed point, g_0. If g_0 is in the convergent region of the logistic map $x_{n+1} = cx_n(1 - x_n)$,

$$\mathrm{C} = \{x | (c - 1)/2c < x < (c + 1)/2c\} \tag{4.1}$$

the basin of entrainment to g_0 is

$$\mathrm{BE}(g_0) = \{x | 1 - g_0 - c^{-1} < x < g_0 + c^{-1}\}, \tag{4.2}$$

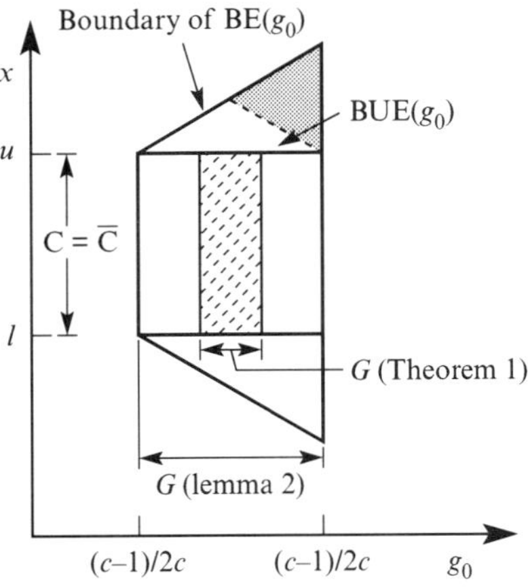

Figure 4 The upper and lower boundaries of the basin of entrainment of x as a function of the possible fixed-point goals, g_0, for the logistic maps, $x_{n+1} = cx_n(1 - x_n)$. Theorem 1 only ensures entrainment in the striped region, whereas Lemma 2 yields the general result. The shaded region indicates non-uniform entrainment.

whose upper and lower boundaries are illustrated in Figure 4. Outside $\mathrm{BE}(g_0)$ the solutions tend to the attractor at $x = -\infty$. While Lemma 2 is restricted in a number of ways, it is general enough to predict (4.2). Indeed both Lemma 2 and Theorem 2, refine the analysis done by Jackson and Hübler, by noting the distinction between the $\mathrm{BE}(g_0)$ and $\mathrm{BUE}(g_0)$. The latter basin of uniform entrainment has an upper boundary which differs from $\mathrm{BE}(g_0)$ only in the region indicated in Figure 4. That is, if (x_0, g_0) lies in the shaded (not striped) region, then $|x_1 - g_0| > |x_0 - g_0|$, whereas $|x_n - g_0| > |x_{n+1} - g_0|$ for $n \geq 1$, so the convergence is not uniform.

Theorem 1, which is based on the mean-value theorem, is more restrictive in its prediction. The goal region G it allows, and its ensured basin of entrainment is illustrated by the striped region in Figure 4. Note that in the logistic case, $\overline{\mathrm{C}}(E) = \mathrm{C}(E)$; that is the additional requirement for $\overline{\mathrm{C}}(E)$ in terms of $\frac{1}{2}(u - l)$ is automatically satisfied by the logistic map. This shows that Theorem 1 is quite conservative, but its advantage appears to be in its extension to higher dimensions, as will be discussed in a subsequent study.

Next consider Figure 5 [19] illustrating the region of stability for period-two goal dynamics $\{g_k\} = (g_0, g_1, g_0, g_1, \ldots)$, namely where $|\partial E(g_0)/\partial g_0)(\partial E(g_1)/\partial g_1)| < 1$. In this region there is some type of basin of entrainment. What is the nature of these basins of entrainment? The study of Jackson and Hübler explicitly showed that these basins of entrainment are often disjoint, and possibly even fractal in character, even for the simple logistic map. If one studies this for the present period-two goal, it is found that the disjoint basins of entrainment occur in three of the 'arms' of the stable region. These are illustrated in Figure 5. The other basins which intersperse this basin of entrainment may be other bounded periodic solutions, or unbounded solutions $(x_n \to -\infty)$. Thus goal dynamics in these regions cannot generally be reliably achieved, using only 'macroscopic' initial-state information, x_0, about the system.

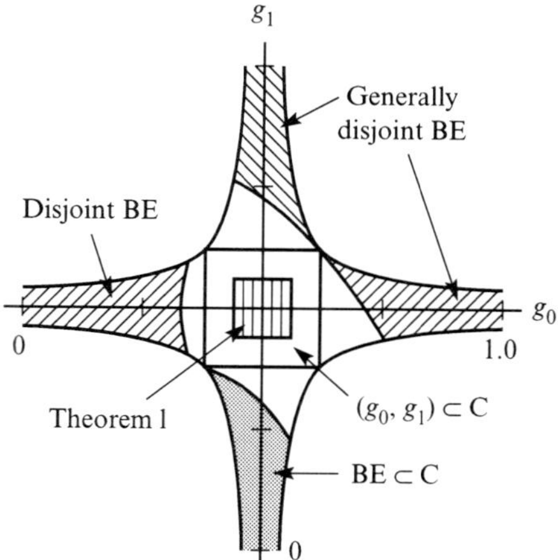

Figure 5 The goal region for period-two entrainment of the logistic map, $\left|c^2(1 - 2g_0)(1 - 2g_1)\right| < 1$. The basin of entrainment is general disjoint in the striped regions. In the shaded region, the basin of entrainment is smaller than the convergent region. Theorem 1 only applies in the vertically striped region. The square region corresponds to the goals in the convergent region, which is therefore sufficient, but not necessary, to ensure connected basins of entrainment.

It is interesting to note that in the fourth 'arm' ($g_0 \cong 0.5$, $g_1 \to 0$), the basin of entrainment does not become disjoint, but it becomes vanishingly small. We, of course, would like to have a 'substantial' basin of entrainment – say at least as large as C. The undesirable region of (g_0, g_1), where BE $\subset$ C, is indicated in Figure 5.

What do the results of Section 3 give in this case? Theorem 1 is very conservative in its character. It says that if $(g_0, g_1) \in \{x | \frac{7}{16} < x < \frac{9}{16}\}$ then the basin of entrainment satisfies BE $\supset$ C. This region of (g_0, g_1) is vertically striped in Figure 5. This is clearly only a small portion of the actual region. Neither Lemma 2, nor Theorem 2 say anything about this case, since they are limited to fixed-point goal dynamics.

In contrast with the study of Jackson and Hübler, which concentrated on the periodic entrainment of the chaotic logistic dynamics ($c = 4$), we now demonstrate the topological freedom which exists for both the goal dynamics and experimental dynamics. All of the goal dynamics is restricted only by the condition $\{g_k\} \subset$ C, (4.1). None of the theorems of Section 3 address this general situation. They do, however, illustrate that frequently the basin of entrainment of a set $\{g_k\}$ is at least as large as C. As explained in Section 3, this cannot be generally true, but the sets $\{g_k\}$ for which it is not true are frequently 'artificial', 'contrived', or 'unnatural' in character. In any case, the following results are only numerical examples.

Figure 6 shows x_k versus k for the experimental system and g_k versus k for the goal dynamics. The goal dynamics is confined to the convergent region (4.1). The vertical lines show when the search is initiated (to determine when $x_k \in$ C) and when the control is terminated. In the search, it

may take anywhere from $k = 0$ to $k = 6$ before the condition $x_k \in$ C is satisfied; then the control is initiated ($k = 0$ and $S = 1$ in (2.7)). The figures show several initiations simply to illustrate the reliability of the entrainment. Figure 6(a) shows entraining a period-eight experimental dynamics to a trigonometric period-30 goal. The transition is thus order → order. Figure 6(b) shows the entrainment of chaotic motion to periodic motion (chaos → order). Figure 6(c) illustrates the entrainment of one form of chaotic motion to another type of chaos (chaos → chaos). Note that the entrainment is rapid and precise, provided that the dynamic equations are known (see the appendix). Finally, Figure 6(d) shows the entrainment of a period-eight dynamics to a chaotic goal dynamics (order → chaos). This topological freedom is implicit in Theorem 1, since only a goal region constrains the goal dynamics.

In order to pursue the study of the MAS discussed in Section 1, we next consider maps with several attractors and convergent regions.

5. Gaussian-map dynamics: global basins of entrainment

In this section we introduce a Gaussian-related map which, although similar to the logistic map for $x \gtrsim 0$, has the advantage that it is a global attractor for all $x \in \mathbb{R}$. In particular, it replaces the logistic second attractor at $x = -\infty$, with one which is in a finite region $x < 0$. In addition to having two attractors, each basin of attraction has two convergent regions, one which is infinite in extent This latter region gives rise to a new, and very important effect; namely goal-dynamic regions which have a global basin of entrainment, BE $= \mathbb{R}$. For such goals, the control can be initiated at any time because the system is always in its basin of entrainment.

To explore these ideas we first introduce an anti-symmetric Gaussian-related map, $E(x) \equiv H(x)$,

$$x_{n+1} = H(x_n), \quad x \in \mathbb{R}, \tag{5.1}$$

$$H(x; r) = r\alpha x\, e^{-2x^2}, \quad \alpha = \tfrac{1}{2}e^{1/2}, \tag{5.2}$$

which has a maximum at $x = \tfrac{1}{2}$, and α has been selected so that $H(\tfrac{1}{2}) = 1$ if the control parameter $r = 4$. These choices have been made simply to make $H(x)$ similar to the logistic map in the region $0 \leq x \leq \tfrac{1}{2}$. Obviously other choices could be made, and in fact might be experimentally more relevant (e.g. Refs. [2, 32]), but the present study is heuristically motivated. The function $H(x; 4)$ is illustrated in Figure 7, where the vertical lines indicate the boundaries of convergent regions, $(C_1^{\pm}, C_2^{\pm})$. In the autonomous case, (5.1), the dynamics of (5.2) in $x > 0$ and $x < 0$ are identical and separate from each other.

When controls are introduced,

$$x_{n+1} = H(x_n) - g_{n+1} + H(g_n)S_n, \quad n \geq 0, \tag{5.3}$$

the dynamics in these two regions of phase space ($\mathbb{R}^{\pm}$) are generally not isolated. Of particular interest are the infinite

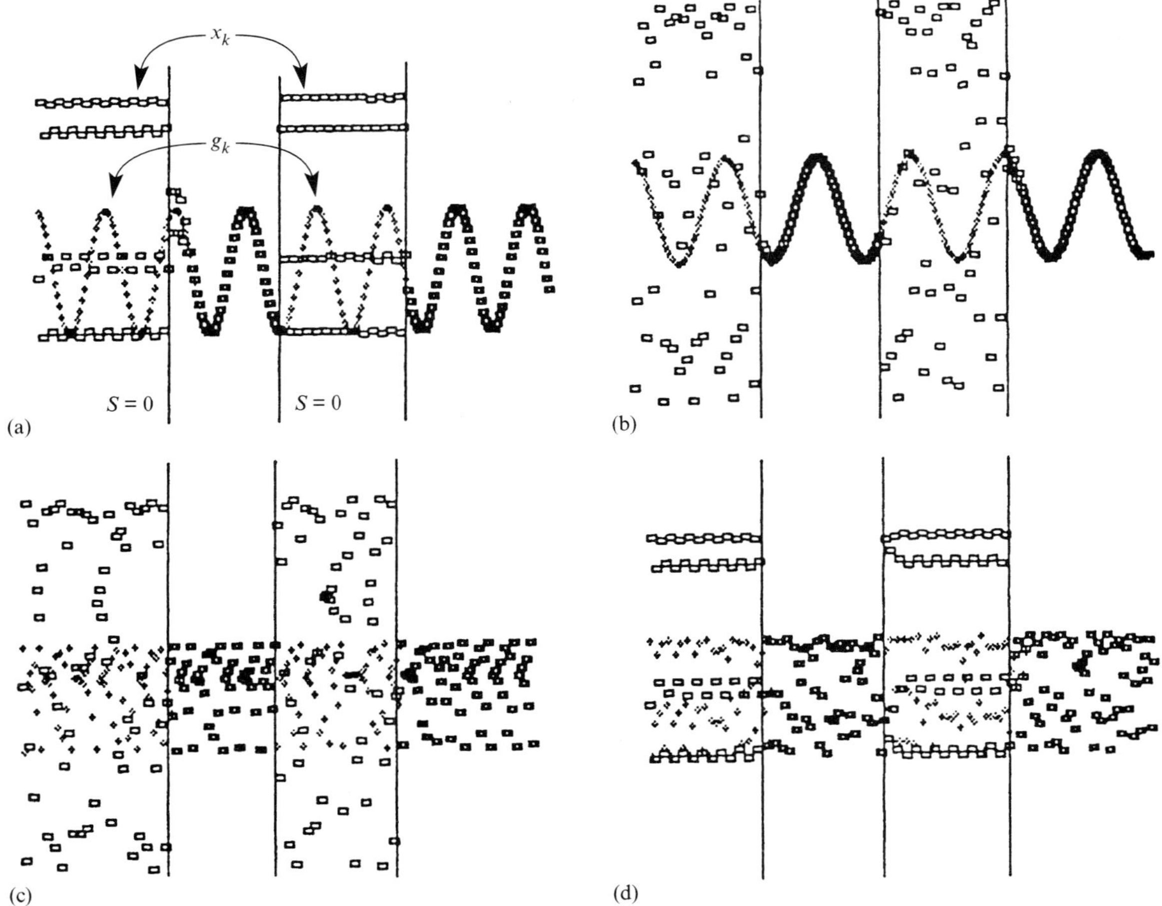

Figure 6 The logistic map is used to illustrate the topological generality of the system's dynamics (x_k: larger circles) and the goal dynamics (g_k) to which it can be entrained as a function of time (k). All goal dynamics are in the convergent region, (4.1). The search is initiated at the first and third vertical lines, whereas the control is terminated ($S = 0$) at the second line. They illustrate the system's transition (a) order $\rightarrow$ order, (b) chaos $\rightarrow$ order, (c) chaos $\rightarrow$ chaos, and (d) order $\rightarrow$ chaos.

convergent regions $C_2^\pm$. Consider for example C_2^+,

$$C_2^+ = \{x \mid |\partial E/\partial x| < 1, x > x_l\}. \tag{5.4}$$

For any $g_0 \in C_2^+$, $|E(x) - E(g_0)| < |x - g_0|$ for all $x > g_0$, so $u = +\infty$ in Theorem 2. Moreover, if g_0 is sufficiently large, then this inequality also holds for all $x < g_0$, in which case $l = -\infty$ in Theorem 2. For example, if $r = 4$, one finds the $x_l \cong 1.156$ in (5.3), and $l = -\infty$ if $g_0 \gtrsim 1.55$. In these cases the entire phase space is the basin of uniform entrainment, $\mathrm{BUE}(g_0) = \mathbb{R}$.

For smaller values of $q_0 > x_l$, there is a largest root of $|E(l) - E(g_0)| = |l - g_0|$ ($l < g_0$) (see Figure 8, where $g_0 = 1.35$). If $x_n \lesssim l$ then $|x_{n+1} - g_0| > |x_n - g_0|$ and any entrainment is no longer uniform. Nonetheless, in the case (5.2) it is not difficult to see that $|x_{n+2} - g_0| < |x_n - g_0|$, so that there is a 'two-step' type of uniform entrainment. This sort of result is typical of the types of details one encounters with special cases, $E(x)$, and which are difficult to formulate with any generality (i.e., without making many special statements on

the global character of $E(x)$). Thus, in this Gaussian example, if $g_0 \in C_2^\pm$ the basin of entrainment is global, $\mathrm{BE}(g_0) = \mathbb{R}$, but it is generally only 'two-step-uniform', until $x_n \in C_2^\pm$, after which it becomes uniformly entrained.

Systems with infinite convergent regions are quite common, and may be characteristic of all systems with global attractors. This can be true of both maps and flows (systems of first-order ODE) in $\mathbb{R}^n$, as will be shown in subsequent studies. If, as in the present Gaussian case, goal dynamics in these convergent regions have global basins of entrainment, it may be possible to use these goals to cause the system to transfer from one attractor to another.

To illustrate this point, consider the asymmetric generalization of (5.2)

$$H(x; r, a) = r \alpha x e^{-2x^2 + ax} \tag{5.5}$$

with an asymmetry parameter a. $H(x; 4, \frac{1}{2})$ is shown in Figure 9. Because of the asymmetry, the dynamics (5.1) is no longer the same in the regions $x < 0$ and $x > 0$. Figure 10

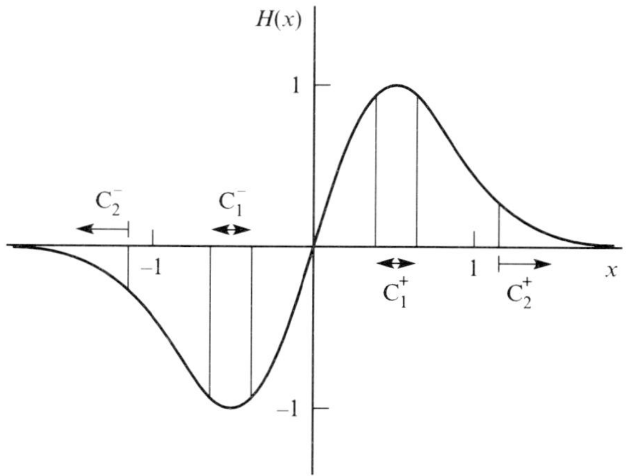

Figure 7 The Gaussian-map function, (5.2), when $r = 4$. This has two attractors, $A^\pm \subset \mathbb{R}^\pm$, each with two convergent regions, $C_k^\pm$ ($k = 1,2$). The two convergent regions $C_2^\pm$ are infinite in extent.

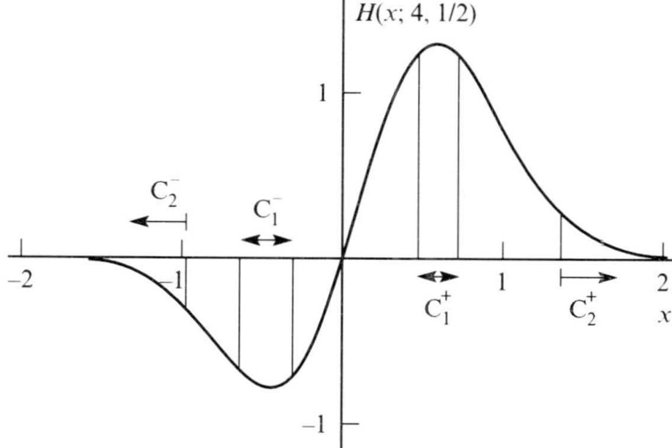

Figure 9 The asymmetric map function, (5.5), which has attractors in $\mathbb{R}^\pm$ that have topologically distinct dynamics.

shows the attractors for (5.5) over the range $2 \le r \le 8$. For $r \le 5.5$ the attractor in $\mathbb{R}^-$ is periodic, whereas a variety of dynamics could occur in $\mathbb{R}^+$. Thus, if $r = 4$, the attractor A^+ is chaotic, whereas A^- has period-two. If the system is in BA^+ (attracted to A^+), we could use a goal dynamics which is a simple fixed point in C_2^-, $g_0 \in C_2^-$; we can then turn on the control at any time ($S_n = 1$ in (5.3)), because g_0 has a global basin of entrainment, and $x_n \to g_0$ in a non-uniform fashion. Anytime that $x_k \in BA^-$ (i.e., $x_k < 0$) we can turn off the control ($S_n = 0$, $\forall n > k$), and the system will tend toward the attractor A^-. Therefore, in the above system we could transfer the system permanently between any of the attractors with the same value of r in Figure 9, using a control only for a limited time. Moreover this could be accomplished at any time, regardless of the state of the system. Examples of this type of control will be given in Section 6 (see Figure 13).

This illustrates the potential power of such global basins of entrainment. However it will be shown in the following

section that the transferring between attractors cannot always be accomplished this easily, even when there are global basins of entrainment. Figure 11 schematically illustrates possible sets and regions of the autonomous system (2.1) and (2.2) discussed up to this point. The goal sets and their basins of entrainment, associated with the controlled systems, (2.7), are not illustrated. Figure 11 shows two attractors, (A_1, A_2), their basins of attraction (BA_1, BA_2) and some possible convergent regions, $C_i(k)$, for each attractor A_k ($k = 1,2$). The invariant boundary set between BA_1 and BA_2 is shown to be 'complicated' in the lower region, and a new type of convergent region, $C_\infty(1,2)$, is also indicated. In the above Gaussian example only $C_1(k)$, $C_\infty(k)$ ($k = 1, 2$) were present. We will see in the next section an example of $C_\infty(1,2)$, and the other convergent regions. These will illustrate the usefulness of other goal dynamics besides the entrainment goals considered so far. To obtain a more general type of control of complex systems, we will next introduce the concept of migration goals.

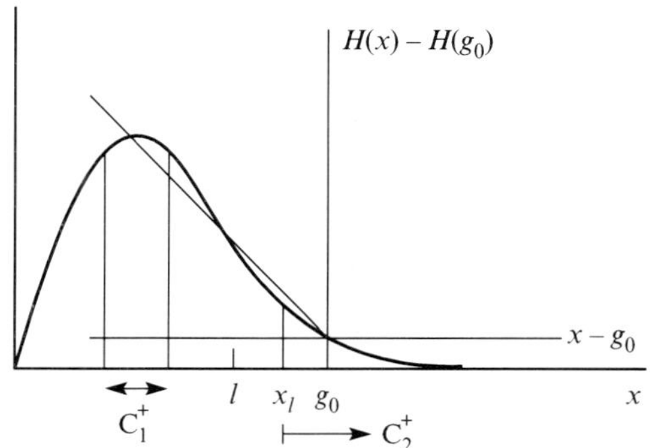

Figure 8 Graphical illustration of non-uniform global entrainment when the goal g_0 is near the lower boundary of the infinite convergent region C_2^+.

6. Migration-goal dynamics

In this section we introduce a second type of goal dynamic, distinct from the entrainment-goals,

$$\text{e-goals} \equiv \{g_k\}$$

$$\text{such that } \lim_{n \to \infty} |x_n - g_n| = 0,$$

$$\{g_k | k = 0, 1, \ldots\} \subset G_l \subset C_l. \tag{6.1}$$

The second condition, $\{g_k\} \subset C_l$, is necessary if the sets $\{g_k\}$ are allowed to be arbitrary (topologically) within a region G_l. The purpose of the second type of goal dynamics is to take a system from one convergent region to another, and therefore will be referred to as migration-goals:

$$\text{m-goals} \equiv M_{ij}$$

$$= \{g_k | k = 0, \ldots, N;$$

$$g_0 \subset C_i, g_N \subset C_j, j \ne i\}$$

$$\text{so } M_{ij} \cap C_i \ne \varnothing, \ M_{ij} \cap C_j \ne \varnothing \ (i \ne j). \tag{6.2}$$

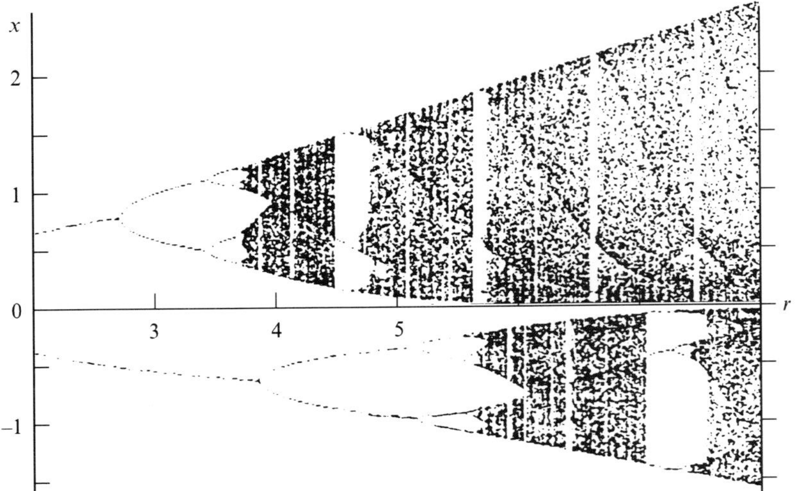

Figure 10 The attractors of the autonomous system, (5.1) and (5.5) when $a = \frac{1}{2}$ and $2 \leq r \leq 8$. The infinite convergent regions, $C_2^{\pm}$ (Figure 9), can be used to transfer the system between any two attractors with the same r.

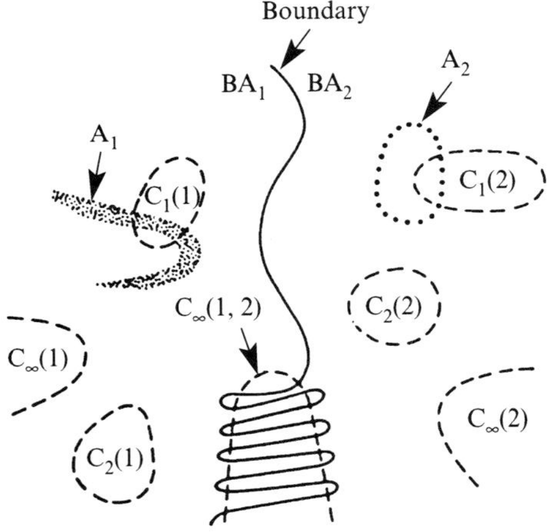

Figure 11 A schematic illustration of two attractors A_1 (chaotic), A_2 (periodic) and finite convergent regions $C_i(k)$ ($i = 1,2$) in their basins of attraction, BA_k ($k = 1,2$). Infinite convergent regions $C_\infty(k)$ are also illustrated. The convergent region $C_\infty(1,2)$ is discussed in Section 6.

Thus an m-goal is a finite-set (finite-time) control set, whose purpose is to transfer the experimental system from a convergent region C_i to C_j ($i \neq j$), again using control

$$x_{n+1} = E(x) + g_{n+1} - E(g_n),$$

$$n = 0, \ldots, N - 1. \tag{6.3}$$

Whether a particular m-goal set M_{ij}, which is inherently unstable, can in fact reliably produce this desired transfer from C_i to C_j involves a number of considerations (e.g., optimal 'velocity', noise limitations, etc.). The present study will simply illustrate this possibility, and note some potentially important applications, without doing justice to the many subtleties involved in its application. These will have to be explored in future studies.

As noted in the introduction, many complex systems have a number of attractors, A_k ($k = 1,2, \ldots$), each with their basins of attraction, BA_k. In these basins of attractors there may be a number of convergent regions, $C_l(k)$ ($l = 1,2, \ldots$) as illustrated in Figure 11. Thus the Gaussian maps of Section 5 had two attractors, each with two convergent regions ($k = 1,2$; $l = 1,2$). The asymmetric case (5.5) involved attractors with very different types of dynamics (e.g., periodic, intermittent, chaotic, etc.), one of which may be most beneficial to the system. Thus chaos may cause a system to self-destruct, so that a periodic dynamics might be more 'healthy'. Conversely, as in the case of epileptic seizures, regular dynamics may be the 'unhealthy' state, whereas a form of chaos may be the healthy state of a system; e.g., allowing it to 'search' and respond to a changing environment (e.g., Refs. [26,30]). It is clear that there are many other situations where, if a system is disturbed into one attractor dynamics by some outside influence, it might be beneficially transferred to another attractor dynamics. There are also a number of other reasons one might want to transfer a system from one convergent region to another (as distinct from one attractor to another). As an example, this would allow one to study the controlled behavior of a system in distinct regions of its phase space. Such knowledge can be used to improve the dynamic models of the experimental system (e.g., see Appendix A; also Ref. [1]. Thus migratory controls could be of both practical and basic importance in the future.

The usefulness of m-goals is usually only in conjunction with some e-goals at both ends of the M_{ij} set. A possible scenario might go like the following (see Figure 11):

1. The system is in BA_1, attracted to a chaotic attractor A_1.
2. A simple fixed-point e-goal, $g_0 \in C_1(1)$, is initiated when $x_0 \in BE(g_0)$.

3. If we know that x_0 is in the basin of uniform entrainment, BUE(g_0), so that $|x_n - g_0| \leq \varepsilon^n |x_0 - g_0|$ then after a finite time, $n = T$, we can ensure that $|x_T - g_0| < \varepsilon^T |x_0 - g_0|$. At this time we can attempt to migrate the system from $C_1(1)$ to (say) $C_2(1)$.

4. During this migration, many of the g_k may be outside any convergent region, so that the system will begin to diverge from the m-goal set. In order for the m-goal set to be effective, it is important for $\varepsilon^T |x_0 - g_0|$ in (3) to be 'sufficiently small'.

5. If the system reaches $C_2(1)$, the control can be transferred to another e-goal set (e.g., a new fixed point $g_0 \subset C_2(1)$), so that the system can be re-entrained. This might be necessary in the presence of noise (see the example below).

6. Once a 'sufficient degree' of entrainment is achieved, a second m-goal set might transfer the system from BA_1 to BA_2.

7. Once in BA_2 it might simply be 'released' (no control), and let the autonomous system dynamics take it to A_2, or it might be transferred to one of the convergent regions $C_2(2)$, and reentrained (e.g., to explore the accuracy of the model of the experimental system).

The above scenario clearly has numerous variations, and it appears that many of these can be accomplished rather easily if the noise is not too large. The m-goals which are needed depend not on the regions $C_j(k)$, but on the basins of entrainment of the e-goals within these regions. Thus, if $g_{0j} \in C_j$ ($j = 1,2$) are two fixed-point e-goals in two convergent regions, it might happen that $g_{01} \in BE(g_{02})$. In this case the m-goal set $\{g_{01}, g_{02}\}$ is sure to transfer the system from C_1 to C_2, provided that $[x_T - g_{01}]$, in step (3) above, is sufficiently small, to ensure that $x_T \in BE(g_{02})$ (of course, one then shifts to the e-goal set $\{g_k\} = g_{02}$ to keep the system in C_2). In general, of course, it will not happen that $g_{02} \in BE(g_{01})$, and not only will some m-goal states be outside any C_j, but also the system may be subjected to environmental noise. The analysis of all of these factors will be important for the future. Here we merely exhibit some simple examples.

It was shown in the last section that sometimes a system can be transferred from one attractor to another using a convergent region whose goal dynamics, $G \subset C$, have a global basin of entrainment. We will illustrate in this section that even regions with this global property do not necessarily ensure a reliable method for making a transfer $A_i \rightarrow A_j$. A simple example of this can be given using the map

$$E(x) = F(x) \equiv H^2(x; 4, 0), \quad x \geq 0 \qquad (6.4)$$

where $H(x; r, a)$ is given by (5.5). The notation $H^2(x) \equiv H(H(x))$ is used in (6.4). For simplicity we restrict the dynamics to $x \geq 0$, and illustrate $F(x)$ in this region in Figure 12. In $\mathbb{R}^+$, $F(x)$ has four convergent regions, C_i ($i = 1, \ldots, 4$); since the diagonal curve intersects $F(x)$ where $|\partial F/\partial x| < 1$, in C_2 and C_3, these regions each have a stable period-one attractor which we denote as A_2 and A_3. The region C_4 is infinite and, following the analysis of Section 5, any goal region $G \subset C_4$ has a global basin of entrainment (including $\mathbb{R}^-$).

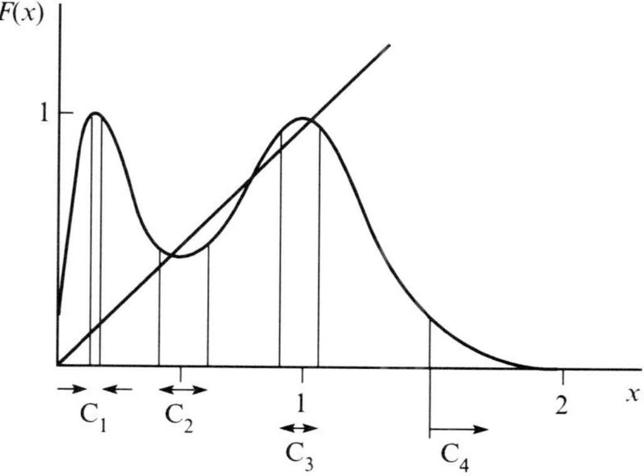

Figure 12 The map function, (6.4), which has four convergent regions in $\mathbb{R}^+$. The two attractors $A_2 \in C_2$ and $A_3 \in C_3$ are stable, as shown by the diagonal line. C_4 is an infinite convergent region, so $g_0 \in C_4$ has a global basin of entrainment, but C_4 intersects both BA_2 and BA_3 (in contrast to (5.5)).

The small convergent region C_1 has no attractor, but the basin of entrainment of any $G_1 \subset C_1$ contains the attractor $A_3 \subset C_3$ but not the closer attractor $A_2 \subset C_2(BE(G_1) \cap A_2 = \emptyset; BE(G_1) \cap A_3 \neq \emptyset)$. This can be readily seen by using the graphical method indicated in Figure 3. Therefore the system can be taken from $A_3 \rightarrow G_1$ using any e-goal in G_1. However neither the transfer $A_2 \rightarrow G_1$ nor the more important transfers $A_2 \rightarrow A_3$ or $A_3 \rightarrow A_2$ can be accomplished using only e-goals (i.e., $BE(G_2) \cap A_3 = \emptyset; BE(G_3) \cap A_2 = \emptyset$).

It might appear reasonable to transfer the system between A_2 and A_3 using an entrainment goal $G_4 \subset C_4$, since this goal has a global basin of entrainment. Moreover it is not difficult to see that $BA_k \cap C_4 \neq \emptyset$ for $k = 2, 3$. Hence, once the system is entrained by the goal G_4, we could terminate the control, and let the system tend autonomously to the desired attractor. This was what could be done in the example of Section 5. The difficulty now is that both BA_2 and BA_3 intersect C_4 (see $C_\infty(1,2)$ in Figure 11, for a schematic example in $\mathbb{R}^2$). Thus, when the control is terminated, it may not be possible (using only limited experimental accuracy) to determine whether the system will tend to A_2 or to A_3.

This sensitivity is illustrated in Figure 13, showing x_k and g_k versus k. The boundaries of the four convergent regions are indicated by horizontal lines. The goal set has been taken to be a trigonometric function in C_4, $g_k = 1.95 + 0.3 \sin(0.3k)$, whereas the system (larger circles) initially is in A_3. When the control is transferred on ($S = 1$ in (2.7)), indicated by the first vertical line, $x_k \rightarrow g_k$ in a non-uniform fashion (note the initial large value of x_k following the switch-on). The system is rapidly entrained, then the control is terminated (second vertical line) at a larger value of x_k, and $x_k \rightarrow A_3$ so no transfer of attractors has occurred. A second attempt, where the control is ended at a smaller value of x_k is likewise unsuccessful. However, a third attempt at an intermediate value of x_k produces the desired

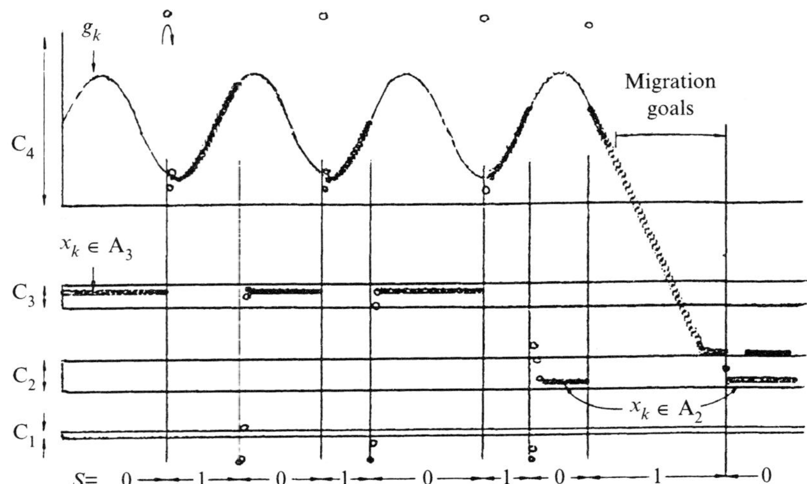

Figure 13 Entrainments to $\{g_k\} \in C_4$, with control terminations occurring when $x_k \in BA_3$ (twice) and $x_k \in BA_2$, illustrating unreliable transfer $A_2 \leftrightarrow A_3$. Entrainment followed by a migration-goal, producing a reliable transfer $A_3 \rightarrow A_2$.

transfer to A_2. This illustrates the disjoint character of the basins of attraction BA_2 and BA_3 in C_4 (again see $C_\infty(1,2)$ in Figure 11). The only certain way to affect $A_2 \leftrightarrow A_3$, using the e-goal G_4, is to follow the entrainment with the migratory sets M_{43} or M_{42}. Figure 13 illustrates the use of a migratory set from $G_4 \rightarrow BA_2$. When the system in A_2 is entrained to the periodic set in C_4 it is soon transferred to an m-goal, $g_k = g_0 - 0.03k$, until g_k reaches near C_2, where g_k is then made constant. Technically this was not a migratory set M_{42} because it did not terminate in C_2, but it is sufficient to enter BA_2, as can be seen by the final termination of the control in Figure 13.

The influence of noise is illustrated by replacing (6.3) with

$$x_{n+1} = E(x_n) + g_{n+1} - E(g_n) + N(\mathrm{RND} - 0.5), \quad (6.5)$$

where RND generates uniformly distributed random numbers between zero and one. Figure 14 shows a number of migrations in the presence of large noise ($N = 5 \times 10^{-3}$). Initially $x_k \in A_3$ and $g_k \in C_2$. If the control is initiated ($S = 1$), it only produces an undesired period-two response from the system. With the control off ($S = 0$), g_k is transferred to the lower boundary of C_3 and the control is then initiated ($S = 1$). The 'near-entrainment' in this case (see step (4) above) is not adequate to overcome the large noise, when a migration goal, M_{32}, is used. The system is scattered from this unstable m-goal, and returns to A_3. Again with $S = 0$, the goal set is returned further inside C_3, and the system is near-entrained ($S = 1$). Now the near-entrainment is adequate for the same migration set, M_{32} to (barely) transfer $A_3 \rightarrow A_2$. Finally a second entrainment to $g_0 \in C_2$ is made, and an attempt is made to transfer the system to C_1. Now the instability throws the system into the region $\mathbb{R}^-$ (below the figure). It could be recovered from

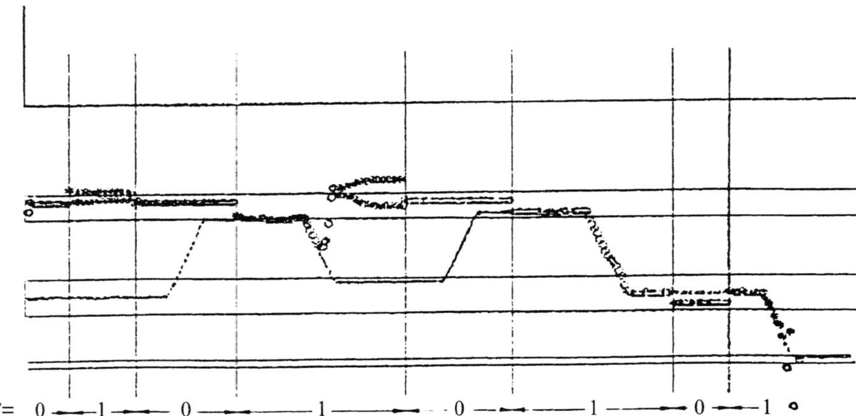

Figure 14 Successful and failed migration attempts ($A_3 \rightarrow A_2$, $A_2 \rightarrow C_1$) of the system in the presence of noise, (6.5). See text for details.

there by using any e-goal in C_4, as explained above. A number of issues are not explored in this illustration, such as the optimum velocity of the m-goal set for a given noise level (in the present examples, $g_{k+1} = g_k - 0.03$). These, and other important issues, will require further research.

7. Conclusion

The present study explores applications of a control method, (2.7), (2.9), which can be used on systems which have dynamic attractors, A_k. It is believed that all such systems have at least one convergent region, C_l, in the basin of attraction, BA_k, of each attractor A_k. The control method then employs goal dynamics which are related to these convergent regions. Of particular importance is the fact that this study begins the exploration of controls that might be used on multiple attractor systems (MAS). MAS are of basic importance in most truly complex systems, which can exhibit more than one dynamic response to environmental stimuli.

When controls are imposed on a MAS it can be done for a variety of reasons. Among these objectives are:

1. To cause the system to tend to some desired (goal) behavior for all future times, using a single sustained control. By this is meant that the control action is quantitatively restricted to one convergent region of the system's phase space. This type of control is represented by the entrainment of a system's dynamics, $x(t)$, to some goal dynamics, $g(t)$

$$\lim_{t \to +\infty} \left| x(t) - g(t) \right| = 0,$$

where $g(t)$ can have any topological character, but is restricted dynamically and to a region, G, contained in some convergent region, C_k. The dynamic (not topological) limitations are difficult to predict, but are related to the rapid nature of the goal dynamics relative to the damping rates, $\mu(x)$ and $\lambda(x)$ in (2.3) and (2.4). These limitations are system-specific, and will be illustrated in subsequent studies. The control is only initiated ($S = 1$ in (2.7)) when the system enters the basin of entrainment of G, $BE(G)$. From the study of many examples, it is conjectured that there is always a $G \subset C_k$, such that $BE(G) \cap A_k \neq \varnothing$. In other words, when the system is near an attractor A_k, it is possible to find a goal region, G in some C_k, such that the basin of entrainment to G intersects A_k. If this is not so then the control cannot be initiated (unless the system is disturbed away from A_k; e.g., using noise). The topological freedom of both the experimental dynamics and the goal dynamics in this entrainment was illustrated in Section 4.

2. To cause the system to tend to some ultimate behavior for all times, using a sequence of (topologically) distinct goals for finite periods of time, ending with a sustained control, (1). This type of control, which was illustrated in Section 6, is represented by a sequence of goals:
   (a) An entrainment goal is selected involving some convergent region, C_i; this goal is then terminated

once the system is (near-entrained) within some accuracy

$$\left| x(t) - g(t) \right| \leq \varepsilon_e.$$

   (b) A migration-goal set is used, M_{ij}, (6.2), to transfer the system from the convergent region C_i to C_j ($i \neq j$).
   (c) An entrainment goal within C_j is selected, and the system is either permanently entrained within C_j, or else (a) and (b) are repeated, transferring the system to another region, C_k.

The reason why such a sequence might be necessary is because every migratory control is unstable, so a re-entrainment at intermediary regions (e.g., C_j) might be required.

As discussed in Section 6, one of the reasons one might want to entrain the system in all of the predicted convergent regions, is to check the accuracy of the theoretical model of the experimental system (e.g., (6.3) and (6.4)). Another aspect of model accuracy is also established in the appendix.

3. The third and perhaps most important objective, involves interacting with the system for only a finite time, but producing a permanent change in its dynamics. This can be accomplished in a MAS by causing the system to transfer from one attractor, A_i, to another, A_j: $A_i \to A_j$ ($i \neq j$). To do this, there are two distinct possibilities. The simplest is to use a goal, g_0, in BA_j, which has a basin of entrainment that intersects $A_i (BE(g_0) \cap A_i \neq \varnothing)$, such as illustrated in Section 5. In particular, a global basin of entrainment would obviously serve nicely for this purpose. Once this control is initiated the system will tend to $g_0 \subset BA_j$ and when $x(t) \subset BA_j$ the control can be terminated ($S = 0$ in (2.7)). The system will then autonomously tend to A_j, and a permanent transfer $A_i \to A_j$ will have been accomplished.

The second possibility is that it may be difficult to determine, given only 'macroscopic' information, whether the g_0 which satisfies $BE(g_0) \cap A_i \neq \varnothing$ also satisfies $g_0 \subset BA_j$. In this case, even if $BE(g_0)$ is a global attractor, x_k (at any given time) may either be in BA_i or BA_j (or possibly some other BA_k), which makes the method unreliable. This was illustrated in Section 6. In that case a series of migration goals $M_{ia}, M_{ab}, \ldots, M_{xj}$ may be used to make the transfer $A_i \to A_j$. Several m-goals, M_{ij}, were illustrated in Section 6.

The application of the above concepts to maps and ODE in higher dimensions will soon be reported [18, 20]. These examples will extend the elementary examples in $\mathbb{R}^n$, given in Section 2, to many of the classic examples of maps and flows.

The control method which is proposed here is not the only method by which complex systems can be controlled, but it has the advantage of general reliability, and great flexibility. Thus, Hübler's method can also be used with goal dynamics, $g(t)$, which are not always confined to convergent regions, provided that all of their Lyapunov exponents have negative real parts (e.g., (2.10) in the case of one-dimensional maps). Examples of this were given in

Jackson and Hübler [19]. As shown there, the difficulty is that such controls may yield fractal basins of entrainment, or other reliability complications, depending on the particular system. Nonetheless, such controls may be very useful for particular systems (e.g., see the limit cycle examples in ref. [17]).

Other control methods have also been recently suggested by Ott, Grebogi and Yorke [29], by Huberman and Lumer [13], and by Sinha, Ramaswamy and Rao [33]. In contrast with the present method, these methods are all based on the ability to change the control parameters, c, of a system $\dot{x} = F(x; c)$ ($x \in \mathbb{R}^n$), and represent an interesting alternative to the present approach. However the topological freedom of the present method, the no-feedback aspects of this control, the macroscopic basins of entrainment, and the ability to transfer MAS between attractors in a reliable fashion, adds considerable flexibility to the present control options.

Acknowledgement

This work was supported by the Department of Physics and Beckman Institute at the University of Illinois at Urbana-Champaign.

Appendix. Near-entrainment with an approximate model but accurate initial-condition knowledge

Numerical studies by Chang, Kodogeorgiou, Hübler and Jackson [1] have shown that near-entrainment can be achieved even when the experimental function, $E(x)$, is not known precisely, provided the initial error $\varepsilon_i \equiv |x_0 - g_0|$ is small. Here we present a simple analytic proof of this result, under suitable restrictions on ε_i and the model error ε_m.

Assume that

$$x_{n+1} = M(x_n), \quad x \in \mathbb{R} \tag{A.1}$$

is the best model equation which is known for the experimental system, and that the maximum error over $C(E)$, (2.3), is

$$\max_{C(E)} |E(x) - M(x)| \equiv \varepsilon_m. \tag{A.2}$$

If $M(x)$ is used to govern the system, the dynamics is given by

$$x_{n+1} = E(x_n) + g_n - M(g_n), \quad n \geq 0 \tag{A.3}$$

rather than (2.7) and (2.9). Now assume that the governing set $\{g_k\}$ and model satisfy

$$\max_{\{g_k\}} |\partial M(g_k)/\partial g_k| \equiv \mu_1 < \delta < 1, \tag{A.4}$$

and let $\Delta_n \equiv |x_n - g_n|$. (A.3) yields

$$\Delta_{n+1} < |E(x_n) - M(x_n) - M(g_n)|$$
$$< \varepsilon_m + |x_n - g_n) \, \partial M(g_n)/\partial g_n$$
$$+ \tfrac{1}{2}(x_n - g_n)^2 (\partial^2 M/\partial x^2)_{mn}|,$$

where the last derivative is evaluated at some mean point between x_n and g_n. If we denote

$$\max_{C(E)} |\partial^2 M/\partial x^2| = \mu_2, \tag{A.5}$$

the last equation yields

$$\Delta_{n+1} < \varepsilon_m + \Delta_n \mu_1 + \tfrac{1}{2}\Delta_n^2 \mu_2$$
$$= \varepsilon_m + \delta\Delta_n \big[(\mu_1/\delta) + \tfrac{1}{2}\Delta_n(\mu_2/\delta)\big] \tag{A.6}$$

and

$$\mu_1/\delta + \tfrac{1}{2}\Delta_n(\mu_2/\delta) < 1 \quad \text{if } \Delta_n < 2(\delta - \mu_1)/\mu_2. \tag{A.7}$$

We assume now that the initial error satisfies

$$\Delta_0 \equiv |x_0 - g_0| \equiv \varepsilon_i < 2(\delta - \mu_1)/\mu_2, \tag{A.8}$$

in which case, (A.7) and (A.6) yield

$$\Delta_1 < \varepsilon_m + \delta\Delta_0. \tag{A.9}$$

We must now place a bound on the model error, ε_m, to ensure near-entrainment. We require that Δ_1 satisfy the inequality (A.7), or

$$\varepsilon_m + \delta\Delta_0 < 2(\delta - \mu_1)/\mu_2$$

and, using (A.8), this is assured if

$$\varepsilon_m < 2(1 - \delta)(\delta - \mu_1)/\mu_2. \tag{A.10}$$

Since (A.7) is then satisfied for $n = 1$, (A.6) yields $\Delta_2 < \varepsilon_m + \delta\Delta_1$, and by induction, using (A.10), we can conclude that

$$\Delta_{n+1} < \varepsilon_m \sum_{k=0}^{n} \delta^k + \delta^{n+1}\Delta_0,$$

proving that near-entrainment results

$$\lim_{k \to \infty} |x_k - g_k| < \varepsilon_m/(1 - \delta). \tag{A.11}$$

Subject to the conditions, (A.4), (A.8), and (A.10). This linear dependence of the near-entrainment ε_m was discovered numerically by Hübler some time ago, and a result similar to (A.11) was obtained by A. Kodogeorgiou under the assumption that Δ_n remains small enough to ensure that the linear term in the Taylor expansion sufficed (see ref. [1]). The present result simply formalizes these results and insights, by obtaining the above sufficient conditions, (A.4), (A.8), and (A.10).

References

[1] K. Chang, A. Kodogeorgiou, A. Hübler and E.A. Jackson, General resonance spectroscopy, in: *SIAM Conference on Dynamical Systems*, May, 1990.

[2] K. Coffman, W.D. McCormick and H.L. Swinney, Multiplicity in a chemical reaction with one-dimensional dynamics, *Phys. Rev. Lett.* **56** (1986) 999–1002.

[3] D. Coles, Transition in circular couette flow, *J. Fluid Mech.* **21** (1965) 385–425.

[4] P.R. Fenstermacher, H.L. Swinney and J.P. Gollub, Dynamical instabilities and transition to chaotic Taylor vortex flow, *J. Fluid Mech.* **94** (1979) 103–128.

[5] R.R. Gantmacher, *The Theory of Matrices* (Chelsea, 1959).

[6] R. Georgii, Control of nonlinear continuous systems based on poincaré maps, in: *SIAM Conference on Dynamical Systems*, May 1990.

[7] L. Glass, *Physica D* **40** (1989) 299–310.

[8] L. Glass and M.C. Mackey, *The Rhythms of Life* (Princeton Univ. Press, Princeton, NJ, 1988).

[9] L. Glass, A. Shrier and J. Bélair, Chaotic cardiac rhythms, in: *Chaos*, ed. A.V. Holden (Princeton Univ. Press, Princeton, NJ, 1986) pp. 237–256.

[10] P. Grassberger and I. Procaccia, Dimensions and entropies of strange attractors from a fluctuating dynamics approach, *Physica D* **13** (1984) 34–54.

[11] M.R. Guevara, G. Ward, A. Shrier and L. Glass, in: *Computers in Cardiology* (IEEE Comput. Soc., Silver Springs, 1984) pp. 167–170.

[12] S.M. Hammel, C.K.R.T. Jones and J.V. Moloney, J. *Opt. Soc. Am. B* **5** (1988) 552–564.

[13] B.A. Huberman and E. Lumer, Dynamics of adaptive systems, *IEEE Trans. Circuits Syst.* **CAS-37** (1990) 547–555.

[14] A.W. Hübler, Dissertation, Department of Physics, Technical University of Munich (November, 1987).

[15] A. Hübler and E. Lüscher, Resonant stimulation and control of non-linear oscillators, *Naturwissenschaft* **76** (1989) 67.

[16] E.A. Jackson, Understanding complex systems, in: *Perspectives of Nonlinear Dynamics*, Vol. 2 (Cambridge Univ. Press, Cambridge, 1990).

[17] E.A. Jackson, Controls of dynamic flows with attractors, Report CCSR-90–15, Center for Complex Systems Research, Beckman Institute, University of Illinois at Urbana-Champaign (1990).

[18] E.A. Jackson, The entrainment of migration controls of multiple-attractor systems, *Phys. Lett. A* (1990) 478–484.

[19] E.A. Jackson and A. Hübler, Periodic entrainment of chaotic logistic map dynamics, *Physica D* **44** (1990) 407–420.

[20] E.A. Jackson and A.E. Kodogeorgiou, Entrainment and migration controls of the Hénon and Ikeda maps, in preparation.

[21] A. Kodogeorgiou and A. Hübler, Optimal control of catastrophes, in: *SIAM Conference on Dynamical System*, May 1990.

[22] C. Kurrer, Dissertation, Department of Physics, Technical University of Munich (August 1989).

[23] E. Lüscher and A. Hübler, Resonant stimulation of complex systems, *Helv. Phys. Acta* **62** (1989) 543.

[24] L.A. Lugiato, Theory of optical bistability, in: *Progress in Optics*, 21, ed. E. Wolf (North-Holland, Amsterdam 1984) pp. 71–211.

[25] J.M. Nese, Quantifying local predictability in phase space, *Physica D* **35** (1989) 237–250.

[26] J.S. Nicolis, Chaotic dynamics in biological information processing: A heuristic outline, in: *Chaos in Biological Systems*, eds. H. Degn, A.V. Holden and L.F. Olsen (Plenum Press, New York, 1987) pp. 221–232.

[27] J. Nicolis, G. Meyer-Kress and G. Haubs, *Z. Naturforch.* **38a** (1983) 1157.

[28] H. Nijmeijer and A. van der Schaft, *Nonlinear Dynamical Control Systems* (Springer, Berlin, 1990).

[29] E. Ott, C. Grebogi and J.A. Yorke, Controlling chaos, *Phys. Rev. Lett.* **64** (1990) 1196–1199.

[30] R. Pool, Is it healthy to be chaotic?, *Science* **243** (1989) 604.

[31] R. Shermer, Control of the dynamics of shock waves and complicated flows by aperiodic perturbations, in: *SIAM Conference on Dynamical Systems*, May 1990.

[32] R.M. Simoyi, A. Wolf and M.L. Swinney, One-dimensional dynamics in a multicomponent chemical reaction, *Phys. Rev. Lett.* **49** (1982) 245.

[33] S. Sinha, R. Ramaswamy and J.S. Rao, Adaptive control in nonlinear dynamics, *Physica D* **43** (1990) 118–128.

[34] A.T. Winfree, *When Time Breaks Down* (Princeton Univ. Press, Princeton, NJ, 1987).

Paper 9
Synchronization in chaotic systems

Louis M. Pecora and Thomas L. Carroll
Code 6341, Naval Research Laboratory, Washington, D.C. 20375, USA

Received 20 December 1989

Certain subsystems of nonlinear, chaotic systems can be made to synchronize by linking them with common signals. The criterion for this is the sign of the sub-Lyapunov exponents. We apply these ideas to a real set of synchronizing chaotic circuits.

Chaotic systems would seem to be dynamical systems that defy synchronization [1]. Two identical autonomous chaotic systems started at nearly the same initial points in phase space have trajectories which quickly become uncorrelated, even though each maps out the same attractor in phase space. It is thus a practical impossibility to construct identical, chaotic, synchronized systems in the laboratory.

In this paper we describe the linking of two chaotic systems with a common signal or signals. We show that when the signs of the Lyapunov exponents for the subsystems are all negative the systems will synchronize. By synchronize we mean that the trajectories of one of the systems will converge to the same values as the other and they will remain in step with each other. The synchronization appears to be structurally stable.

We apply these ideas to several well-known systems [2] (e.g., Lorenz and Rössler) as well as the construction of a real set of chaotic synchronizing circuits.

The capability of synchronization is not obvious in nonlinear systems. We derive the results for flows (differential equations), but only a slight variation is needed to use them for iterated maps. Consider an autonomous n-dimensional dynamical system,

$$\dot{u} = f(u). \tag{1}$$

Divide the system, arbitrarily, into two subsystems [$u = (v,w)$],

$$\dot{v} = g(v, w), \quad \dot{w} = h(v, w), \tag{2}$$

where $v = (u_1, \ldots, u_m)$, $g = (f_1(u), \ldots, f_m(u))$, $w = (u_{m+1}, \ldots, u_n)$, and $h = (f_{m+1}(u), \ldots, f_n(u))$.

Now create a new subsystem w' identical to the w system, substitute the set of variables v for the corresponding v' in the function h, and augment Eqs. (2) with this new system, giving

$$\dot{v} = g(v, w), \quad \dot{w} = h(v, w), \quad \dot{w}' = h(v, w'). \tag{3}$$

Examine the difference, $\Delta w = w' - w$. The subsystem components w and w' will synchronize only if $\Delta w \to 0$ as $t \to \infty$. In the infinitesimal limit this leads to the variational equations for the subsystem,

$$\xi = D_w h(v(t), w(t))\xi, \tag{4}$$

where $D_w h$ is the Jacobian of the w subsystem vector field with respect to w only. The behavior of Eq. (4) or its matrix version [9] depends on the Lyapunov exponents of the w subsystem. We refer to these as sub-Lyapunov exponents. We now have the following theorem: The subsystems w and w' will synchronize only if the sub-Lyapunov exponents are all negative.

The above theorem is a necessary, but not sufficient, condition for synchronization. It says nothing about the set of 'initial conditions' in w' which will synchronize with w. We do not mention here any results regarding these sets of points. They are under investigation and will be reported elsewhere.

Taking a broader view, one can think of the $v = (v_1, \ldots, v_m)$ components as being driving variables and the $w' = (w'_{m+1}, \ldots, w'_n)$ as being responding variables. We take just such a view in our application to a chaotic electronic circuit, below.

It is natural to ask how the synchronization is affected by differences in parameters between the w and w' systems which would be found in real applications. Let μ be a vector of the parameters of the y subsystem and μ' of the w' subsystem, so that $h = h(v,w,\mu)$, for example. If the w subsystem were one dimensional, then for small Δw and small $\Delta\mu = \mu' - \mu$,

$$\Delta\dot{w} \approx h_w \Delta w + h_\mu \Delta\mu, \tag{5}$$

where h_w and h_μ are the derivatives of h. Roughly, if h_w and h_μ are nearly constant in time, the solution of this will follow the form

$$\Delta w(t) = \left[\Delta w(0) - \frac{h_\mu}{h_w}\right]e^{h_w t} + \frac{h_\mu}{h_w}. \tag{6}$$

If $h_w < 0$, the difference between w and w' will level off at some constant value. Although this is a simple one-dimensional approximation, it turns out to be the case for all systems we have investigated numerically, even when the differences in parameters are rather large (~10%–20%).

The phenomenon of synchronization is reminiscent of the 'slaving principle' of Haken [10]. Haken applied his principle mostly to systems near singularities, like bifurcations, showing that the degrees of freedom of the system for which the eigenvalue of the linear part of the vector field were ≥ 0 determined the behavior of all other variables associated with negative eigenvalues. Just as the Lyapunov exponent is the generalization of the Jacobian for stability studies, our use of the sub-Lyapunov exponents appears to be a generalization of concepts like Haken's slaving.

We have tested these ideas on several models, including several two-dimensional maps. Here we present the results for the Rössler [4] and Lorenz [3] attractors which are typical for all our systems [2].

We found that in the Rössler system it was possible to use the y component to drive an (x', z') response Rössler system and attain synchronization with the (x, z) components of the driving system. Figure 1 shows three-dimensional views of the drive and response systems for a particular set of parameters in the chaotic regime. One can see that although the response system starts far away from the drive values it soon spirals into the same type of attractor where it remains in synchronization with the drive-system attractor. Table 1 shows the sub-Lyapunov exponents [11] of various configurations of drive and response for the Rössler system. Note that only the y drive configuration will synchronize.

Table 1 also shows the sub-Lyapunov exponents for the Lorenz system in the chaotic regime. In this case, synchronization will occur for either x or y driving. Figure 2(a) shows a plot of time versus log of the differences $y' - y$ and $z' - z$ for the Lorenz attractor. The convergences to synchronization are consistent with the values in Table 1.

Figure 2(b) shows the results for the same situation, but with a slight change in the parameters of the response system. As expected from the simple one-dimensional argument above, the differences level off. The systems

Table 1 A listing of the various subsystems and driving components for the Lorenz and Rössler systems and their sub-Lyapunov exponents

System	Drive	Response	Sub-Lyapunov exponents
Rössler	x	(y,z)	$(+0.2, -8.89)$
$a = 0.2, b = 0.2$	y	(x,z)	$(-0.056, -8.81)$
$c = 9.0$	z	(x,y)	$(+0.1, +0.1)$
Lorenz	x	(y,z)	$(-1.81, -1.86)$
$\sigma = 10, b = \frac{8}{3}$	y	(x,z)	$(-2.67, -9.99)$
$r = 60.0$	z	(x,y)	$(+0.0108, -11.01)$

partially synchronize in that y' and z' stay within some neighborhood of y and z as they proceed around the attractor.

We have investigated all the above phenomena in other models [2] and have found similar results.

We used a modified version of an electronic chaotic circuit by Newcomb and Sathyan [6] to test these ideas on a real system. The drive circuit consists of an unstable second-degree oscillator coupled to a hysteritic circuit which continually shifts the center of the unstable focus causing the system to be reinjected into the region near one of two unstable focii. This keeps the motion bounded and chaotic in certain parameter regimes. This is a three-dimensional dynamical system. The response circuit was chosen to be a subcircuit in which the hysteritic circuitry was mostly cut off, so the drive signal came from a point just at the cutoff. The details of the circuits and these experiments will be given elsewhere.

The equations of motion for the model of the drive circuit

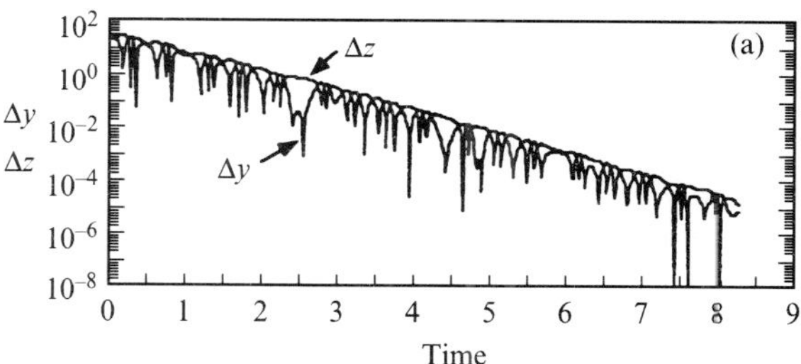

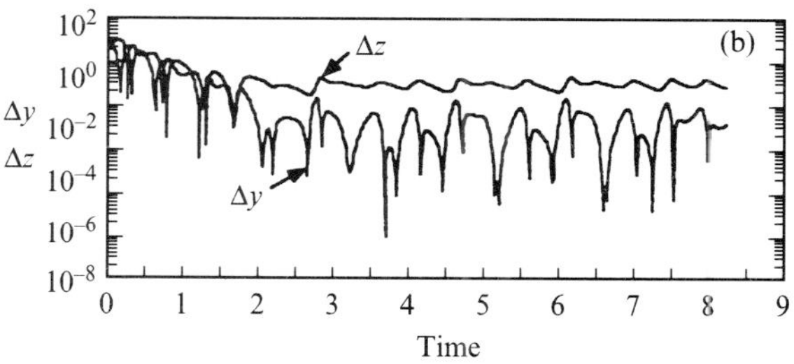

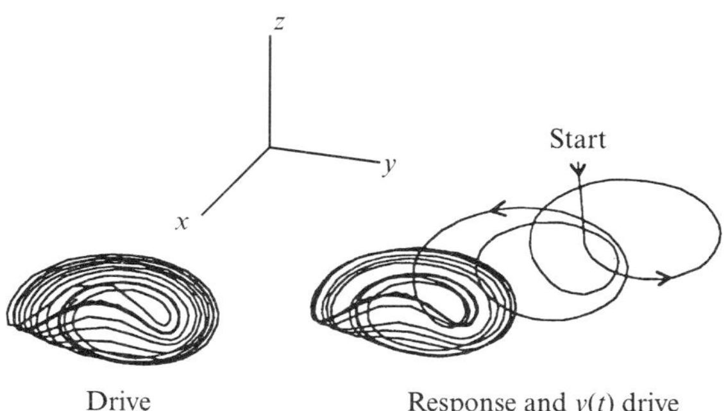

Figure 1 The attractors for the Rössler drive system and the $(x'–z')$ response system and $y(t)$ drive variable.

Figure 2 The differences $y' - y$ and $z' - z$ between the response variables and their drive counterparts for the Lorenz system for (a) when parameters are the same for both systems and (b) when the parameters differ by 5%.

can be written in terms of the above oscillator-hysteresis description (see Ref. 12 for a description of modeling hysteresis). These must be transformed so that the drive signal, x_3, is explicitly shown. This gives

$$\dot{x}_1 = x_2 + \gamma x_1 + c(\alpha x_3 - \beta x_1),$$

$$\dot{x}_2 = -\omega_2 x_1 - \delta_2 x_2,$$

$$\varepsilon \dot{x}_3 = \alpha^{-1}\{[1 - (\alpha x_3 - \beta x_1)^2](sx_1 - r + \alpha x_3 - \beta x_1) - \delta_3 \alpha x_3 - \beta x_1 - \beta x_2 - \beta \gamma x_1 - \beta c(\alpha x_3 - \beta x_1)\}. \tag{7}$$

The equations for x_1 and x_2 model the response circuit as well. For the chaotic regime the circuit settings dictate that $\gamma = 0.2$, $c = 2.2$, $\alpha = 6.6$, $\beta = 7.9$, $\delta_2 = 0.01$, $\omega_2 = 10$, $s = 1.667$, and $r = 0.0$. The sub-Lyapunov exponents can be calculated directly since the Jacobian for Eqs. (7) is a constant in the x_1 and x_2 variables. The exponents are -16.587 and -0.603, implying synchronization will occur.

The circuit itself runs in the realm of a few kHz. We find that the response synchronizes with the drive within about 2 ms which is consistent with the above sub-Lyapunov exponents whose units are inverse milleseconds. Figure 3

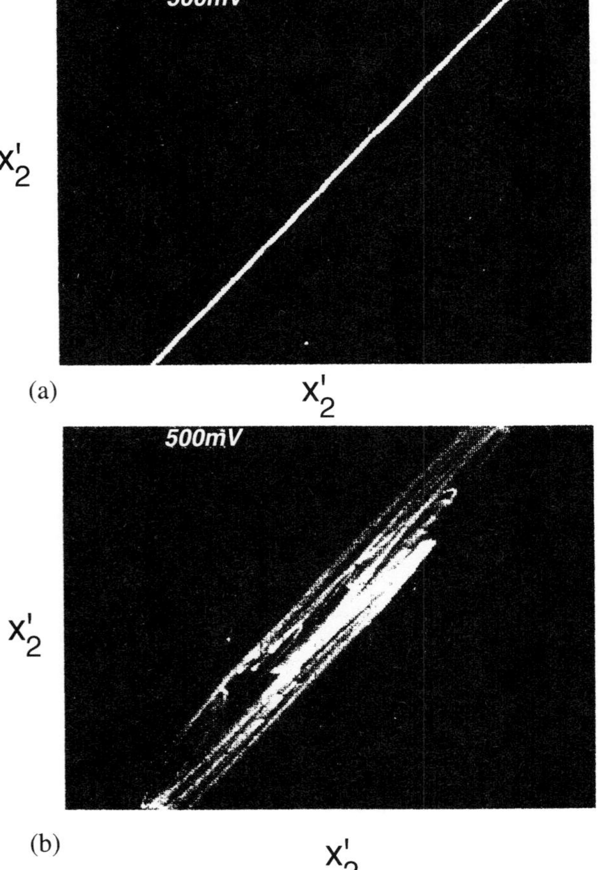

Figure 3 Oscilloscope traces of the response voltage x_2' vs its drive counterpart voltage x_2 for (a) circuit parameters the same and (b) circuit parameters different by 50%.

shows oscilloscope traces of the variable x_2 versus its response counterpart x_2' for the synchronizing circuits for two different parameter values. The parameter varied was a resistor in the response circuit which effectively changed α and β. In Fig. 3(b) $\alpha = 9.9$ and $\beta = 10.4$. The values for the driving circuit remained unchanged. This shows changes ($\sim50\%$) of the circuit parameters effect synchronization greatly. Even though the sub-Lyapunov exponents in the latter cases both remain negative, synchronization is degraded.

At this point much more remains to be done (theoretically and experimentally) on synchronizing systems. All of the systems studied so far have been low dimensional with one positive Lyapunov exponent. Can synchronization be accomplished in the case of two or more positive exponents, but with only one drive? Can one predict which components will synchronize based on the structure of the center, unstable, and stable manifolds? Despite these and other open questions, we would like to offer some speculations.

The ability to design synchronizing systems in non-linear and, especially, chaotic systems may open interesting opportunities for applications of chaos to communications, exploiting the unique features of chaotic signals. One now has the capability of having two remote systems with many internal signals behaving chaotically yet still synchronized with each other through the one linking drive signal.

Recent interesting results [13, 14] suggest the possibility of extending the synchronization concept to that of a metaphor for some neural processes. Freeman has suggested that one should view the brain response as an attractor. The process of synchronization can be viewed as a response system that 'knows' what state (attractor) to go to when driven (stimulated) by a particular signal. It would be interesting to see whether this dynamical view could supplant the more 'fixed-point' view of neural nets [15, 16].

We would like to acknowledge useful conversations with R.W. Newcomb and the continued encouragement of A.C. Ehrlich, S. Wolf, M. Melich, and W. Meyers. One of us (T.L.C.) was supported on an Office of Naval Technology Postdoctoral Associateship.

Notes

[1] Y.S. Tang, A.I. Mecs, and L.O. Chua, *IEEE Trans. Circuits* **30**, 620 (1983).

[2] References to 'all systems' in this paper include the Lorent (Ref. 3), Rössler (Ref. 4), scroll (Ref. 5), Newcomb hysteresis (Ref. 6), three-mode spin system (Ref. 7), and laster emulation (Ref. 8) systems. We hope to report on these results in the future.

[3] J. Gukenheimer and P. Holmes, *Nonlinear Oscillations Dynamical Systems, and Bifurcations of Vector Fields* (Springer-Verlag, New York, 1983), pp. 92–102.

[4] O.E. Rössler, *Phys. Lett.* **57A**, 397 (1976).

[5] T. Matsumoto, L.O. Chua, and M. Komuro, *IEEE Trans. Circuits Syst.* **32**, 798 (1985).

[6] R.W. Newcomb and S. Sathyan, *IEEE Trans. Circuits Syst.* **30**, 54 (1983).

[7] T.L. Carroll, L.M. Pecora, and F.J. Rachford, *Phys. Rev. A* **40**, 377 (1989).

[8] F. Mitschke and N. Flüggen, *Appl. Phys. B* **35**, 59 (1984).

[9] J. Gukenheimer and P. Holmes, *Nonlinear Oscillations, Dynamical Systems, and Bifurcations of Vector Fields* (Springer-Verlag, New York, 1983), p. 25.

[10] H. Haken, *Synergetics* (Springer-Verlag, Berlin, 1977); *Advanced Synergetics* (Springer-Verlag, Berlin, 1983).

[11] Lyapunov exponents were calculated by using the technique suggested by J.-P. Eckmann and D. Ruell [*Rev. Mod. Phys.* **57**, 617 (1985)] employing QR decompositions of the fundamental solution matrix of the equation of motion at points along the trajectory.

[12] O.E. Rössler, Z. *Naturforsch.* **38a**, 788 (1983)

[13] C. Skarda and W.J. Freeman, *Behav. Brain Sci.* **10**, 161 (1987), and the commentaries following the article.

[14] A. Garfinkel, *Am. J. Physiol.* **245**, R455 (1983).

[15] *Proceedings of the IEEE First Annual International Conference on Neural Networks, San Diego 1987*, edited by M. Caudil and C. Butler (IEEE, New York, 1987).

[16] C. Skarda and W.J. Freeman, *Behav. Brain Sci.* **10**, 170 (1987).

Paper 10
Predictable chaos in slightly perturbed unpredictable chaotic systems

K. Pyragas[1]
Physical Institute, University of Tübingen, W-7400 Tübingen, Germany
[1]*Alexander von Humboldt Fellow on leave from the Institute of Semiconductor Physics, Goštauto 11, 2600 Vilnius, Lithuania.*

Received 26 February 1993; revised manuscript received 8 June 1993; accepted for publication 13 August 1993
Communicated by A.R. Bishop

A method for stabilizing aperiodic orbits of a strange attractor is suggested. It enables the transformation of an unpredictable chaos into a predictable one by synchronizing the current behavior of a chaotic system with its past behavior. This is achieved by a *small* self-controlling feedback perturbation using the past output signal of the system, recorded previously in a memory. An experimental realization of the method is very simple. It does not require any computer analysis of the system behavior, and can be carried out by a purely analogous technique.

1. Introduction

It is well known that the prediction of the long-term behavior of chaotic systems is practically impossible, although these systems can be described by strongly determined dynamic models. Lorenz was the first to run into this problem when investigating the simple dynamic model consisting of three nonlinear ordinary differential Equations [1]. The actual source of unpredictability is the property of a nonlinear system to separate initially close trajectories by an exponential law. Since, in practice, one can only fix the initial conditions of the system with finite accuracy, the errors increase exponentially fast. The characteristic time of reliable prediction is determined by the reciprocal of the maximal positive Lyapunov exponent of the system. Lorenz called this sensitive dependence on initial conditions the 'butterfly effect', because the outcome of his equations, which describe in a crude sense the problem of weather forecasting, could be changed by a butterfly flapping its wings.

In spite of this fundamental difficulty, many investigations in the field of dynamic chaos are devoted to the development of forecasting methods [2–7]. These are based on building mathematical models directly from experimental data. The short-term prediction is then obtained as a solution of these models. It is common for all methods of forecasting to assume that the investigator (forecaster) is a passive subject, who cannot act on the system. The aim of this paper is to show that using only a *small* external perturbation of a special form, one can synchronize the current behavior of the system with its past behavior recorded previously in a memory. As a result, a reliable prediction becomes possible for any length of time.

The method suggested is based on stabilizing aperiodic orbits of the strange attractor. It represents a connection of two ideas, namely, the controlling chaos [8] suggested by Ott, Grebogi and Yorke (OGY), and the synchronization of chaos [9,10] suggested by Pecora and Carroll. OGY have suggested a method to stabilize the unstable *periodic* orbits of the strange attractor by using only a *small* feedback perturbation. The idea of Pecora and Carroll is based on synchronizing *aperiodic* orbits of two *strongly* coupled chaotic systems. Here we demonstrate the possibility of the stabilization of *aperiodic* orbits by a *small* feedback perturbation.

The subject of controlling chaotic systems has recently received a fair amount of attention of both theoretical [11–17] and experimental workers [18–22]. The standard methods of stabilizing periodic orbits are discrete in time since they deal with the Poincaré map of the system. The controlled perturbation is usually applied to the system one time per period. Recently we have proposed two methods of permanent control by a small self-controlling feedback [17]. They are noise resistant and can be realized in an experiment by an analogous technique. Here we extend the ideas of these methods for the case of stabilizing aperiodic orbits.

The paper is organized as follows. The method and its illustration for the Rössler [23], Lorenz [1], and Duffing [24] systems are presented in Section 2. In Section 3, the problem of stabilizing aperiodic orbits is reduced to the problem of synchronizing two identical chaotic systems. This permits the use of the conditional Lyapunov exponents, introduced by Pecora and Carroll as a criterion of stabilization. The influence of restricting the perturbation on the system transient dynamics is considered in Section 4, and the conclusions are presented in Section 5.

2. Method

Let us consider the chaotic system that can be simulated by a set of ordinary differential equations [17],

$$\dot{y} = P(y, x) + F(t), \quad \dot{x} + Q(y, x). \tag{1}$$

We imagine that Eqs. (1) are unknown, but some scalar variable $y(t)$ can be measured as a system output. The vector $x(t)$ describes the remaining variables of the system that are not available or are not of interest for observation. $F(t)$ is an external perturbation fed to the system input. Here we assume, for simplicity, that the input signal $F(t)$ disturbs only the first equation corresponding to the output variable. A more complicated multi-variable perturbation will be considered in Section 3. The block diagram of the method is presented in Figure 1. The experiment is carried out in two stages. In the first, preparatory, stage an appropriate segment of the output signal $y_{ap}(t)$ of the unpertubed system has to be singled out and recorded in a memory. In the second stage, the system can be forced to repeat exactly the recorded signal by using a small feedback perturbation of the form

$$F(t) = K[y_{ap}(t) - y(t)]. \tag{2}$$

Here K is an experimentally adjustable weight of the perturbation. The perturbation has to be introduced into the system as a negative feedback ($K > 0$). The important feature of this perturbation is that it vanishes when the output signal coincides with the signal recorded in a memory, $F(t) = 0$ at $y(t) = y_{ap}(t)$. Therefore, it does not change the solution of the system corresponding to the segment of the aperiodic signal $y_{ap}(t)$. The perturbation performs the function of self-control, since it always tends to attract the current trajectory $y(t)$ of the system to the desired aperiodic orbit $y_{ap}(t)$. At a sufficiently large weight K, it can stabilize this trajectory. When the stabilization is achieved $y(t) \approx y_{ap}(t)$, and the perturbation becomes very small.

The results of such a stabilization for the Rössler, Lorenz,

and Duffing systems are shown in Figure 2. After switching on the control, the perturbation is at first large, but then rapidly decreases to a very small value.[2] After this transient

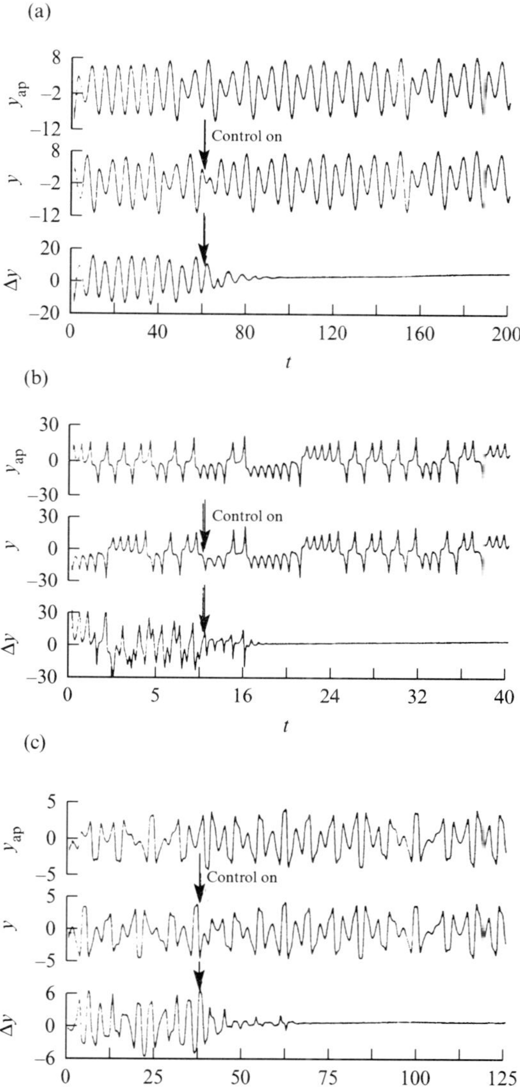

Figure 2 Segments of 'recorded' aperiodic output signals $y_{ap}(t)$ and the dynamics of the output signals $y(t)$ and the differences $\Delta y(t) = y_{ap} - y$, (a) for the Rössler system: $\dot{x} = -y - z$, $\dot{y} = x + 0.2y + K[y_{ap}(t) - y]$, $\dot{z} = 0.2 + z\,(x - 5.7)$, $K = 0.4$, (b) for the Lorenz system: $\dot{x} = 10\,(y - z)$, $\dot{y} = -xz + 28x - y + K[y_{ap}(t) - y]$, $\dot{z} = xy - \frac{8}{3}z$, $K = 4$, and (c) for the nonautonomous Duffing oscillator: $\dot{x} = y$, $\dot{y} = x - x^3 - dy + a\cos(\omega t) + K[y_{ap}(t) - y]$, $a = 2.5$, $\omega = 1$, $d = 0.02$, $K = 0.4$. The arrows show the moment of switching on the perturbation.

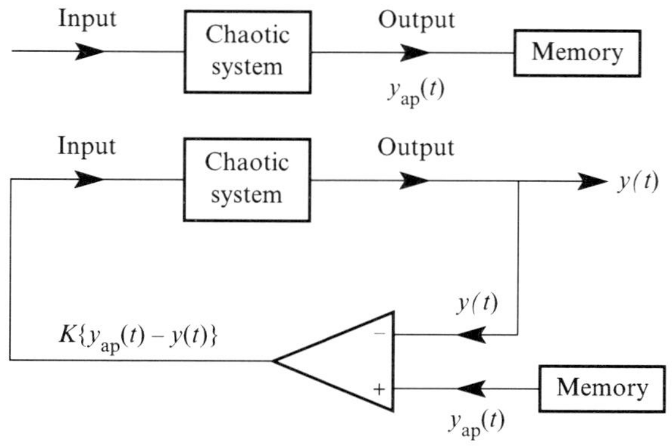

Figure 1 Block diagram of the method.

[2]We have tried many different initial conditions for the trajectories $y(t)$ and $y_{ap}(t)$. The stabilization has been achieved for all systems considered independent of these conditions. Therefore, we ignore the possibility of other basins of attraction for now. Should this problem arise for some systems, we hope that it can be solved by restriction of the perturbation [17].

process, the system begins to repeat exactly its previous behavior corresponding to the recorded signal $y_{ap}(t)$. Therefore, the *small* feedback perturbation in the form of the difference between the output signal and the signal recorded in a memory forces the system to behave chaotically, however, makes it absolutely predictable. The resulting behavior depends, within certain limits, on our desire. The point is that any one of the different segments $y_{ap}(t)$ can be stabilized,[3] and the choice can be made to achieve the best system performance among those segments.

In a real experiment, the control will be negatively affected by at least two factors: fluctuation noise and gradual deviation of the system parameters from their initial values. These factors lead to the finite amplitude of the perturbation in a post-transient regime. Figure 3 illustrates the influence of both factors on the dispersion $\langle F^2(t)\rangle$ of the perturbation for the Duffing system. The amplitude of the perturbation decreases linearly with the decrease of the noise amplitude, as well as with the decrease of the parameter deviation. If both factors are small, the stabilization of the aperiodic orbit can be achieved with a very small perturbation $F(t)$, and the experiment can be performed with a small external signal.

3. Linear analysis

In order to illustrate the law by which the perturbed system approaches the desired aperiodic orbit, Figure 4

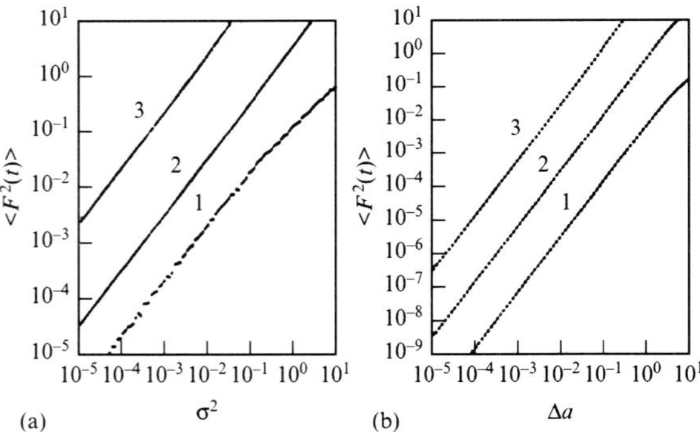

(a)

(b)

Figure 3 Dispersion $\langle F^2(t)\rangle$ of the perturbation of the Duffing oscillator versus (a) dispersion σ^2 of the external noise and (b) versus deviation Δa of the amplitude of the external force at three different values of K: $K = 0.5$ (1), 5 (2) and 50 (3). The noise has been simulated by adding to the right-hand sides of the Duffing equations random functions independent of each other, having the mean value 0, and the mean squared value σ^2. To simulate the deviation of the amplitude a, we calculated at first an unperturbed aperiodic orbit $y_{ap}(t)$ at the fixed initial value $a = a_0 = 2.5$. Then the dynamics of the perturbed system has been calculated with the changed value $a = a_0 + \Delta a$, but with the old function $y_{ap}(t)$ corresponding to $a = a_0$.

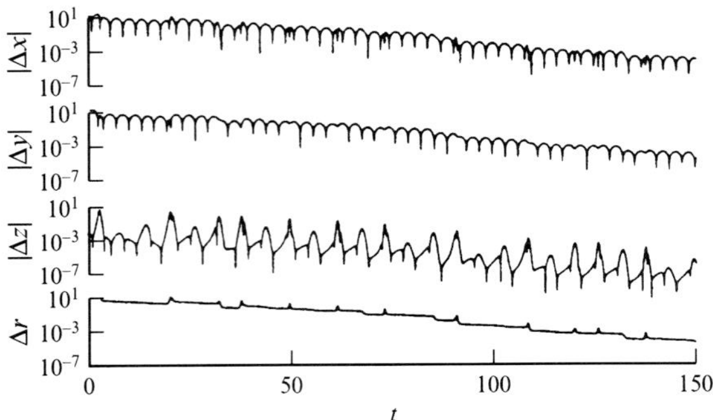

Figure 4 Dynamics of the differences $\Delta x = x_{ap} - x$, $\Delta y = y_{ap} - y$, $\Delta z = z_{ap} - z$, and $\Delta r = \sqrt{(\Delta x)^2 + (\Delta y)^2 + (\Delta z)^2}$ of the Rössler system. $K = 0.3$.

shows the dynamics for the Rössler system in a half logarithmic scale. As is evident from the figure, an asymptotic behavior follows an exponential law. The characteristic exponent depends neither on the initial conditions of the desired aperiodic orbit nor on the current initial conditions of the system corresponding to the moment of switching on the perturbation. This is illustrated in Figure 5 for the Lorenz system. The features above permit the introduction of the Lyapunov exponents as characteristics of the linear behavior of system (1) close to the desired aperiodic orbit. The use of the Lyapunov exponents is possible in spite of the fact that system (1) is disturbed by a chaotic external signal.

The theory of systems driven with chaotic signals, including the introduction of the Lyapunov exponents for such systems, has been developed recently by Pecora and Carroll [9,10]. Our problem can be reduced to that considered in this theory, and we can simply use its results. The theory deals with a compound autonomous dynamic system, which can be divided into two one-way coupled subsystems. By one-way coupling is meant that the behavior of one (response) system is dependent on the behavior of another (drive) system, but the other is not

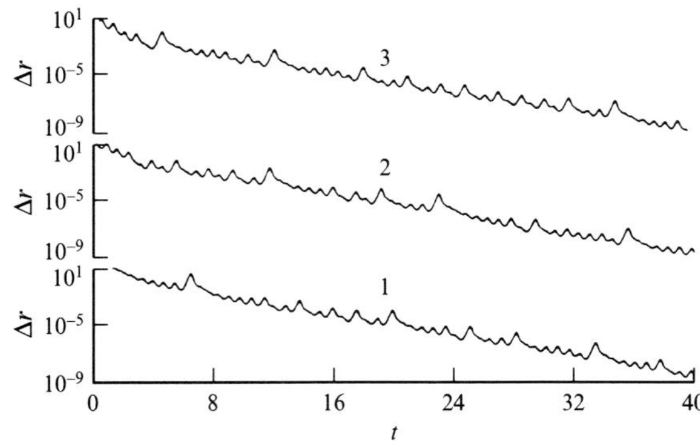

Figure 5 Dynamics of the difference Δr of the Lorenz system at three different arbitrary chosen initial conditions for the signals $y_{ap}(t)$ and $y(t)$. $K = 4$.

---

[3]The length of the segment has to be larger than the characteristic length of the transient process.

influenced by the behavior of the first. To characterize the stability of the response system, Pecora and Carroll introduced the conditional Lyapunov exponents, the characteristics of the variational equations of the response system. The name 'conditional' has been used because these equations depend on the variables of the drive system. It has been shown that the response system synchronizes with the drive system if all conditional Lyapunov exponents are negative.

Our method can be analyzed by the above theory since the nonautonomous system presented in Figure 1 can be reduced to a compound autonomous system consisting of two one-way coupled subsystems. Indeed, a memory element used in the second stage of the method (Figure 1) to generate a past output signal can be replaced by an additional, identical chaotic system (Figure 6), which, starting at the appropriate initial conditions, can generate an exactly aperiodic signal recorded in the memory. As a result, the two-stage experiment presented in Figure 1 can be replaced by the physically equivalent one-stage experiment presented in Figure 6, and the initial problem reduces to the problem of synchronizing two coupled, identical chaotic systems. Mathematically, this problem can be presented as follows,[4]

$$\text{drive:} \quad \dot{y}_{ap} = P(y_{ap}, \boldsymbol{x}_{ap}),$$

$$\dot{\boldsymbol{x}}_{ap} = \boldsymbol{Q}(y_{ap}, \boldsymbol{x}_{ap}),$$

$$\text{response:} \quad \dot{y} = P(y, \boldsymbol{x}) + K(y_{ap} - y),$$

$$\dot{\boldsymbol{x}} = \boldsymbol{Q}(y, \boldsymbol{x}). \tag{3}$$

The conditional Lyapunov exponents $\lambda(K)$ are defined by variational equations of the response system:

$$\delta\dot{y} = \delta y \frac{\partial}{\partial y} P(y_{ap}, \boldsymbol{x}_{ap}) + \delta\boldsymbol{x} \frac{\partial}{\partial\boldsymbol{x}} P(y_{ap}, \boldsymbol{x}_{ap}) - K\delta y,$$

$$\delta\dot{\boldsymbol{x}} = \delta y \frac{\partial}{\partial y} \boldsymbol{Q}(y_{ap}, \boldsymbol{x}_{ap}) + \delta\boldsymbol{x} \frac{\partial}{\partial\boldsymbol{x}} \boldsymbol{Q}(y_{ap}, \boldsymbol{x}_{ap}). \tag{4}$$

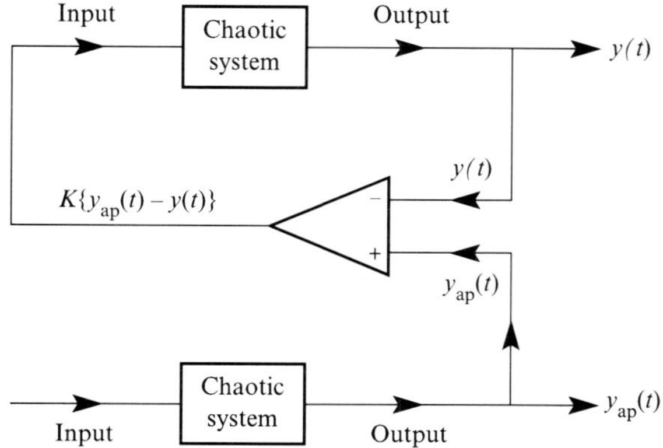

Figure 6 Block diagram of a physically equivalent system to that presented in Figure 1. The memory element is replaced by an additional, identical chaotic system.

Here $\delta y = y - y_{ap}$, $\delta\boldsymbol{x} = \boldsymbol{x} - \boldsymbol{x}_{ap}$ define the deviations of the response system from the aperiodic orbit, determined by the drive system. Equations (4) differ from the variational equations defining the usual Lyapunov exponents of the unperturbed ($K = 0$) system (1) by the term $-K\delta y$. At $K = 0$, the conditional Lyapunov exponents coincide with the usual Lyapunov exponents of the unperturbed system. With the increase of K, this term results in a decrease of $\lambda(K)$ and the inversion of the sign of the initially positive Lyapunov exponents. Figure 7 shows the dependence of the maximal conditional Lyapunov exponents of K for the Rössler, Lorenz, and Duffing systems. The Lyapunov exponents are shown for all possible cases of a one-variable control and also for a multi-variable control. To explain these different cases, let us represent system (1) in a symmetrical form,

$$\dot{\xi}_i = \Phi_i(\boldsymbol{\xi}) + K_i[(\xi_{ap})_i - \xi_i], \quad i = 1, 2, \ldots, m. \tag{5}$$

Here $\boldsymbol{\xi} \equiv \{\xi_1, \xi_2, \ldots, \xi_m\} \equiv \{y, \boldsymbol{x}\}$ is the complete vector of the dynamic variables, $\boldsymbol{\Phi} = \{P, \boldsymbol{Q}\}$, and the vector $\boldsymbol{\xi}_{ap}$ represents the aperiodic orbit of the unperturbed system (5) that we intend to stabilize. The one-variable control by the variable ξ_i corresponds to $K_i = 0$ for all $i \neq j$, and $K_j \equiv K$. The multi-variable control considered in Figure 7 represents the simplest case of such a control, when all K_i are equal: $K_1 = K_2 = \cdots = K_m \equiv K$.

The negative values of the maximal conditional Lyapunov exponent $\lambda(K) < 0$ define the intervals of K corresponding to the synchronization. In other words, these values of K define the operating range of our method. The boundaries of these intervals, K_{min} and K_{max}, correspond to the minimal and maximal thresholds of the synchronization, $\lambda(K_{min}) = \lambda(K_{max}) = 0$. As can be seen from the figure, a one-variable control depends on the choice of the controlled variable. For example, control of the Rössler system is most efficient by the y variable, since it leads to the maximal interval of K corresponding to the synchronization. The control by the z variable of this system is less efficient, since synchronization is possible only in a small interval of the parameter K. Some chaotic systems can have more than one interval of synchronization. An example is the Lorenz system, which in the case of control by the z variable has two of such insulated intervals.

In a one-variable control, all characteristics $\lambda(K)$ have minima at some $K = K_{op}$, which leads to the maximal rate of synchronization, and, therefore, provides the optimal control. The existence of an optimal value of K can be understood as follows. The weight K of the perturbation has to be sufficiently large, in order to compensate the divergence of the trajectories close to the desired aperiodic orbit. A rather large K is not efficient since the perturbation disturbs only one equation of the system, corresponding to the output variable. For large K, the changes of this variable are very fast, and the remaining variables have no time to follow these changes. Therefore, one can conclude that the minimum in $\lambda(K)$ is related to the nonsymmetrical nature of a one-variable control. The calculation of $\lambda(K)$ in the case

[4]This presentation is also more convenient than (1), (2) for computer simulation. The difficulty with the application of higher-order Runge–Kutta methods to system (1), (2) is related to the fact that these methods require knowledge of the external signal $y_{ap}(t)$ values at the moments sited inside the integration intervals. This difficulty does not occur for system (3) since it is autonomous.

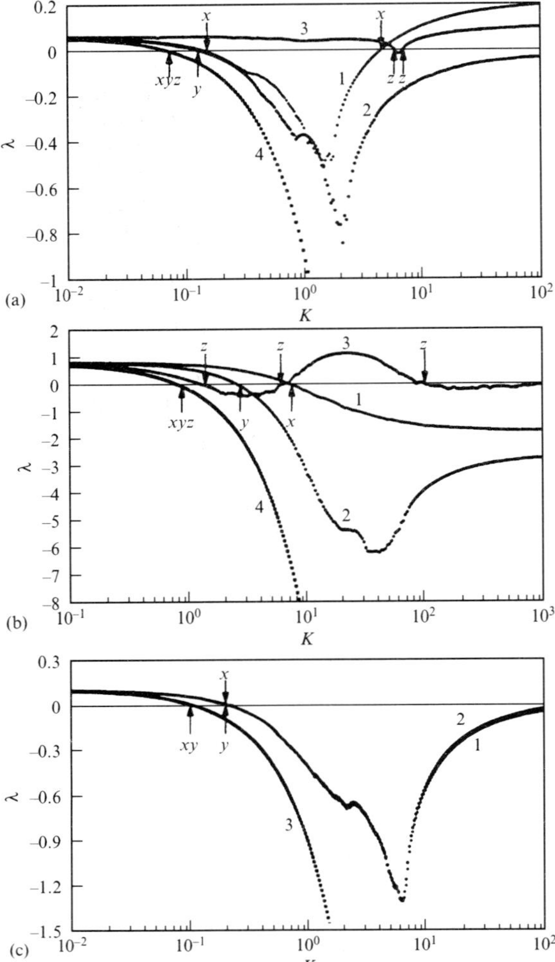

Figure 7 Maximal conditional Lyapunov exponents λ versus weight K of the perturbation for the (a) Rössler (b) Lorenz, and (c) Duffing systems. Curves 1, 2, and 3 in diagrams (a) and (b) correspond to a one-variable control by the x, y, and z variable, respectively. Curve 4 corresponds to a multi-variable control: $K_x = K_y = K_z \equiv K$. In diagram (c) curves 1 and 2 correspond to a one-variable control by the x and y variables, respectively, and curve 3 corresponds to a multi-variable control: $K_x = K_y \equiv K$. The arrows, marked by corresponding controlled variables, show the thresholds of synchronization.

of multi-variable control supports this statement. This control leads to monotonically decreasing characteristics $\lambda(K)$ at any K, for all systems considered in this paper (Figure 7). Therefore, a multi-variable control is more efficient. It leads to a smaller threshold K_{min} and to a faster rate of synchronization at any fixed $K > K_{min}$, as compared with those in a one-variable control.

The following question becomes important for an experimental application of the method: What kind of chaotic systems can be synchronized with their past

behavior by a one-variable control? Our answer to this question is based on the assumption that the stabilization of the response system can be achieved only if the perturbation has a sufficiently large number of degrees of freedom to suppress the divergence of the system flow in all expanded directions. More precisely, we assume that the minimal number of controlled variables has to be equal to the number of positive Lyapunov exponents of the system. All models considered up to now support this assumption. They all have only one positive Lyapunov exponent and they all can be synchronized by a one-variable control. To check this assumption for a more complicated system, we have considered the hyperchaos equations [25] with two positive Lyapunov exponents. The dependence of the conditional Lyapunov exponents on K for different types of control is shown in Figure 8. It is impossible to synchronize this system by a one-variable control: the maximal $\lambda(K)$ is positive at any K for all dynamic variables. However, it is possible to invert the sign of one out of two initially positive Lyapunov exponents. This is illustrated in the figure for the case of control by the y variable. Although synchronization is impossible here, the flow of the dynamic system close to the desired aperiodic orbit diverges now only in one unstable direction. One can say that close to this orbit the control turns hyperchaos into chaos. Applying the perturbation to two equations of the system, one can invert the sign of both positive Lyapunov exponents. This is illustrated in the figure for the case of a two-variable control by the y and w variables. Therefore, this model also supports the above assumption.

4. Restriction of the perturbation

Let us discuss now the transient process. The initial amplitude of the perturbation depends on the distance

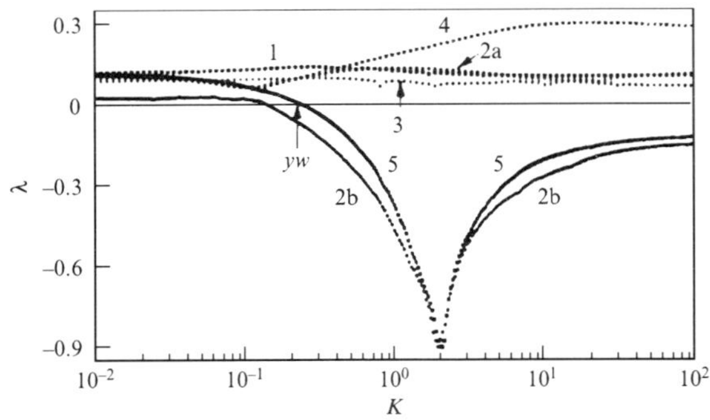

Figure 8 Conditional Lyapunov exponents versus K for a hyperchaos system [25]: $\dot{x} = -y - z$, $\dot{y} = x + 0.25y + w$, $\dot{z} = 3 + xz$, $\dot{w} = -0.5z + 0.05w$. Curves 1, 2a, 3, and 4 correspond to the maximal Lyapunov exponents of a one-variable control by the x, y, z, and w variables, respectively. Curve 2b shows the second largest Lyapunov exponent in the case of control by the y variable. Curve 5 corresponds to a two-variable control by the y and w variables simultaneously: $K_y = K_w \equiv K$.

between the states of the response and the drive systems at the moment of switching on the perturbation. In the typical case, this distance is not short, and the perturbation has a large initial amplitude. Large initial values of the perturbation can be undesired or inaccessible for some experimental situations. Here, as well as in our previous paper [17], we consider the restricted perturbation of the form

$$F(t) = -F_0, \qquad KD(t) \leq -F_0,$$
$$= KD(t), \quad -F_0 < KD(t) < F_0,$$
$$= F_0, \qquad KD(t) \geq F_0. \tag{6}$$

Here $F_0 > 0$ is the saturating value of the perturbation, and $D(t) = y_{ap}(t) - y(t)$. Saturation can be achieved by introducing some nonlinear element into the feedback circuit. In proximity to the recorded signal, $y(t) \approx y_{ap}(t)$, both perturbations (2) and (6) are working identically, but they are leading to different transients. Figure 9 illustrates the influence of the restriction on the system dynamics. Here the perturbation is always small including the transient process, however, the duration of this process, on average, is now much longer. The control is not sufficiently efficient until the state of the response system does not come close to the state of the drive system.

For small F_0, the average time of the transient $\langle \tau_0 \rangle$ can be estimated as follows. The probability of the repetition of the state of the dynamic system with some accuracy ε is proportional to the correlation integral $C_m(\varepsilon)$ that scales as $C_m(\varepsilon) \propto \varepsilon^d$ [26]. Here d is the correlation dimension of the strange attractor. The efficient control leading to the synchronization is possible only if the difference ε of the states is of the order of the amplitude of the perturbation, $\varepsilon \propto F_0$. Therefore, the average time $\langle \tau_0 \rangle$ of the transient increases with the decrease of F_0 by a power law,

$$\langle \tau_0 \rangle \propto C_m^{-1}(\varepsilon) \propto F_0^{-d}, \quad F_0 \ll K\Delta y_{max}. \tag{7}$$

Here Δy_{max} is the size of the strange attractor in the y direction.

For large $F_0 > K\Delta y_{max}$, the perturbation does not achieve the saturating value F_0, and the system behaves in the same manner as if without any restriction. The average length of

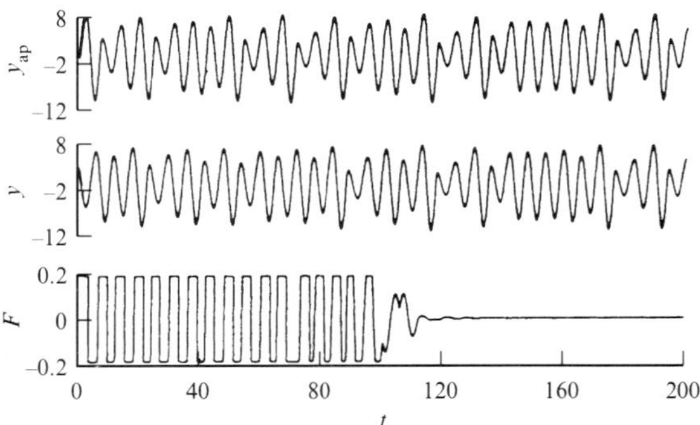

Figure 9 The same as in Figure 2(a), but for the case of a restricted perturbation. $F_0 = 0.2$. The perturbation is switched on at $t = 0$.

the transient $\langle \tau_0 \rangle$ now is proportional to the reciprocal of the maximal conditional Lyapunov exponent of the system,

$$\langle \tau_0 \rangle = |\lambda(K)|^{-1}, \quad F_0 \geq K\Delta y_{max}. \tag{8}$$

Figure 10 confirms the above relations for the Lorenz system. The exponent defined from the slope of the linear dependence $\log(\langle \tau_0 \rangle)$ versus $\log F_0$ at small values of F_0 is in good agreement with the correlation dimension of the Lorenz system, $d = 2.05$. The saturating values of $\langle \tau_0 \rangle$ at large F_0 agree with the reciprocal of the Lyapunov exponent defined in Figure 7(b).

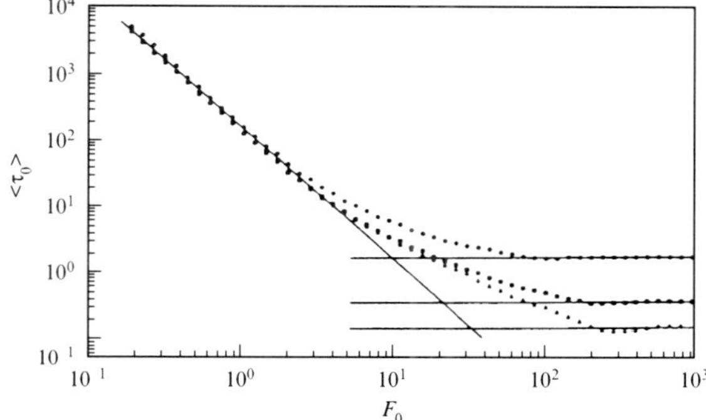

Figure 10 The average time $\langle \tau_0 \rangle$, needed to achieve synchronization of the Lorenz system, versus maximally allowed amplitude of the perturbation F_0 for three different values of K: $K = 4$ (1), 10 (2) and 40 (3).

5. Conclusions

We have shown that the current behavior of a chaotic system can be synchronized with its past behavior, recorded previously in a memory. This is achieved by a small self-controlling feedback perturbation in the form of the difference between the current and past output signals. As a result, the system behavior becomes absolutely predictable. This behavior can be changed, within certain limits, according to our desire by choosing different intervals of the past output signal. An experimental application of the method does not require any computer analysis of the system behavior. It can be easily carried out by a purely analogous technique. The operating range of the method can be determined from the variational equations of the perturbed system. The method works if the maximal conditional Lyapunov exponent of the perturbed system is negative. To stabilize the chaos of higher order, multivariable control has to be used. The minimal number of controlled variables has to be equal to the number of positive Lyapunov exponents of the unperturbed system.

The large initial values of the perturbation corresponding to the transient dynamics can be avoided by some restriction of the perturbation. However, the transient process becomes longer in this case. The average length of the transient increases with the decrease of the maximally allowed amplitude of the perturbation by a power law with the characteristic exponent being equal to the dimension of the strange attractor.

Acknowledgement

I thank R.P. Huebener, A. Kittel, R. Richter and J. Peinke for stimulating discussion during the work, J. Parisi for a critical reading of the manuscript and many useful suggestions, and O.E. Rössler for a useful discussion of the results. The research was supported by the Alexander von Humboldt Foundation.

References

[1] E.N. Lorenz, *J. Atmos. Sci.* **20** (1963) 130.

[2] J.P. Crutchfield and B.S. McNamara, *Complex Syst.* **1** (1987) 417.

[3] J.D. Farmer and J.J. Sidorowich, *Phys. Rev. Lett.* **59** (1987) 845.

[4] M. Casdagli, *Physica D* **35** (1989) 335.

[5] G. Sugihara and R.M. May, *Nature* **344** (1990) 734.

[6] A.I. Mess, Int. J. Bifurc. *Chaos* **1** (1991) 777.

[7] A. Mess, K. Aihara, M. Adachi, K. Judd, T. Ikeguchi and G. Matsumoto, *Phys. Lett. A* **169** (1992) 41.

[8] E. Ott, C. Grebogi and J.A. Yorke, *Phys. Rev. Lett.* **64** (1990) 1196.

[9] L.M. Pecora and T.L. Carroll, *Phys. Rev. Lett.* **64** (1990) 821.

[10] L.M. Pecora and T.L. Carroll, *Phys. Rev. A* **44** (1991) 2374.

[11] N.J. Mehta and R.M. Henderson, *Phys. Rev. A* (1991) 4861.

[12] B. Peng, V. Petrov and K. Showalter, *J. Phys. Chem.* **95** (1991) 4957.

[13] U. Dressler and G. Nitsche, *Phys. Rev. Lett.* **68** (1992) 1.

[14] T.L. Carroll, I. Triandof, I. Schwartz and L. Pecora, *Phys. Rev. A* **46** (1992) 6189.

[15] D. Auerbach, C. Grebogi, E. Ott and J.A. Yorke, *Phys. Rev. Lett.* **69** (1992) 3479.

[16] T. Shinbrot, C. Grebogi, E. Ott and J.A. Yorke, *Phys. Lett. A* **169** (1992) 349.

[17] K. Pyragas, Phys. Lett. A 170 (1992) 421.

[18] W.L. Ditto, S.N. Rauseo and M.I. Spano, *Phys. Rev. Lett.* **65** (1990) 3211.

[19] A. Azevedo and S.M. Rezende, *Phys. Rev. Lett.* **66** (1991) 1342.

[20] E.R. Hunt, *Phys. Rev. Lett.* **67** (1991) 1953.

[21] J. Singer, Y-Z. Wang and H.H. Bau, *Phys. Rev. Lett.* **66** (1991) 1123.

[22] R. Roy, T.W. Murphy Jr., T.D. Maier, Z. Gills and E.R. Hunt, *Phys. Rev. Lett.* **68** (1992) 1259.

[23] O.E. Rössler, *Phys. Lett. A* **57** (1976) 397.

[24] J. Guckenheimer and P. Holmes, *Nonlinear Oscillations, Dynamical Systems, and Bifurcations of Vector Fields* (Springer, Berlin, 1983) pp. 82–91.

[25] O.E. Rössler, *Phys. Lett. A* **71** (1979) 155.

[26] P. Grassberger and I. Procaccia, *Phys. Rev. Lett.* **50** (1983) 346.

Paper 11
Circuit implementation of synchronized chaos with applications to communications

Kevin M. Cuomo and Alan V. Oppenheim
Research Laboratory of Electronics, Massachusetts Institute of Technology, Cambridge, Massachusetts 02139, USA

Received 21 January 1993

An analog circuit implementation of the chaotic Lorenz system is described and used to demonstrate two possible approaches to private communications based on synchronized chaotic systems.

In 1990 Pecora and Carroll [1] reported that certain chaotic systems possess a self-synchronization property. A chaotic system is self-synchronizing if it can be decomposed into subsystems: a drive system and a stable response subsystem that synchronize when coupled with a common drive signal [1–3]. They showed numerically that synchronization occurs if all of the Lyapunov exponents for the response subsystems are negative. For some synchronizing chaotic systems the ability to synchronize is robust. For example, the Lorenz system is decomposable into two separate response subsystems that will each synchronize to the drive system when started from any initial condition. As discussed in [4–6], the combination of synchronization and unpredictability from purely deterministic systems leads to some potentially interesting communications applications. In this Letter, we focus on the synchronizing properties of the Lorenz system, the implementation of the Lorenz system as an analog circuit, and the potential for utilizing the Lorenz circuit for various communications applications. It should be stressed that the applications indicated are very preliminary and presented primarily to suggest and illustrate possible directions.

The Lorenz system [7] is given by

$$\dot{x} = \sigma(y - x),$$
$$\dot{y} = rx - y - xz, \tag{1}$$
$$\dot{z} = xy - bz,$$

where σ, r, and b are parameters. As shown by Pecora and Carroll an interesting property of (1) is that it is decomposable into two stable subsystems. Specifically, a stable (x_1, z_1) response subsystem can be defined by

$$\dot{x}_1 = \sigma(y - x_1),$$
$$\dot{z}_1 = x_1 y - bz_1 \tag{2}$$

and a second stable (y_2, z_2) response subsystem by

$$\dot{y}_2 = rx - y_2 - xz_2,$$
$$\dot{z}_2 = xy_2 - bz_2. \tag{3}$$

Equation (1) can be interpreted as the drive system since its dynamics are independent of the response subsystems. Equations (2) and (3) represent dynamical response systems which are driven by the drive signals $y(t)$ and $x(t)$, respectively. The eigenvalues of the Jacobian matrix for the (x_1, z_1) subsystem are both negative and thus $|x_1 - x|$ and $|z_1 - z| \to 0$ as $t \to \infty$. Also, it can be shown numerically that the Lyapunov exponents of the (y_2, z_2) subsystem are both negative and thus $|y_2 - y|$ and $|z_2 - z| \to 0$ as $t \to \infty$.

As we show below, the two response subsystems can be used together to regenerate the full-dimensional dynamics which are evolving at the drive system. Specifically, if the input signal to the (y_2, z_2) subsystem is $x(t)$, then the output $y_2(t)$ can be used to drive the (x_1, z_1) subsystem and subsequently generate a 'new' $x(t)$ in addition to having obtained, through synchronization, $y(t)$ and $z(t)$. It is important to recognize that the two response subsystems given by Eqs. (2) and (3) can be combined into a single system having a three-dimensional state space. This produces a full-dimensional response system which is structurally similar to the drive system (1). Further discussion of this result is given below in the context of the circuit implementations.

A direct implementation of Eq. (1) with an electronic circuit presents several difficulties. For example, the state variables in Eq. (1) occupy a wide dynamic range with values that exceed reasonable power supply limits. However, this difficulty can be eliminated by a simple transformation of variables. Specifically, we define new variables by $u = x/10$, $v = y/10$, and $w = z/20$. With this scaling, the Lorenz equations are transformed to

$$\dot{u} = \sigma(v - u),$$
$$\dot{v} = ru - v - 20uw, \tag{4}$$
$$\dot{w} = 5uv - bw.$$

This system, which we refer to as the transmitter, can be more easily implemented with an electronic circuit because the state variables all have similar dynamic range and circuit voltages remain well within the range of typical power supply limits.

An analog circuit implementation of the circuit Eqs. (4) is shown in Figure 1. The operational amplifiers (1–8) and associated circuitry perform the operations of addition, subtraction, and integration. Analog multipliers implement the nonlinear terms in the circuit equations. We emphasize that our circuit implementation of (4) is exact, and that the coefficients σ, r, and b can be independently varied by adjusting the corresponding resistors R_5, R_{11}, and R_{18}. In addition, the circuit time scale can be easily adjusted by changing the values of the three capacitors, C_1, C_2, and C_3, by a common factor. We have chosen component values (resistors (kΩ): R_1, R_2, R_3, R_4, R_6, R_7, R_{13}, R_{14}, R_{16}, R_{17}, R_{19} = 100; R_5, R_{10} = 49.9; R_8 = 200; R_9, R_{12} = 10; R_{11} = 63.4; R_{15} = 40.2; R_{18} = 66.5; R_{20} = 158; capacitors (pF): C_1, C_2, C_3 = 500; op-amps (1–8): LF353 multipliers: AD632AD) which result in the coefficients σ = 16, r = 45.6, and b = 4.

To illustrate the chaotic behavior of the transmitter circuit, an analog-to-digital (A/D) data recording system was used to sample the appropriate circuit outputs at a 48 kHz rate and with 16-bit resolution. Figure 2(a) shows

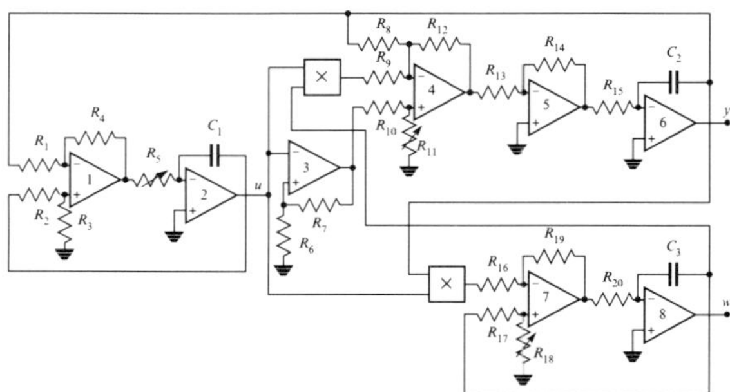

Figure 1 Lorenz-based chaotic circuit.

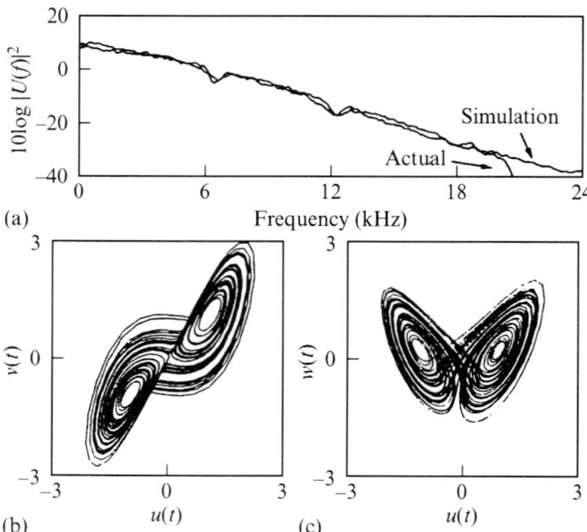

(a)

(b)

(c)

Figure 2 Circuit data: (a) averaged power spectrum of $u(t)$; (b) chaotic attractor projected onto the uv plane; (c) chaotic attractor projected onto the uw plane.

the averaged power spectrum of the circuit wave form $u(t)$. The power spectrum is broadband which is typical of a chaotic signal. Figure 2(a) also shows a power spectrum obtained from a numerical simulation of the circuit equations. As we see, the performance of the circuit and the simulation are consistent. Figures 2(b) and 2(c) show the circuit's chaotic attractor projected onto the uv plane and uw plane, respectively. These data were obtained from the circuit using the stereo recording capability of the A/D system to simultaneously sample the x-axis and y-axis signals at a 48 kHz rate and with 16-bit resolution. A more detailed analysis of the transmitter circuit is given in [6].

A full-dimensional response system which will synchronize to the chaotic signals at the transmitter (4) is given by

$$\dot{u}_r = \sigma(v_r - u_r),$$

$$\dot{v}_r = ru - v_r - 20uw_r, \qquad (5)$$

$$\dot{w}_r = 5uv_r - bw_r.$$

We refer to this system as the receiver in light of some potential communications applications. We denote the transmitter state variables collectively by the vector $\mathbf{d} = (u, v, w)$ and the receiver variables by the vector $\mathbf{r} = (u_r, v_r, w_r)$ when convenient.

By defining the dynamical errors by $\mathbf{e} = \mathbf{d} - \mathbf{r}$, it is straightforward to show that synchronization in the Lorenz system is a result of stable error dynamics between the transmitter and receiver. Assuming that the transmitter and receiver coefficients are identical, a set of equations which govern the error dynamics are given by

$$\dot{e}_1 = \sigma(e_2 - e_1),$$

$$\dot{e}_2 = -e_2 - 20u(t)e_3,$$

$$\dot{e}_3 = 5u(t)e_2 - be_3.$$

The error dynamics are globally asymptotically stable at the origin provided that $\sigma, b > 0$. This result follows by considering the three-dimensional Lyapunov function defined by $E(\mathbf{e},t) = \frac{1}{2}(1/\sigma)e_1^2 + e_2^2 + 4e_3^2]$. The time rate of change of $E(\mathbf{e},t)$ along trajectories is given by

$$\dot{E}(\mathbf{e},t) = (1/\sigma)e_1\dot{e}_1 + e_2\dot{e}_2 + 4e_3\dot{e}_3$$

$$= -(e_1 - \tfrac{1}{2}e_2)^2 - \tfrac{3}{4}e_2^2 - 4be_3^2,$$

which shows that $E(\mathbf{e},t)$ decreases for all $\mathbf{e} \neq 0$. As $E(\mathbf{e},t)$ goes to zero synchronization occurs. Note that the transmitter and receiver need not be operating chaotically for synchronization to occur. In [8], a similar Lyapunov argument is given for the synchronization of the (y,z) subsystem of the Lorenz equations.

A comparison of the receiver equations (5) with the transmitter equations (4) shows that they are nearly identical, except that the drive signal $u(t)$ replaces the receiver signal $u_r(t)$ in the $(\dot{v}_r, \dot{w}_r)$ equations. This similarity allows the transmitter and receiver circuits to be built in an identical way, which helps to achieve perfect synchronization between the transmitter and receiver. In [6] we discuss and illustrate the synchronization performance of the receiver circuit.

As one illustration of the potential use of synchronized chaotic systems in communications, we describe a system

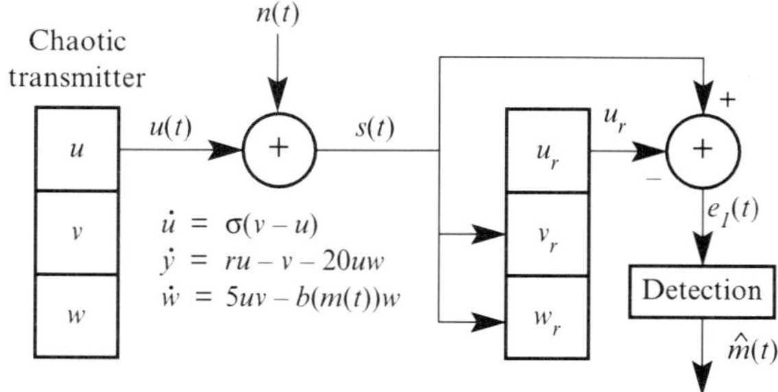

Figure 3 Chaotic communication system.

to transmit and recover binary-valued bit streams [6]. The basic idea is to modulate a transmitter coefficient with the information-bearing wave form and to transmit the chaotic drive signal. At the receiver, the coefficient modulation will produce a synchronization error between the received drive signal and the receiver's regenerated drive signal with an error signal amplitude that depends on the modulation. Using the synchronization error the modulation can be detected.

The modulation/detection process is illustrated in Figure 3. In this figure, the coefficient b of the transmitter Equations (4) is modulated by the information-bearing wave form, $m(t)$. For purposes of demonstrating the technique, we use a square wave for $m(t)$ as illustrated in Figure 4(a). The square wave produces a variation in the transmitter coefficient b with the zero-bit and one-bit

coefficients corresponding to $b(0) = 4$ and $b(1) = 4.4$, respectively. In [6] we show that the averaged power spectrum of the drive signal with and without the embedded square wave present are very similar. Figure 4(b) shows the synchronization error power, $e_1^2(t)$, at the output of the receiver circuit. The coefficient modulation produces significant synchronization error during a '1' transmission and very little error during a '0' transmission. Figure 4(c) illustrates that the square-wave modulation can be reliably recovered by low pass filtering the synchronization error power wave form and applying a threshold test. This approach has also been shown to work using Chua's circuit [9].

Another potential approach to communications applications is based on signal masking and recovery. In signal masking, a noiselike masking signal is added at the transmitter to the information-bearing signal $m(t)$ and at the receiver the masking is removed. In our system, the basic idea is to use the received signal to regenerate the masking signal at the receiver and subtract it from the received signal to recover $m(t)$. This can be done with the synchronizing receiver circuit since the ability to synchronize is robust, i.e., is not highly sensitive to perturbations in the drive signal and thus can be done with the masked signal. It is interesting to note that this idea is not restricted to just the Lorenz circuit but has wider potential; for example, Kocarev et al. [10] have also demonstrated our signal masking concept in [4,5] using Chua's circuit. While there are many possible variations, consider, for example, a transmitted signal of the form $s(t) = u(t) + m(t)$. It is assumed that for masking, the power level of $m(t)$ is significantly lower than that of $u(t)$. The dynamical system implemented at the receiver is

$$\dot{u}_r = 16(v_r - u_r),$$

$$\dot{v}_r = 45.6s(t) - v_r - 20s(t)w_r,$$

$$\dot{w}_r = 5s(t)v_r - 4w_r.$$

If the receiver has synchronized with $s(t)$ as the drive, then $u_r(t) \simeq u(t)$ and consequently $m(t)$ is recovered as $\hat{m}(t) = s(t) - u_r(t)$. Figure 5 illustrates the approach.

Using the transmitter and receiver circuits, we demonstrate the performance of this system in Figure 6 with a segment of speech from the sentence 'He has the bluest eyes.' As indicated in Figure 7 the power spectra of the chaotic masking signal, $u(t)$, and the speech are highly

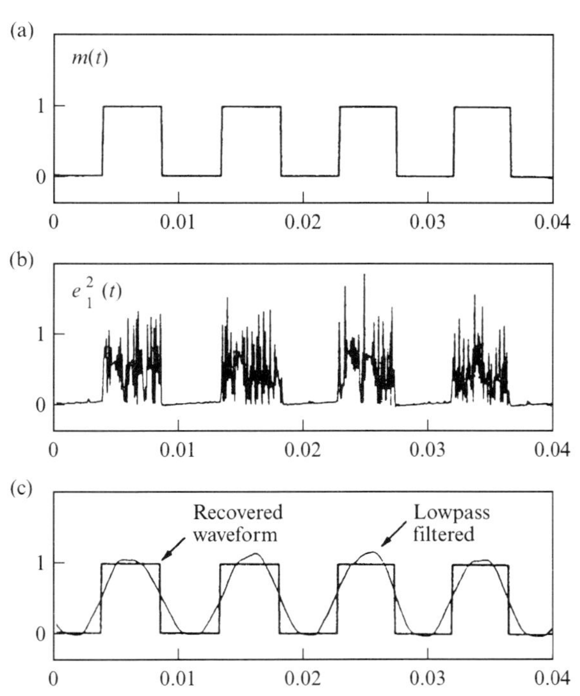

Figure 4 Circuit data: (a) modulation wave form; (b) synchronization error power; (c) recovered wave form.

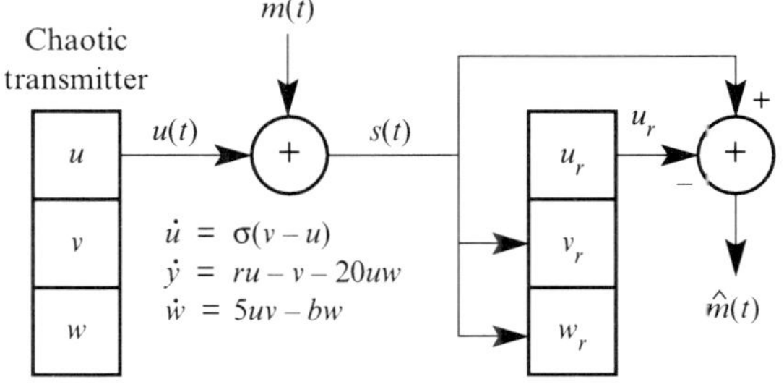

Figure 5 Chaotic signal masking system.

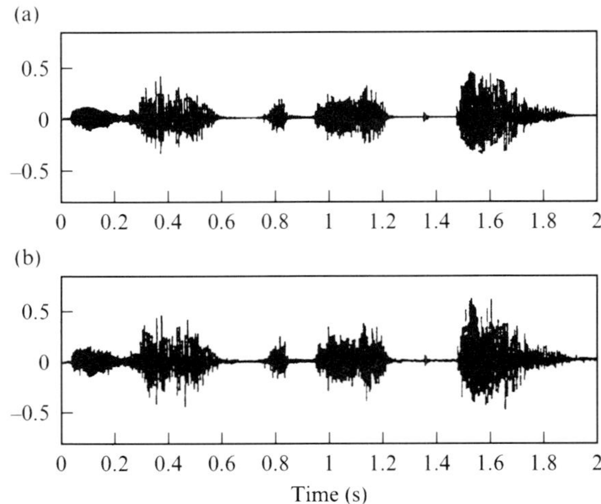

(a)

(b)

Time (s)

Figure 6 Circuit data: speech wave forms. (a) Original; (b) recovered.

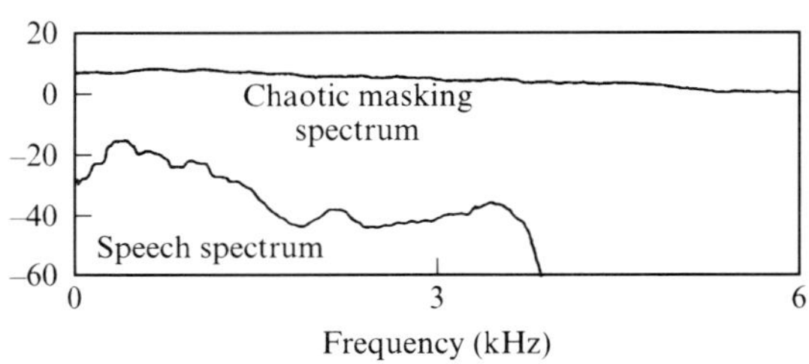

Frequency (kHz)

Figure 7 Circuit data: power spectra of chaotic masking and speech signals.

overlapping with an average signal-to-masking ratio of approximately −20 dB. Figures 6(a) and 6(b) show the original speech, $m(t)$, and the recovered speech signal, $\hat{m}(t)$, respectively. Clearly, the speech signal has been recovered and in informal listening tests is of reasonable quality.

We thank S. Isabelle and S. Strogatz for helpful discussions. This work was sponsored in part by the Air Force Office of Scientific Research under Grant No. AFOSR-91-0034-A, in part by a subcontract from Lockheed Sanders, Inc., under ONR Contract No. N00014-91-C-0125, and in part by the Defense Advanced Research Projects Agency monitored by the Office of Naval Research under Grant No. N00014-89-J-1489. K.M.C. is supported in part through the MIT/Lincoln Laboratory Staff Associate Program.

References

[1] L.M. Pecora and T.L. Carroll, *Phys. Rev. Lett.* **64**, 821 (1990).

[2] T.L. Carroll and L.M. Pecora, *IEEE Trans. Circuits Syst.* **38**, 453 (1991).

[3] L.M. Pecora and T.L. Carroll, *Phys. Rev. A* **44**, 2374 (1991).

[4] K.M. Cuomo, A.V. Oppenheim, and S.H. Isabelle, MIT Research Laboratory of Electronics TR No. 570, 1992 (unpublished).

[5] A.V. Oppenheim, G.W. Wornell, S.H. Isabelle, and K.M. Cuomo, in *Proceedings of the IEEE International Conference on Acoustics, Speech, and Signal Processing*, 1992 (to be published).

[6] K.M. Cuomo and A.V. Oppenheim, MIT Research Laboratory of Electronics TR No. 575, 1992 (unpublished).

[7] E.N. Lorenz, *J. Atmos. Sci.* **20**, 130 (1963).

[8] R. He and P.G. Vaidya, *Phys. Rev. A* **46**, 7387 (1992).

[9] U. Parlitz, L. Chua, Lj. Kocarev, K. Halle, and A. Shang, *Int. J. Bif. Chaos* **2**, 973 (1992).

[10] Lj. Kocarev, K. Halle, K. Eckert, and L. Chua, *Int. J. Bif. Chaos* **2**, 709 (1992).

Paper 12
Extracting messages masked by chaos

Gabriel Pérez[1] and Hilda A. Cerdeiral[2]

[1]*Departamento de Física Aplicada, Centro de Investigación y de Estudios Avanzados del Instituto Politécnico Nacional, Unidad Mérida, A.P. 73 "Cordemex:, 97310 Mérida, Yucatán, México*
[2]*International Centre for Theoretical Physics, P.O. Box 586, 34100 Trieste, Italy*

Received 11 May 1994

We show how to extract messages that are masked by a chaotic signal in a system of two Lorenz oscillators. This mask removal is done for two different modes of transmission: a digital one, where a parameter of the sender is switched between two values, and an analog mode, where a small amplitude message is added to the carrier signal. We achieve this without using a second Lorenz oscillator as a receiver, and without doing a full reconstruction of the dynamics. This method is robust with respect to transformations that impede the unmasking using a Lorenz receiver, and is not affected by the broadband noise that is inherent to the synchronization process. We also discuss the limitations of this way of extraction for messages in high frequency bands.

There has been some recent interest in the idea of using enslaving chaotic variables [1] as a way of transmitting information [2–7]. The principle used here is that if we have two identical nonlinear low-dimensional dynamical systems, where one of the variables from the first system enslaves the second, this chaotic variable can be used as a carrier for a message. The use of these chaotic carriers in a communication channel is intended, among other reasons, for security [8]. The actual transmitted signal is broadbanded and should look at first sight like some type of noise. It is also expected that, since the carrier is able to synchronize only identical dynamical systems, i.e., identical sets of equations [9] with identical – or at least extremely close – parameters, any eavesdropper will be lost in the infinite maze of possible dynamical models and parameter sets and will not be able to extract the message.

Our purpose in this Letter is to show that this type of masking can be easily removed, at least in some of the proposed implementations [3,4], and that this can be achieved without resorting to a nonlinear receiving system. The origin of these weaknesses in the masking is that efficient message reconstruction requires the existence of a low-dimensional attractor and a fast relaxation of the dynamics to that attractor, at least in the time scales used in the message. This allows a third party to do a *partial* reconstruction of the dynamics, using some return maps. By analyzing the evolution of the signal on the attracting sets of those maps, the message can be extracted. This process does not use at any moment the full reconstruction of the continuous dynamics of the sender [10], a more time-consuming procedure that requires embedding in a space of larger dimensionality than that of the intended receiver.

To show how this unmasking is done, we will use computer simulations of the sender–receiver circuits used

in Ref. [3]. These circuits are built so that their dynamics constitutes a scaled implementation of the Lorenz equations [11]. The equations for the sender are

$$\frac{dx_1}{d\tau} = \sigma(y_1 - x_1), \tag{1}$$

$$\frac{dy_1}{d\tau} = rx_1 - y_1 - x_1z_1, \tag{2}$$

$$\frac{dz_1}{d\tau} = x_1y_1 - bz_1, \tag{3}$$

while for the receiver we have

$$\frac{dx_2}{d\tau} = \sigma(y_2 - x_2), \tag{4}$$

$$\frac{dy_2}{d\tau} = rx_1 - y_2 - x_1z_2, \tag{5}$$

$$\frac{dz_2}{d\tau} = x_2y_2 - bz_2, \tag{6}$$

Here $\tau = Kt$, where K is an overall scale factor. The values used here are those from Ref. [3]: $\sigma = 16.0$, $r = 45.6$, and $b = 4.0$. The scale factor has been set to $K = 1/2505$ [12], so that the time scale agrees with the one used in the figures of the cited reference.

In the second set of equations, the use of x_1 instead of x_2 in Eqs. (5) and (6) has the effect of enslaving the second oscillator to the first. This means that if we start the two oscillators from different initial conditions, but using in both the same set of parameters, the variables in the receiver will soon approach the values of those of the sender. The equation for x_2 in the receiver serves as a check of this enslavement, since the process makes x_2 approach x_1.

The actual transmission of data is implemented in one of

two ways. The first one, digital, changes the parameter b in the equations of the sender between its reference value $b = 4.0$ and a shifted one of $b = 4.4$. The corresponding parameter in the receiver is kept fixed at the reference value, giving as a result that the oscillators synchronize when the parameters are equal and are frustrated when they are different. This frustration is manifested in the squared difference $(x_2 - x_1)^2$, which presents persistent fluctuations when the two b's are different. Examples of how this process works are given in Refs. [3,5].

The second option for transmission of messages is to add a small-amplitude analog message $m(t)$ to the variable $x_1(t)$ we get from the sender, producing a modified drive $s(t) = x_1(t) + m(t)$. This new drive is fed to the receiver. A synchronization of sorts is achieved, but it is far from perfect, since the incoming signal is not exactly a variable of the Lorenz system. However, the frustration in the synchronization process can still be used as a way of recovering the message, using for this purpose the difference $s(t) - x_2(t)$. This process, however, is nontrivial. The difference between drive and response does not exactly reproduce the added message $m(t)$, and the correlation between the two is strongly dependent on the frequencies involved. This happens because, as can easily be seen in any numerical simulation, the error in the synchronization process decays in its own time scale, and not monotonically. Given initially unsynchronized oscillators, for the parameters we are using here, the difference $x_2 - x_1$ decays in a time scale of roughly 10^{-4} s, and presents oscillations with a broad spectrum of frequencies, with a peak around $f_0 \approx 3$ kHz.

This synchronization delay affects the quality of the recovered message. For frequencies comparable with those predominant in the synchronization noise, the reconstructed message gets a large admixture of noise and does not reproduce the message well. For smaller frequencies one finds that the reconstructed message contains a broad spectrum of frequencies above the transmitted one, and that the output approximately reproduces the input message only if we perform a low pass filtering.

For very high frequencies, well above f_0, a different phenomenon occurs. The period of the message is much smaller than the decay time of the synchronization process, and the message and the synchronization–frustration mechanism decouple. This makes the recovered message practically identical to the original one. However, this decoupling happens at such high frequencies that a careful eavesdropper may be able to notice some peaks in an otherwise broad spectrum, this being so because the power spectrum of the Lorenz oscillator is quite low at high frequencies.

Our approach to this problem comes from the discovery by Lorenz [11] that, by following just one of the variables in the set of Equations (1)–(3), one can produce a return map where the dynamics is attracted to an almost 1D set. Following this lead, we constructed the following return map from the $x(t)$ variable in the Lorenz oscillator. Starting from some arbitrary point in time, define t_n as the time when $x(t)$ reaches its nth (local) maximum, and X_n as the value of x at that moment. Similarly, define another return map by setting u_m as the time when $x(t)$ reaches its mth local

minimum, and Y_m as the value of x at that moment. Using these discrete values we can construct the return maps X_{n+1} vs X_n and Y_{m+1} vs Y_m. These two maps have attractors that look almost 1D. Under the transformation $Y \to -Y$ the attractor for the Y map is identical to that of the X map. This is due to the fact that the underlying dynamics is invariant under the transformation $x \to -x$, $y \to -y$, $z \to z$, and therefore the maxima of $x(t)$ and the minima of $-x(t)$ give the same return map.

We will not use these two return maps directly; after some experimentation we have found that we get better results using the linear combinations $A_n = (X_n + Y_n)/2$, $B_n = X_n - Y_n$, $C_n = (X_{n+1} + Y_n)/2$, and $D_n = Y_n - X_{n+1}$. These are simply the average values of a consecutive maximum–minimum pair, and the distance between them. The return maps A_n vs B_n and C_n vs D_n have very simple attractors. Each is given by three smooth almost 1D unconnected segments, and they have the same inversion symmetry as for the X and Y maps, so that the A vs B section is identical to the $-C$ vs $-D$ section. These are shown in Figure 1.

The key to extracting messages from the chaotic mask, in this digital mode, is to recognize that a small change in the parameters of the sender not only frustrates the synchronization but also affects the attractor obtained in the return map (here we will just superimpose the A vs B and $-C$ vs $-D$ return maps). Since the change in parameters is small, the only effect is a shift in the position of the segments of the attractor, while its general form is conserved. Therefore the attractor obtained when there is a message shows splitting, with two close parallel branches appearing where only one segment was found for the unperturbed Lorenz oscillator (see Figure 2).

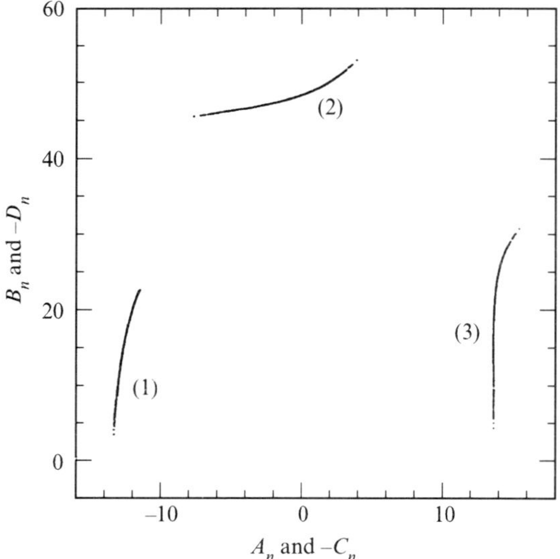

Figure 1 Attractors of the return maps obtained from the maxima and minima of $x(t)$ in the Lorenz oscillator. Here we have superimposed the attractors for the maps A_n vs B_n and $-C_n$ vs $-D_n$, whose definitions are given in the text. The three segments of the attractor are labeled for later convenience.

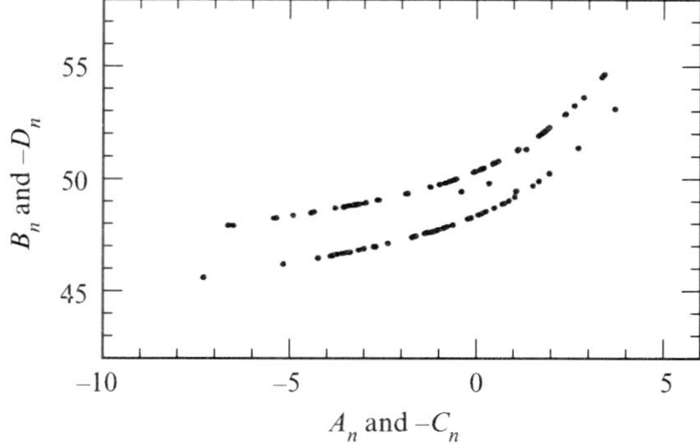

Figure 2 Segment 2 of the attractor of the return map, split by the use of both $b = 4$ and $b = 4.4$ in the generation of the signal. The other two segments of the attractor undergo similar splitting.

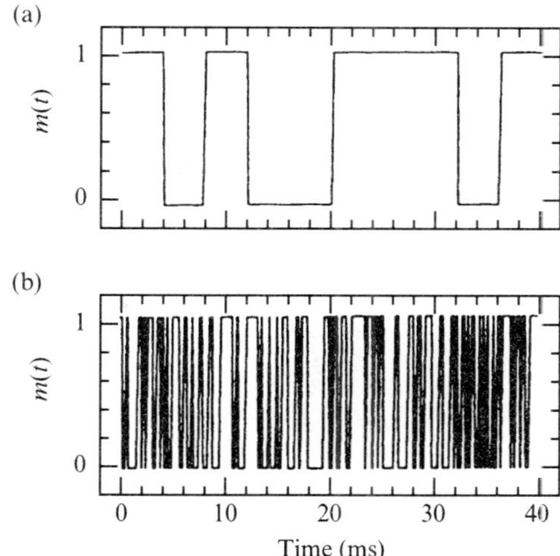

Figure 3 Unmasking of a digital transmission, using the return map. In (a) we show the results obtained with the correct assignment of 0's and 1's to segments 2 and 3 of the attractor. In (b) we show the results of a typical wrong assignment. The correct unmasking gives results identical to the original message, which was the word '1010011101'. Each bit is 4 msec long.

Once we have realized that the switching between the two parameters means also switching between the two parallel branches of the attractor, it is a simple task to go back to the return map and start classifying the points according to which branch of the attractor they fall in. We use only the points that are clearly separated, and assign a 0 or 1 value to each branch of a split segment. Then we read in the time sequences the values t_n and u_n and plot the assigned value vs time. The result will probably be meaningless, since we have done the assignment of 0's and 1's in an arbitrary way. We need to try different assignments and compare the results. The correct one will be that which shows always long sequences of only 0's or only 1's. This is so because the bits of the original message have to be long enough (in time) to overcome the synchronization lag, and that gives $x_1(t)$ in the sender enough time to run through several maxima and minima. Here we have three split segments, which means that we need to try four possible assignments of zeros and ones. (The assignment for the first segment is arbitrary.) In Figure 3 we show the messages extracted using the correct assignment and a typical wrong one.

It is clear from these results that this simple algorithm permits the reconstruction of the message, except for the small ambiguity of deciding which bits are identified as 1's and which as 0's. As a bonus, the out scheme is robust with respect to transformations that affect the mask removal using a Lorenz oscillator, something that we should expect from the amplification of the signal needed for long-distance transmissions. In particular, the simple affine modification $x_1'(t) = ax_1(t) + b$, with a and b constants, is ignored by our scheme, except for an unimportant breaking of the symmetry between the A vs B and C vs D maps. On the other hand, it can completely spoil the synchroniz-ation–frustration process needed for transmission between Lorenz systems.

For the analog mode, the separation of a signal into a small-amplitude message and a carrier is not much more difficult than the extraction of digital messages done before, at least for low frequencies. We use the same principle, i.e., the fact that perturbations of any kind on the carrier signal affect the (quasi) 1D attractors of the return map. For the analog mode of transmission the effect of adding the message $s(t)$ to the carrier $x_1(t)$ is to smear the attractor, turning its three segments into three diffuse stripes. If we superimpose the original 'silent' attractor – which should become apparent in an actual transmission as a denser line that forms during any silences the message happens to have – to these stripes, we find that the broadening is almost symmetric, with equal spreads at both sides.

What we do in order to recover the message from the return maps is measure the distance between the present position of the points in the attractor and the place they should have appeared in the absence of a message, i.e., in the silent attractor, taking into account to which side of it the point has moved. This distance can be reasonably approximated by the closest distance to the silent attractor. In our case, stripes 1 and 3 of the attractor are almost vertical, and stripe 2 is approximately horizontal. Therefore it is enough to take the x distance to the silent attractor for stripes 1 and 3, and the y distance for stripe 2. Once this is done, we need to assign an amplitude factor (including sign) to segments 2 and 3 (segment 1 is assigned $+ 1.0$ by default), and do some trial and error adjustments – mostly for the signs – on these two amplitudes in order to get a meaningful output.

The results of this procedure are quite satisfactory for frequencies below a cutoff $f_c \approx 3000$ Hz, which is the value where the power spectrum for the synchronization noise peaks. For these low frequencies, the carrier has several maxima and minima for each period of the message, allowing for a good reconstruction, with some redundancy to spare, and little noise (see Figure 4). The quality of the recovered message deteriorates as we go through f_c, and for

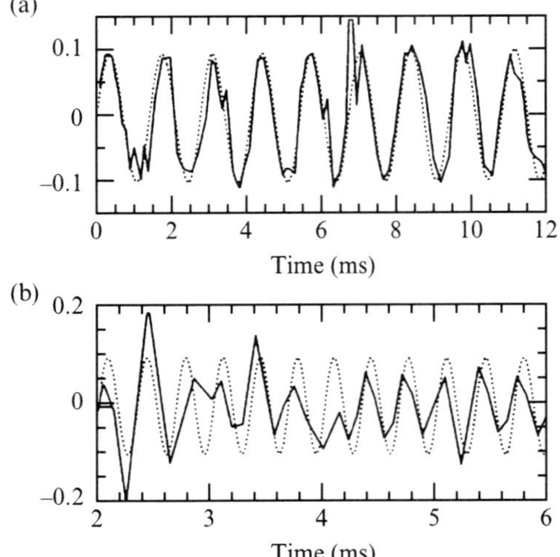

Figure 4 Unmasking of analog messages using the return map. The dotted lines are the original message $m(t) = 0.1$ $\sin(2\pi f t)$, and the solid lines are the reconstructed messages. In (a) we have $f = 750$ Hz, and in (b) $f = 3$ kHz.

frequencies larger than this cutoff the results are very poor. Very often, the output misses completely the oscillations in the message. For frequencies well above the cutoff the return map does not work, since it samples the signal at a rate too low compared to the message's period.

In conclusion, we have shown that it is possible for a diligent eavesdropper to uncover messages transmitted using a Lorenz–Lorenz chaotic pair. The digital mode of transmission is easily and efficiently unmasked, even allowing for signals that have been corrupted, say, by an imperfect amplification process. Given the sharp separation of the two branches of the attractor, this scheme will work even under small noise conditions.

For the masking of analog messages the mask removal will work for frequencies up to a cutoff given roughly by the peak frequency of the noisy synchronization spectrum. For frequencies close to this cutoff both extraction schemes give poor results, while for frequencies much higher than the cutoff our approach to mask removal does not work.

Since the main limiting element in our unmasking algorithm is the existence of a cutoff frequency, it seems that all that is needed to defeat the eavesdropper is to transmit only in the high frequency bands. In this case, however, sender and receiver will be working in a sector where the power spectrum of the Lorenz oscillator is quite low, and it may not be enough to mask the signal.

What we have done here for Lorenz oscillators should also work for transmissions done using other synchronized chaotic systems. In general, dynamics where just one variable is enough to enslave a set of differential equations will have only one positive Lyapunov exponent. Other exponents being negative enough as to ensure fast synchronization, the attractors in 2D return maps will be (quasi) 1D sets, and will necessarily show the effects of any perturbation over the carrier. These effects occasionally may not be clean enough to allow unmasking, but there is always a chance that a good choice of return maps will spoil security for chaotic transmissions.

G.P. wants to thank the International Centre for Theoretical Physics, where this work was started, for their hospitality.

Notes

[1] L.M. Pecora and T.L. Carroll, *Phys. Rev. Lett.* **64**, 821 (1990); Phys. Rev. A **44**, 2374 (1991); T.L. Carroll and L.M. Pecora, *IEEE Trans. Circuits Sys.* **38**, 453 (1991).

[2] A.V. Oppenheim, G.W. Wornell, S.H. Isabelle, and K. Cuomo, in *Proceedings of the IEEE International Conference on Acoustics, Speech, and Signal Processing ICASSP-92* (IEEE, New York, 1992), p. 117; K. Cuomo and A.V. Oppenheim, in *Proceedings of the IEEE International Conference on Acoustics, Speech, and Signal Processing ICASSP-93* (IEEE, New York, 1993).

[3] K. Cuomo and A.V. Oppenheim, *Phys. Rev. Lett.* **71**, 65 (1993)

[4] K. Murali and M. Lakshmanan, *Phys. Rev. E* **48**, R1624 (1993); *Int. J. Bifurcations Chaos* **3**, 1057 (1993).

[5] U. Parliz *et al.*, *Int. J. Bifurcations Chaos* **2**, 973 (1992).

[6] S. Hayes, C. Grebogi, and E. Ott, *Phys. Rev. Lett.* **70**, 3031 (1993).

[7] K.S. Halle *et al.*, *Int. J. Bifurcations Chaos* **3**, 469 (1993).

[8] Comments on the use of these systems for secure communications have appeared in many sources. See, for instance, I. Amato, *Science* **261**, 429 (1993), J.C.G. Lesurf, *Nature (London)* **365**, 604 (1993).

[9] It was found recently that the slave system does not have to be identical to the master. The evolution equations for the slave may contain extra terms, as long as they go to zero when synchronization is achieved. See M. Ding and E. Ott, *Phys. Rev. E* **49**, R945 (1994).

[10] N.H. Packard, J.P. Crutchfield, J.D. Farmer, and R.S. Shaw, *Phys. Rev. Lett.* **45**, 712 (1980); F. Takens, in *Dynamical System and Turbulence*, edited by D.A. Rand and L.S. Young (Springer, Heidelberg, 1981), p. 366; P. Grassberger *et al.*, *Chaos* **3**, 127 (1993).

[11] E. Lorenz, *J. Atmos. Sci.* **20**, 130 (1963).

[12] K. Cuomo (private communication).

Paper 13
General approach for chaotic synchronization with applications to communication

L. Kocarev[1] and U. Parlitz[2]

[1]*Department of Electrical Engineering, St. Cyril and Methodius University, Skopje, P.O. Box 574, Republic of Macedonia*
[2]*Drittes Physikalisches Institut, Universität Göttingen, Bürgerstrasse 42–44, D-37073 Göttingen, Germany*

Received 5 July 1994; revised manuscript received 18 November 1994

A general approach for constructing chaotic synchronized dynamical systems is discussed that is based on a decomposition of given systems into active and passive parts. As a possible application we consider an improved encoding method where the information signal is injected into the dynamical system of the transmitter. Furthermore, we show how to design in a systematic way high-dimensional synchronized systems that may be used for efficient hyperchaotic encoding of information.

Synchronization of periodic signals is a well-known phenomenon in physics, engineering, and many other scientific disciplines. Recently, synchronization of chaos [1] has aroused much interest in light of its potential applications. In particular, the use of chaotic synchronization in communication systems has been investigated by several authors [2–11]. There, an information signal containing a message is transmitted using a chaotic signal as a broadband carrier, and the synchronization is necessary to recover the information at the receiver. Different implementations of this basic idea have been suggested. For example, in Refs. [2–5] the information signal is added to the chaotic signal and in Refs. [2,6] a parametric modulation is used for the transmission of digital signals. Other approaches to use chaos for the purpose of communication include controlling techniques to encode binary information [12] and methods that make use of the quick decay of the correlation function for chaotic signals [13].

In this Letter we discuss a new approach [8–11] for constructing (chaotic) synchronized systems that may be viewed as a generalization of the method introduced by Pecora and Carroll [1]. This approach and two examples for illustration are presented in the first part of the Letter. In the second part we apply a new method for encoding messages using chaotic dynamics [8–11]. In contrast to most of the schemes proposed in the literature until now we consider cases where the information signal drives the dynamical system that is used in the transmitter. The scalar signal which is transmitted from the transmitter to the receiver is a function of the transmitter state variables and the information signal. If the receiver synchronizes with the transmitter, the information signal can be recovered exactly, i.e., without the reconstruction error that typically occurs with other encoding methods based on synchronization [2,3]. Furthermore, it turns out that this modulation technique not only yields a transmission without errors but also a more secure encoding. Finally, in the third part of the Letter we show that the new synchronization method can be used to construct systematically high-dimensional synchronized systems using low-dimensional systems as building blocks. This possibility is, for example, very useful for the design of communication systems that are based on hyperchaotic signals.

The new synchronization method is based on the fact that it is possible to consider more general decompositions of a given dynamical system.

$$\dot{\mathbf{z}} = \mathbf{F}(\mathbf{z}) \tag{1}$$

than the decomposition into subsystems proposed by Pecora and Carroll [1]. Starting from a chaotic autonomous system, for example, one can always formally rewrite it in different ways as a nonautonomous system

$$\dot{\mathbf{x}} = \mathbf{f}(\mathbf{x}, s(t)) \tag{2}$$

with some driving $s(t) = h(\mathbf{x})$ or $\dot{s} = h(\mathbf{x}, s)$. Let

$$\dot{\mathbf{y}} = \mathbf{f}(\mathbf{y}, s(t)) \tag{3}$$

be a copy of the nonautonomous system that is driven by the same signal $s(t)$. If the differential equation for the difference $\mathbf{e} = \mathbf{x} - \mathbf{y}$,

$$\dot{\mathbf{e}} = \mathbf{f}(\mathbf{x}, s) - \mathbf{f}(\mathbf{y}, s) = \mathbf{f}(\mathbf{x}, s) - \mathbf{f}(\mathbf{x} - \mathbf{e}, s), \tag{4}$$

possesses a stable fixed point at $\mathbf{e} = \mathbf{0}$ then there exists for the systems (2) and (3) a synchronized state $\mathbf{x} = \mathbf{y}$ that is stable. This can be proved using stability analysis of the linearized system for small $\mathbf{e}$ or using (global) Lyapunov functions. In general, however, the stability has to be checked numerically using the fact that synchronization occurs if all conditional Lyapunov exponents [1] of the nonautonomous system (2) are negative. In this case system (2) is a passive system that tends to a fixed point when not driven. Therefore, we call the decomposition given by h and $\mathbf{f}$ an *active–passive decomposition* (*APD*) of the original dynamical system (1). The fact that all conditional Lyapunov exponents of (2) are negative does not exclude chaotic solutions.

To illustrate this synchronization scheme we consider different active–passive decompositions of the well-known Lorenz model. In the first example we choose

$$\dot{x}_1 = -10x_1 + s(t),$$
$$\dot{x}_2 = 28x_1 - x_2 - x_1x_3, \tag{5}$$
$$\dot{x}_3 = x_1x_2 - 2.666x_3,$$

with

$$s(t) = h(\mathbf{x}) = 10x_2.$$

To estimate the temporal evolution of the difference $\mathbf{e} = \mathbf{x} - \mathbf{y}$ of the states of the two systems (2) and (3) we note first that the difference $e_1 = x_1 - y_1$ of the first components converges to zero, because $\dot{e}_1 = -10e_1$. Therefore, the remaining two-dimensional system describing the evolution of the differences $e_2 = x_2 - y_2$ and $e_3 = x_3 - y_3$ can for the limit $t \to \infty$ be written as

$$\dot{e}_2 = -e_2 - x_1e_3,$$
$$\dot{e}_3 = x_1e_2 - 2.666e_3,$$

Using the Lyapunov function $L = e_2^2 + e_3^2$ one can show that $\dot{L} = -2(e_2^2 + 2.666e_3^2) < 0$. This means, that the synchronization is globally stable and occurs for all types of driving signals $s(t)$. The conditional Lyapunov exponents of this decomposition are given by $\lambda_1 = -1.805$, $\lambda_2 = -1.861$, and $\lambda_3 = -10$ with respect to the natural logarithm.

Our second example is essentially a decomposition of the Lorenz system into subsystems as suggested by Pecora and Carroll [1]:

$$\dot{x}_1 = 28s(t) - x_1 - s(t)x_2,$$
$$\dot{x}_2 = s(t)x_1 - 2.666x_2, \tag{6}$$

where the function $s(t)$ is now given by the additional differential equation

$$\dot{s} = h(\mathbf{x}, s) = 10(x_1 - s).$$

Using similar arguments as for the first example (5) it is easy to see that the difference $\mathbf{e} = \mathbf{x} - \mathbf{y}$ converges to $\mathbf{0}$ for all $s(t)$ [14]. The conditional Lyapunov exponents are $\lambda_1 = -1.796$ and $\lambda_2 = -1.870$.

This example shows that the Pecora–Carroll method for constructing synchronized systems is included in the more general APD approach. However, our numerical simulations [11] indicate that using the APD any typical chaotic system can be used to implement synchronization using a one-dimensional drive in many ways. The freedom to choose the function h that defines the driving signal therefore leads to a large flexibility in applications. This is different from the synchronization method proposed by Pecora and Carroll [1] where only a finite number of possible couplings exists, that is given by the number of (stable) subsystems of the dynamical system.

What makes the above introduced method for constructing synchronized systems interesting for applications is the fact that in many cases the function $s(t)$ can be rather general. In particular, it may depend not only on the state $\mathbf{x}$ but also on some information signal $i(t)$, i.e., $s = h(\mathbf{x}, i)$ or $\dot{s} = h(\mathbf{x}, s, i)$. This feature can, for example, be used in a communication scheme where $\dot{\mathbf{x}} = \mathbf{f}(\mathbf{x}, s)$ is the (chaotic) dynamical system of the transmitter, $s = h(\mathbf{x}, i)$ is the transmitted and received signal, and $\dot{\mathbf{y}} = \mathbf{f}(\mathbf{y}, s)$ constitutes the receiver. In the case of synchronization, i.e., when $\mathbf{y} \to \mathbf{x}$, the information i can be recovered without error from $s = h(\mathbf{x}, i) = h(\mathbf{y}, i)$ if this equation is uniquely solvable for i.

This encoding method has been implemented experimentally using an APD of Chua's circuit [15]. The transmitter and the receiver are given by two copies of the following system:

$$C_1\frac{dV_{C1}}{dt} = G(V_{C2} - V_{C1}) - g(V_{C1}) - G\alpha(s - V_{C1}),$$
$$C_2\frac{dV_{C2}}{dt} = G(V_{C1} - V_{C2}) + i_L, \tag{7}$$
$$L\frac{di_L}{dt} = V_{C2}.$$

The transmitted signal $s = V_{C1} + i$ is given by the voltage V_{C1} at the capacitor C_1 of the transmitter and the information signal i. The parameters are $C_1 = 10$ nF, $C_2 = 100$ nF, $L = 18$ mH, $G = 1/1700\,\Omega$, $\alpha = 3.4$ and g is a piecewise-linear function defined by $g(V) = m_0V + \frac{1}{2}(m_1 - m_0)\big[|V + B_p| - |V - B_p|\big]$ with $m_0 = -0.409$ mS, $m_1 = -0.756$ mS, and $B_p = 1.08$ V. The circuit diagram and other details of this implementation will be given elsewhere [16]. Figure 1 shows as an example the reconstruction of a triangular information signal $i(t)$. Although the parameters of the two coupled circuits are not exactly the same, the quality of the reconstructed signal (Fig. 1(b)) is already quite good. In general this sensitivity of the synchronization on parameter differences may cause difficulties for typical hardware implementations. On the other hand, however, it is a feature that is very welcome for any private communication. This problem and the influence of additional noise in the transmission channel will be discussed in more detail elsewhere.

Of course, the APDs of the Lorenz system introduced above may also be used for encoding. System (5), for example, can be driven by $s = 10x_2 + ix_3$. Note that the above given proof for the synchronization of this system ($\mathbf{e} \to 0$) holds for all driving signals s. Therefore, $\mathbf{y}$ converges to $\mathbf{x}$ for all information signals i (that, of course, have an effect like dynamic noise and thus modify the underlying attractor). Since the variable x_3 is always positive, the information signal can be recovered exactly as $i_R = (s - 10y_2)/y_3$ as soon as the transient of the synchronization process is over. For the second system (6) the transmitted signal s may be generated by the differential equation $\dot{s} = 10(x_1 - s) + i$. To retrieve the information in the receiver as $i_R = \dot{s} - 10(y_1 - s)$ in this case the derivative $\dot{s}$ has to be computed from s. Of course, other functions h (including linear filters) could also be used to generate the transmitted signal s. The only restriction for the signal s is that it has to be chosen in a way that the transmitter and the receiver remain stable (and chaotic).

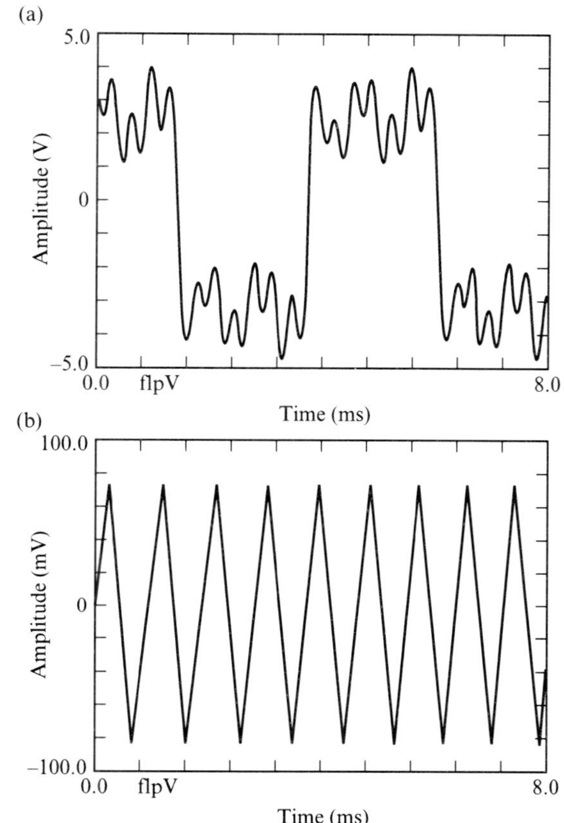

Figure 1 Experimental implementation of a communication scheme using Chua's circuit (7) with driving $s = V_{C1} + i$ and a triangular information signal i. (a) Transmitted signal s. (b) Recovered signal i_R.

The last example shows that the Pecora–Carroll synchronization method can also be used in a more sophisticated way for communication than it was done until now. The communication scheme presented in Refs. [2, 3] is based on Pecora–Carroll synchronization and cascaded subsystems in the receiver that are necessary to recover the information signal. Compared to the method discussed in this Letter the main difference consists in the fact that in Refs. [2, 3] the information is just added to a chaotic carrier but not injected into the dynamical system constituting the transmitter. In this case, the receiver is forced by a sum of the chaotic signal *and* the information signal whereas the transmitter is just driven by the (pure) chaotic signal. Because of this (slightly) different driving **y** does not converge exactly to **x** and the information signal can only be recovered including an error that vanishes in the limit $|i| \to 0$. The specific properties of this inevitable error are investigated in Refs. [4, 5]. If one uses an information signal with a small amplitude $|i|$ in order to minimize this error, however, the information can be destroyed by noise in the transmission channel. Furthermore, this method is not very secure, because it is possible to fit a nonlinear model to the time series given by the transmitted signal s that consists

mainly of the (low-dimensional) chaotic carrier ($|i|$ has to be small to avoid errors). Using this nonlinear model the information signal may then be extracted from s by methods similar to nonlinear noise reduction [17] or other techniques [18]. In contrast, even if we use for our communication scheme just a linear combination like, for example, $s = x_2 + i$, it is difficult to decode the information from the transmitted signal s since the dynamical system of the transmitter is *not autonomous but driven by the information signal* which is in general rather complicated.

The communication method discussed above can also be implemented using high-dimensional systems and transmitted signals that are hyperchaotic. In this case, however, it is in general difficult to find a region in parameter space where hyperchaos exists. Therefore, we propose another method for synthesizing high-dimensional systems in a systematic way, using standard low-dimensional systems with a well-known dynamics. only. For the sake of brevity, only the procedure for designing a six-dimensional transmitter using two three-dimensional systems as building blocks is described in the following. The equations of the transmitter, the transmitted signal, and the receiver are in this case given by

$$\left\{\begin{array}{l} \dot{\mathbf{x}}_1 = \mathbf{f}_1(\mathbf{x}_1, s_{\text{aux}}(t)) \\ s_{\text{aux}} = h_1(\mathbf{x}_1, i(t)) \\ \dot{\mathbf{x}}_2 = \mathbf{f}_2(\mathbf{x}_2, s(t)) \end{array}\right\} \quad \text{transmitter,}$$

$$s = h_2(\mathbf{x}_2, s_{\text{aux}}) \quad \text{transmitted signal,}$$

$$\left\{\begin{array}{l} \dot{\mathbf{y}}_2 = \mathbf{f}_2(\mathbf{y}_2, s(t)) \\ \tilde{s}_{\text{aux}} = h_2^{-1}(\mathbf{y}_2, s(t)) \\ \dot{\mathbf{y}}_1 = \mathbf{f}_1(\mathbf{y}_1, \tilde{s}_{\text{aux}}(t)) \end{array}\right\} \quad \text{receiver,}$$

where we assume that h_1 and h_2 are invertible with respect to i and s_{aux}, respectively. If both pairs of subsystems $(\mathbf{x}_1, \mathbf{y}_1)$ and $(\mathbf{x}_2, \mathbf{y}_2)$ synchronize mutually then at the receiver the information i_R can be recovered as

$$i_R = h_1^{-1}(\mathbf{y}_1, \tilde{s}_{\text{aux}}) = h_1^{-1}(\mathbf{y}_1, h_2^{-1}(\mathbf{y}_2, s)).$$

The generalization to a communication model with an arbitrary number of low-dimensional subsystems is straightforward [11]. The two low-dynamical systems we use in the following numerical example are the Rössler system and the Lorenz system. The transmitter of our communication model is given by

$$\dot{x}_1 = 2 + x_1(x_2 - 4), \qquad \dot{x}_4 = -10x_4 + s,$$
$$\dot{x}_2 = -x_1 - x_3, \qquad \dot{x}_5 = 28x_4 - x_5 - x_4x_6,$$
$$\dot{x}_3 = x_2 - 2.45x_3 + s_{\text{aux}}, \qquad \dot{x}_6 = x_4x_5 - 2.666x_6,$$
$$s_{\text{aux}} = i + 3x_3, \qquad s = 10x_5 + 30s_{\text{aux}}/x_6,$$

where s is the transmitted signal. The receiver reads

$$\dot{y}_4 = -10y_4 + s, \qquad \dot{y}_1 = 2 + y_1(y_2 - 4),$$
$$\dot{y}_5 = 28y_4 - y_5 - y_4y_6, \qquad \dot{y}_2 = -y_1 - y_3,$$
$$\dot{y}_6 = y_4y_5 - 2.666y_6, \qquad \dot{y}_3 = y_2 - 2.45y_3 + \tilde{s}_{\text{aux}},$$
$$\tilde{s}_{\text{aux}} = (s - 10y_5)y_6/30, \qquad i_R = (s - 10y_5)y_6/30 - 3y_3.$$

The parameters of the Rössler and the Lorenz equations are chosen such that both systems have a chaotic attractor. The information signal shown in Figure 2(a) is the spoken word '42' recorded with a sampling rate of 8000 Hz and a resolution of 16 bit. Figure 2(b) shows the transmitted signal s. The transmitter and receiver synchronize, and the difference $|i - i_R|$ between the original information signal i and the reconstructed signal i_R is given in Figure 2(c). The transmitted signal is in this case a sum of two chaotic signals and hyperchaotic. The Lyapunov exponents of the transmitter for $i = 0$ equal $\lambda_1 = 0.710$, $\lambda_2 = 0.177$, $\lambda_3 = 0.000$, $\lambda_4 = -0.006$, $\lambda_5 = -2.59$, and $\lambda_6 = -14.37$. This and other examples of high-dimensional systems with more than one positive Lyapunov exponent [11] show that it is possible to synchronize hyperchaotic systems using a scalar signal, only.

In conclusion, we have presented a general approach to construct a large class of synchronized dynamical systems and discussed its application in a communication scheme where the information can be recovered without errors. Furthermore, we introduced a procedure for synthesizing high-dimensional synchronized systems. In this case the transmitted signal is hyperchaotic and therefore yields a more efficient encoding.

The authors thank Professor W. Lauterborn, M. Wiesenfeldt, R. Mettin, and A. Pikovsky for stimulating discussions, and G. Kirschmann-Schröder, K. Lautscham, and H. Hohmann for support with the photographs and the electronic circuit. L.K. thanks the DAAD for a research grant. This work was also supported in part by the Macedonian Ministery of Science and the DFG (SFB 185).

(a)

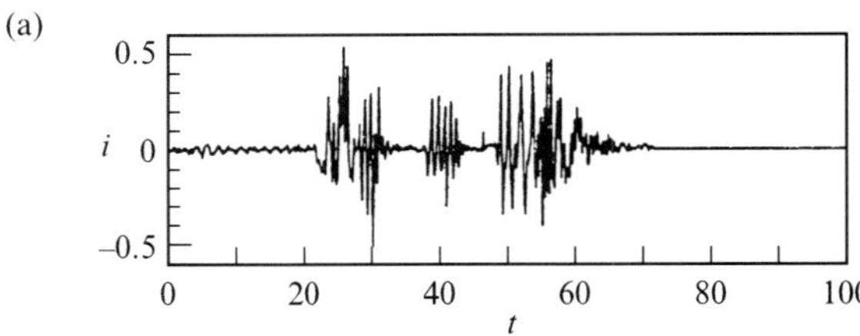

(b)

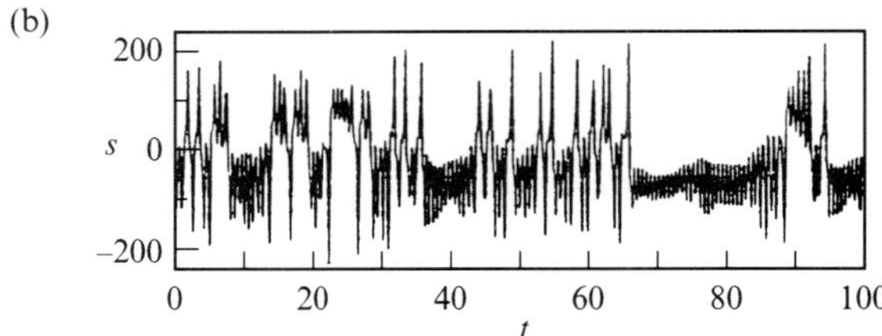

(c)

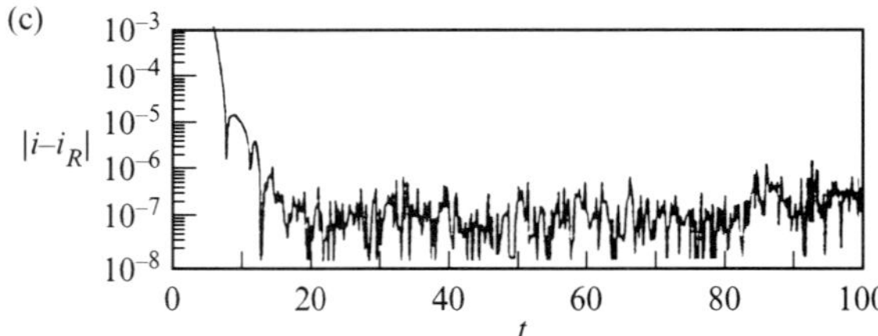

Figure 2 Numerical simulation of a communication scheme using a combination of Rössler and Lorenz systems. (a) Information signal i given by the spoken word '42'. (b) Transmitted signal s. (c) Difference $|i - i_R|$ between the original and the recovered information signal.

References

[1] L. Pecora and T. Carroll, *Phys. Rev. Lett.* **64**, 821–823 (1990); T.L. Carroll and L.M. Pecora, *IEEE Trans. Circuits Syst.* **38**, 453–456 (1991).

[2] K.M. Cuomo and A.V. Oppenheim, *Phys. Rev. Lett.* **71**, 65–68 (1993).

[3] L. Kocarev, K.S. Halle, K. Eckert, L.O. Chua, and U. Parlitz, *Int. J. Bifurcation Chaos* **2**, 709–713 (1992).

[4] R. Lozi and L.O. Chua, *Int. J. Bifurcation Chaos* **3**, 1319–1325 (1993).

[5] K.M. Cuomo, A.V. Oppenheim, and S.H. Strogatz, *Int. J. Bifurcation Chaos* **3**, 1629–1638 (1993).

[6] U. Parlitz, L.O. Chua, L. Kocarev, K.S. Halle, and A. Shang. *Int. J. Bifurcation Chaos* **2**, 973–977 (1992).

[7] K.S. Halle, C.W. Wu. M. Itoh. and L.O. Chua, *Int. J. Bifurcation Chaos* **3**, 469–477 (1993).

[8] C.W. Wu and L. Chua, *Int. J. Bifurcation Chaos* **3**, 1619–1627 (1993).

[9] L. Kocarev and U. Parlitz, in *Proceedings of Nonlinear Dynamics of Electronic Systems*, Krakow, Poland, 1994 (to be published).

[10] L. Kocarev and T. Stojanovski, 'A Model for Secret-Key Cryptography using Chaotic Synchronization,' in *Proceedings of the International Symposium on Information Theory and its Application*, Sydney, 1994 (to be published).

[11] U. Parlitz, L. Kocarev, T. Stojanovski, and H. Preckel (to be published).

[12] S. Hayes, C. Grebogi, and E. Ott, *Phys. Rev. Lett.* **70**, 3031–3034 (1993); S. Hayes, C. Grebogi, E. Ott, and A. Mark, *Phys. Rev. Lett.* **73**, 1781–1784 (1994).

[13] U. Parlitz and S. Ergezinger, *Phys. Lett. A* **188**, 146–150 (1994).

[14] R. He and P.G. Vaidya, *Phys. Rev. A* **46**, 7387–7392 (1992)

[15] M.P. Kennedy, *Frequenz* **46**, 66–80 (1992).

[16] U. Parlitz and L. Kocarev (to be published).

[17] P. Grassberger, R. Hegger, H. Kantz, C. Schaffrath, and T. Schreiber, *Chaos* **3**, 127–141 (1993).

[18] G. Perez and H.A. Cerdeira, *Phys. Rev. Lett.* **74**, 1970–1973 (1995).

Index

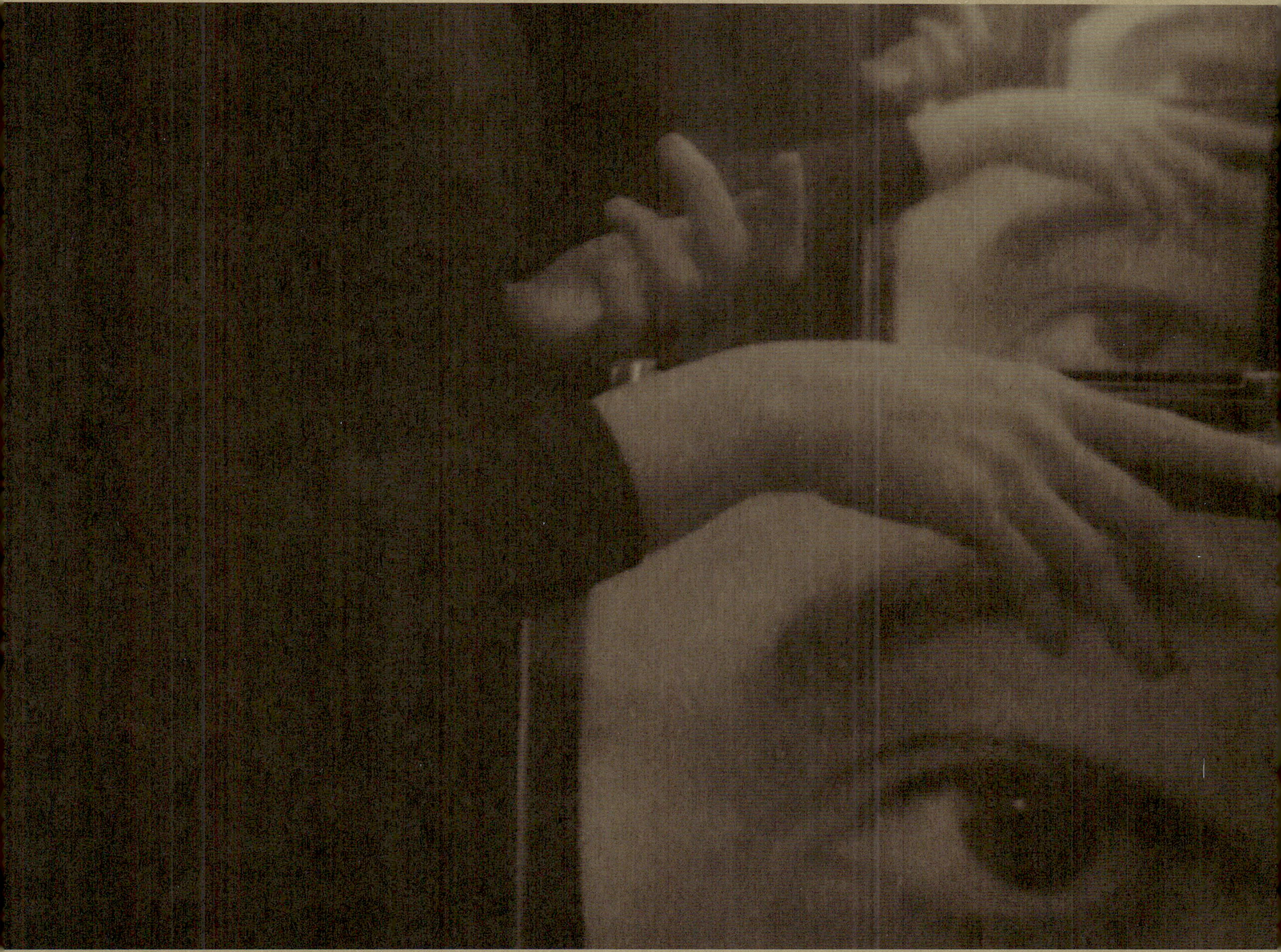

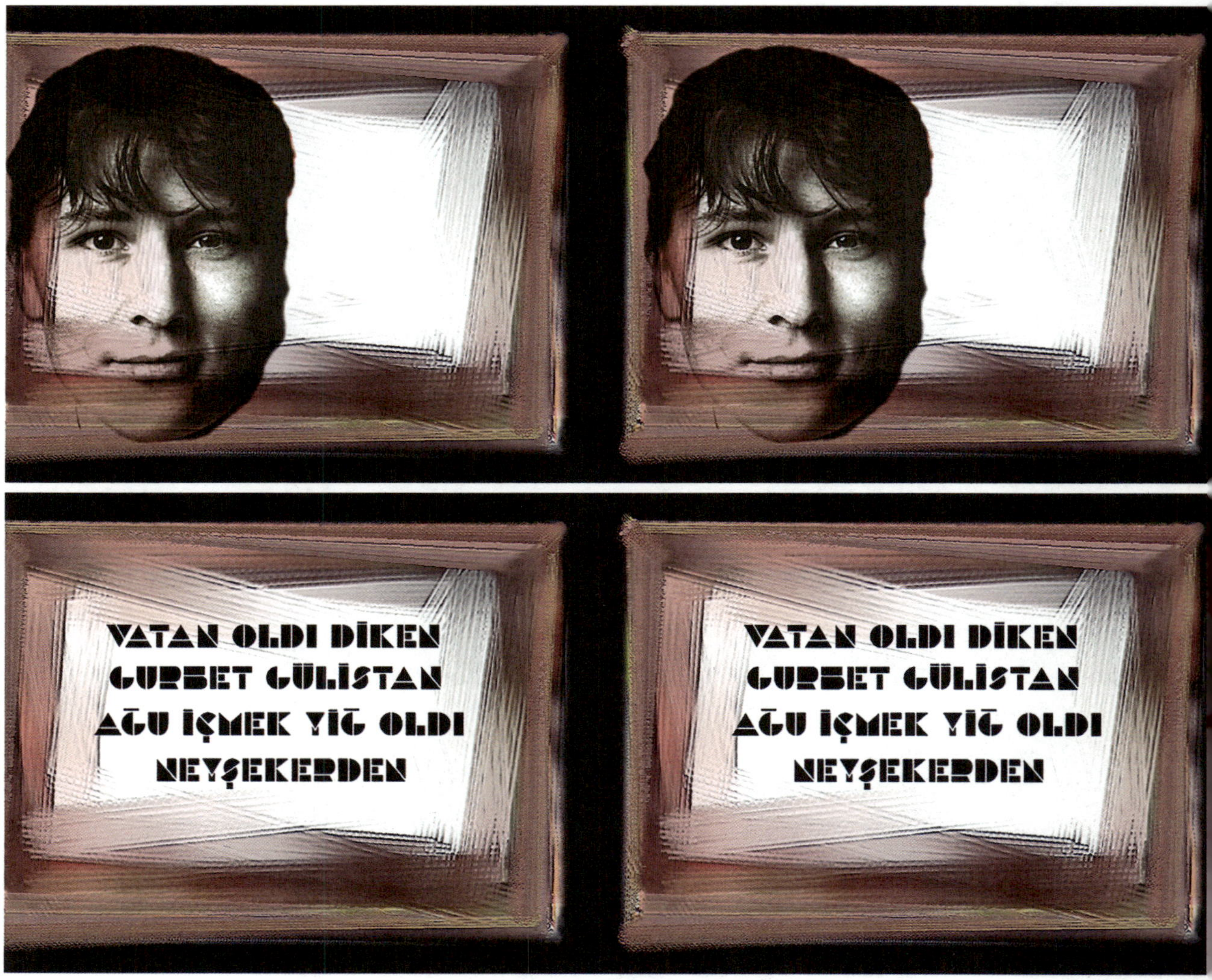

VATAN OLDI DİKEN
GURBET GÜLİSTAN
AĞU İÇMEK YİĞ OLDI
NEYŞEKERDEN

AĞAÇLAR PÜRÇÜNÜ AÇTI
KUŞLAR KILAVUZUN SEÇTİ
YOLUMUZ GURBETE DÜŞTÜ
GARİP DÜŞTÜM ŞİMDENGERÜ
DADALOĞLU'UM DERKİ BANA
DERDİM ARTAR ONDAN YANA
ÖTER BÜLBÜL YANA YANA
GÖZÜM ÇAĞLAR ŞİMDENGERÜ

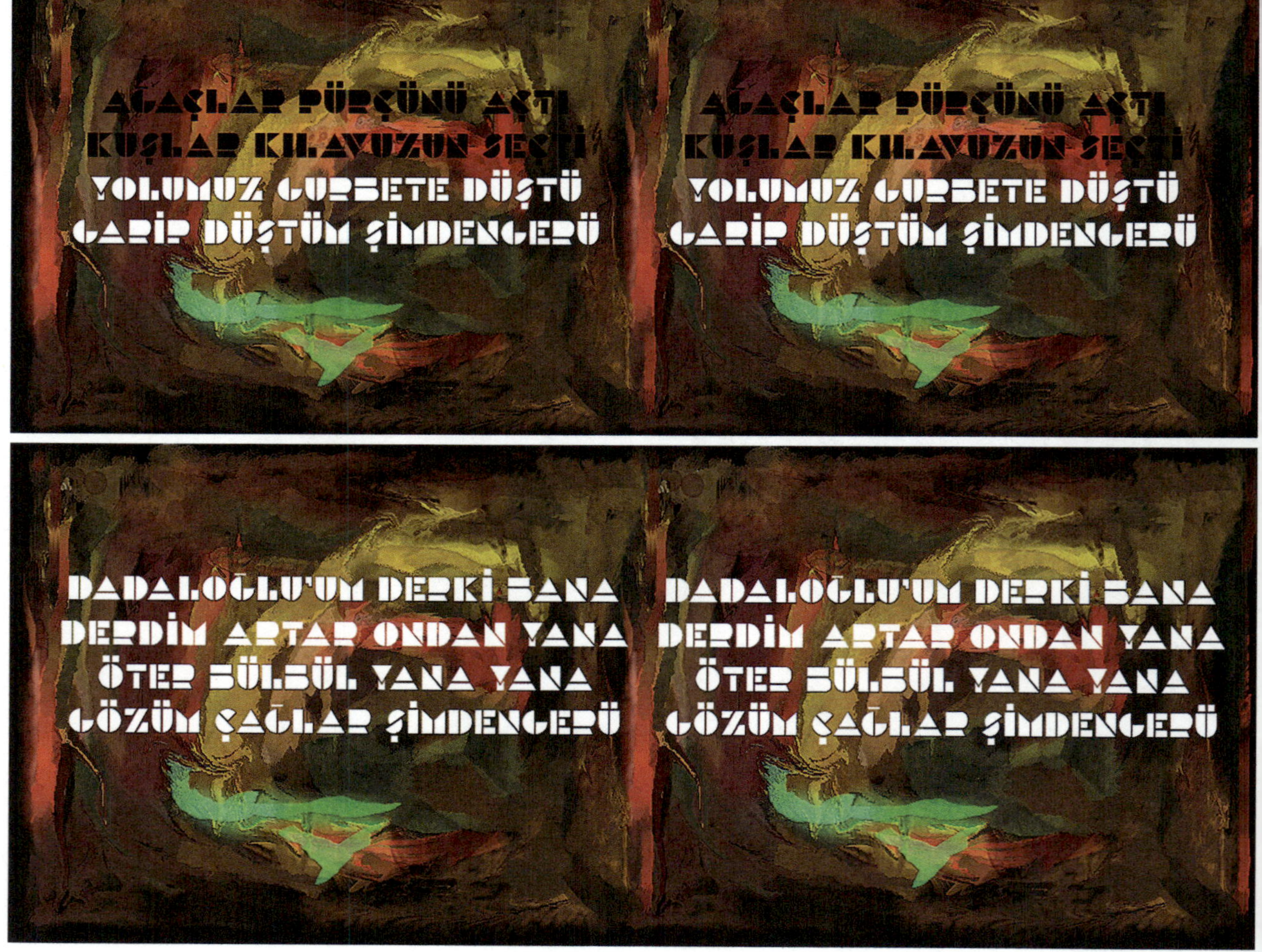

HIZ BİR
HIZ İKİ
HIZ BİR
HIZ İKİ
HIZ ÜÇ
HIZ DÖRT
HIZ ÜÇ
HIZ DÖRT
HIZ BİR
HIZ İKİ
HIZ BİR
HIZ İKİ
YASIYORUZ
YASIYORUZ
HIZ ÜÇ
HIZ DÖRT
HIZ ÜÇ
HIZ DÖRT

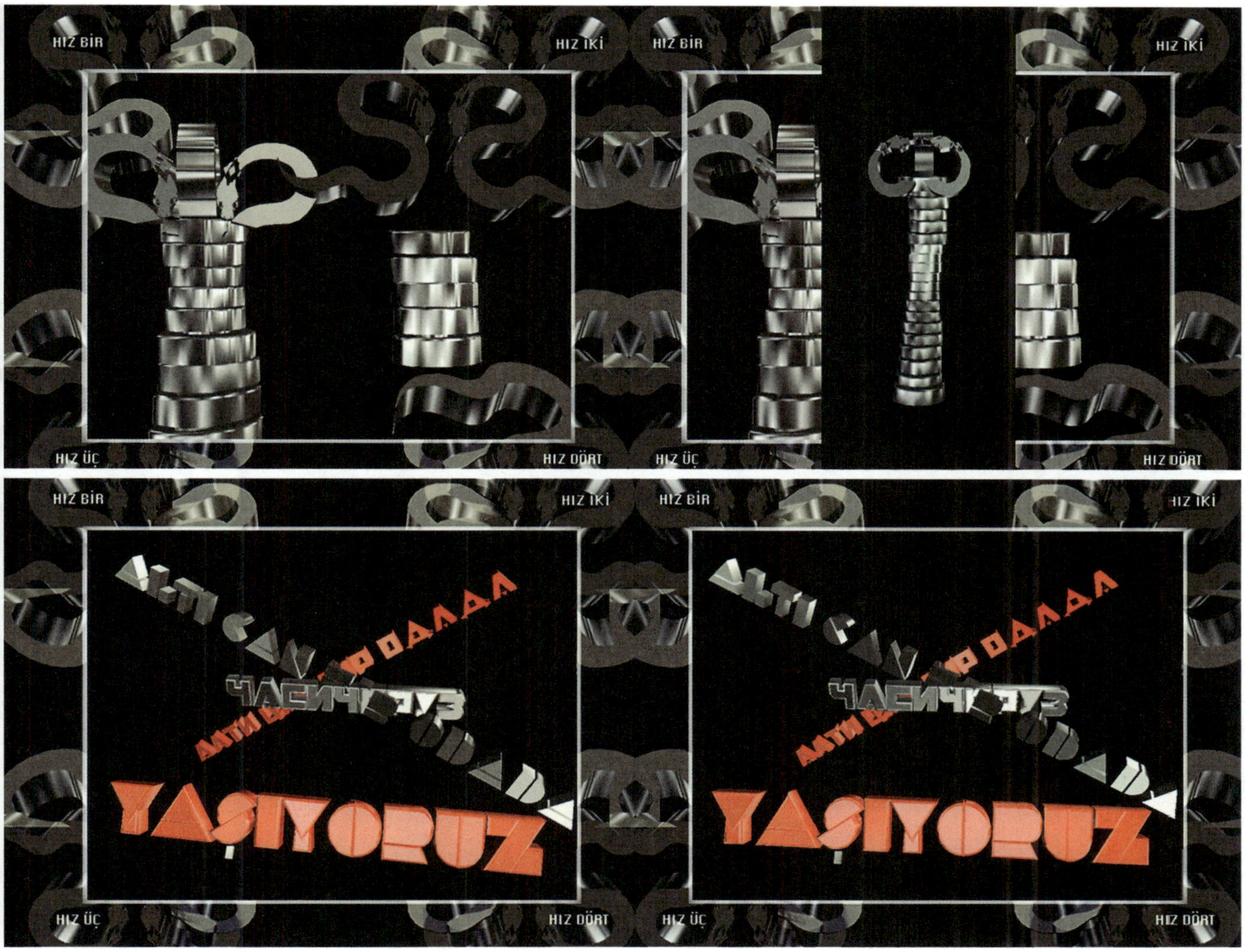

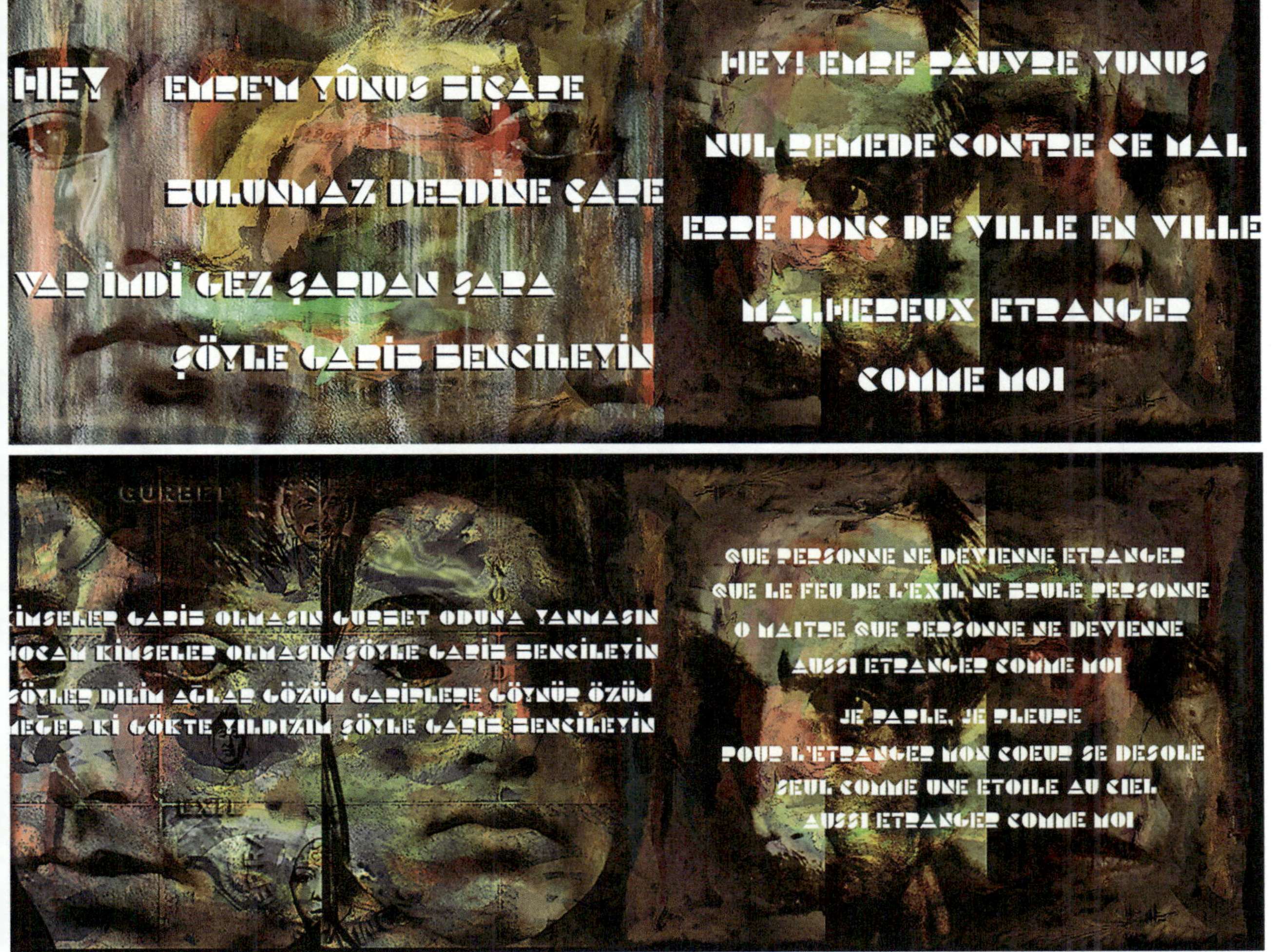

HEY EMRE'M YÛNUS BİÇARE
BULUNMAZ DERDİNE ÇARE
VAR İMDİ GEZ ŞARDAN ŞARA
ŞÖYLE GARİB BENCİLEYİN

HEY! EMRE PAUVRE YUNUS
NUL REMEDE CONTRE CE MAL
ERRE DONC DE VILLE EN VILLE
MALHEREUX ETRANGER
COMME MOI

GURBET
KİMSELER GARİB OLMASIN GURBET ODUNA YANMASIN
HOCAM KİMSELER OLMASIN ŞÖYLE GARİB BENCİLEYİN
ŞÖYLER DİLİM AĞLAR GÖZÜM GARİBLERE GÖYNÜR ÖZÜM
MEGER Kİ GÖKTE YILDIZIM ŞÖYLE GARİB BENCİLEYİN

QUE PERSONNE NE DEVIENNE ETRANGER
QUE LE FEU DE L'EXIL NE BRULE PERSONNE
O MAITRE QUE PERSONNE NE DEVIENNE
AUSSI ETRANGER COMME MOI
JE PARLE, JE PLEURE
POUR L'ETRANGER MON COEUR SE DESOLE
SEUL COMME UNE ETOILE AU CIEL
AUSSI ETRANGER COMME MOI

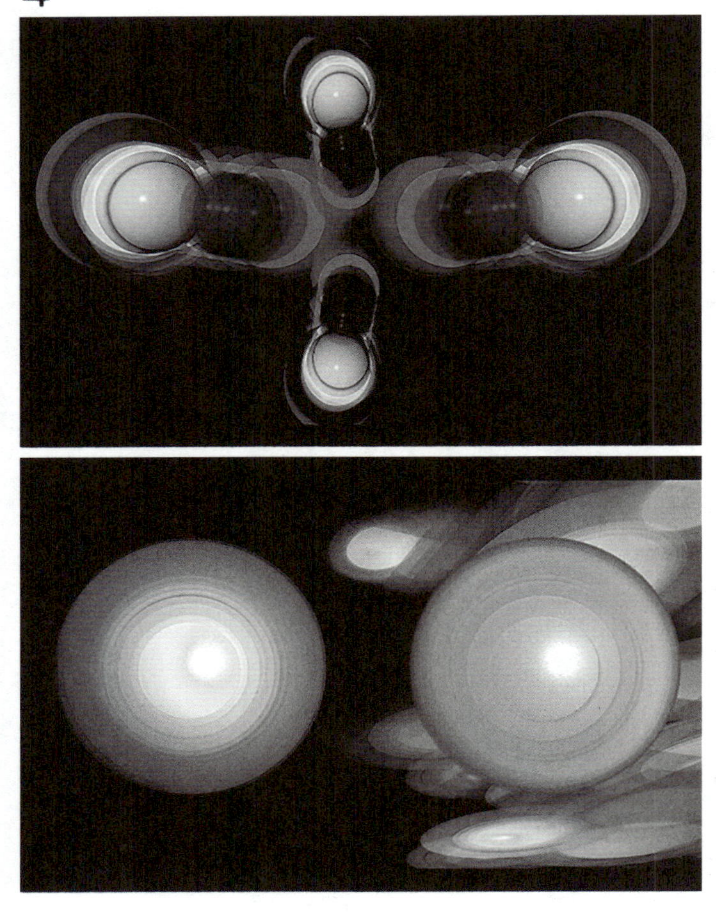

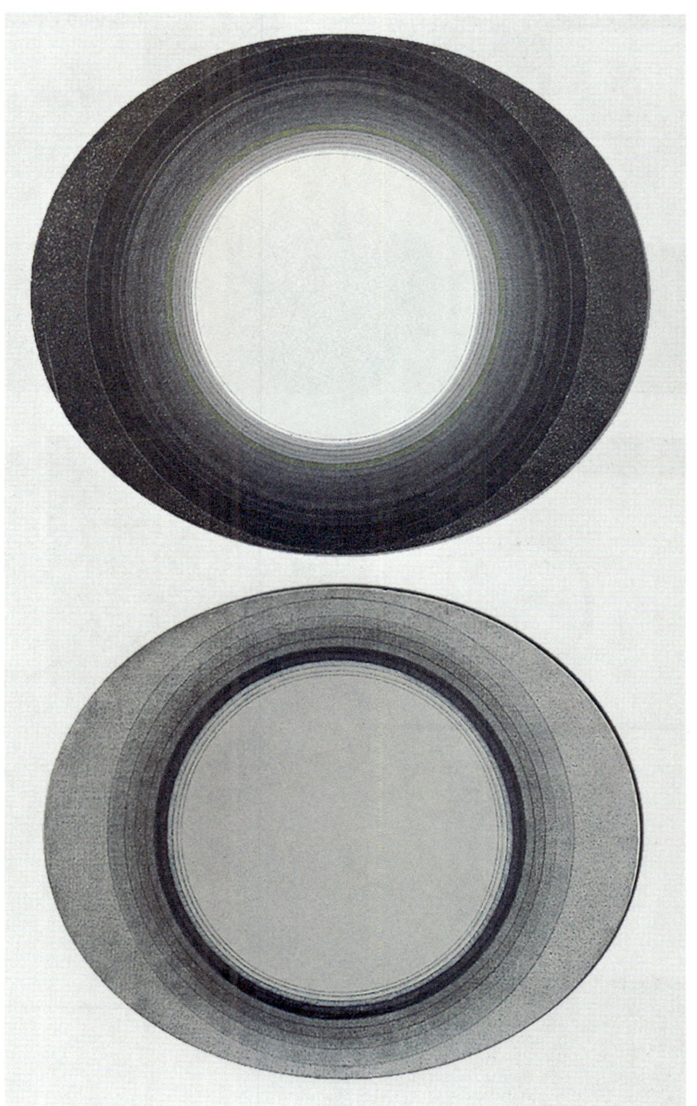

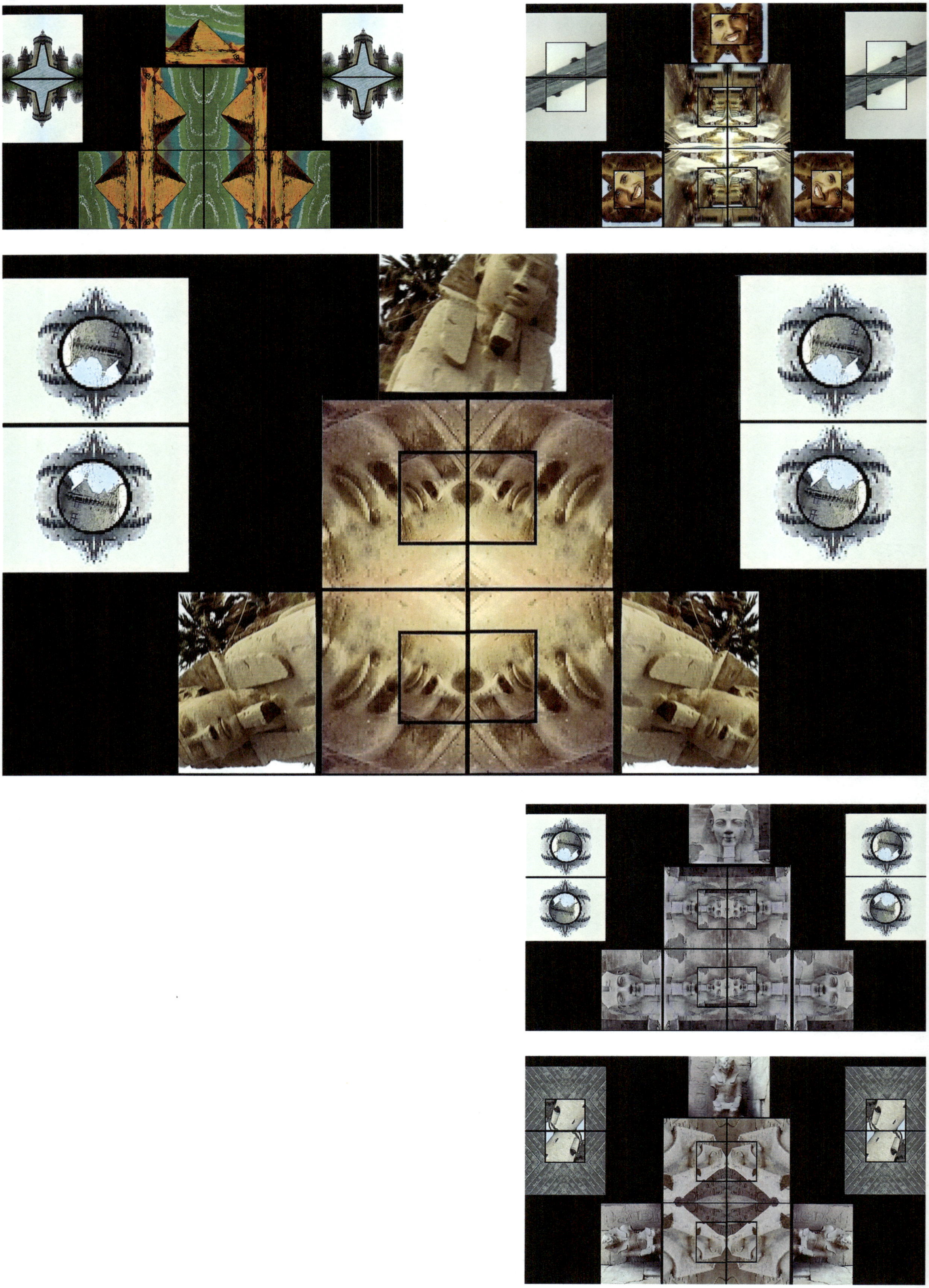

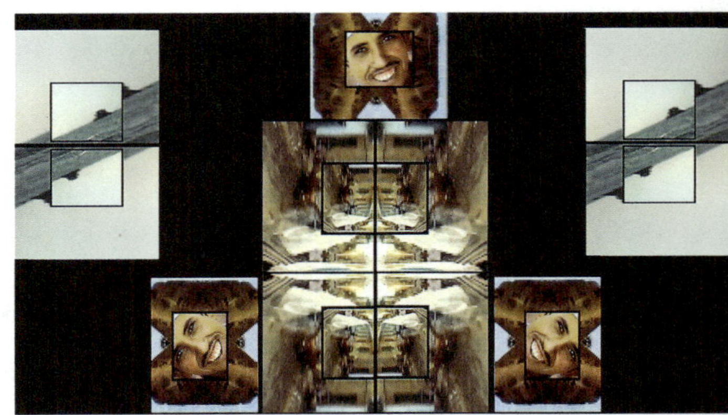
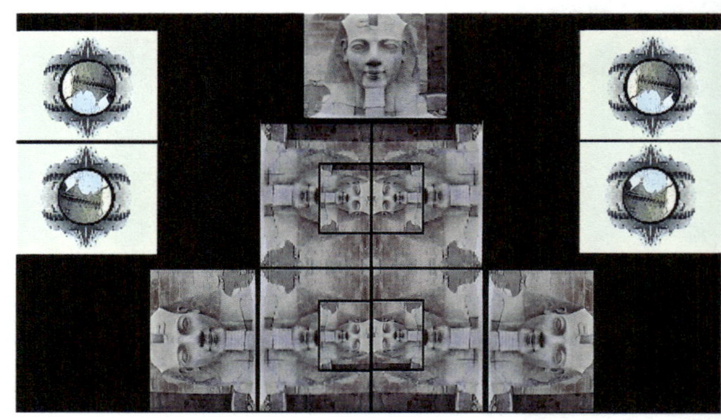

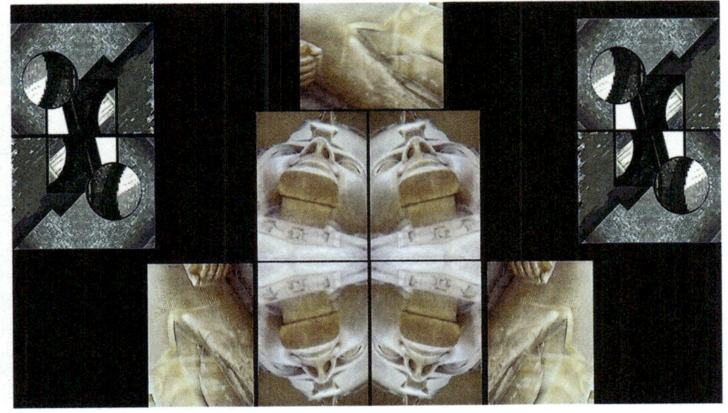

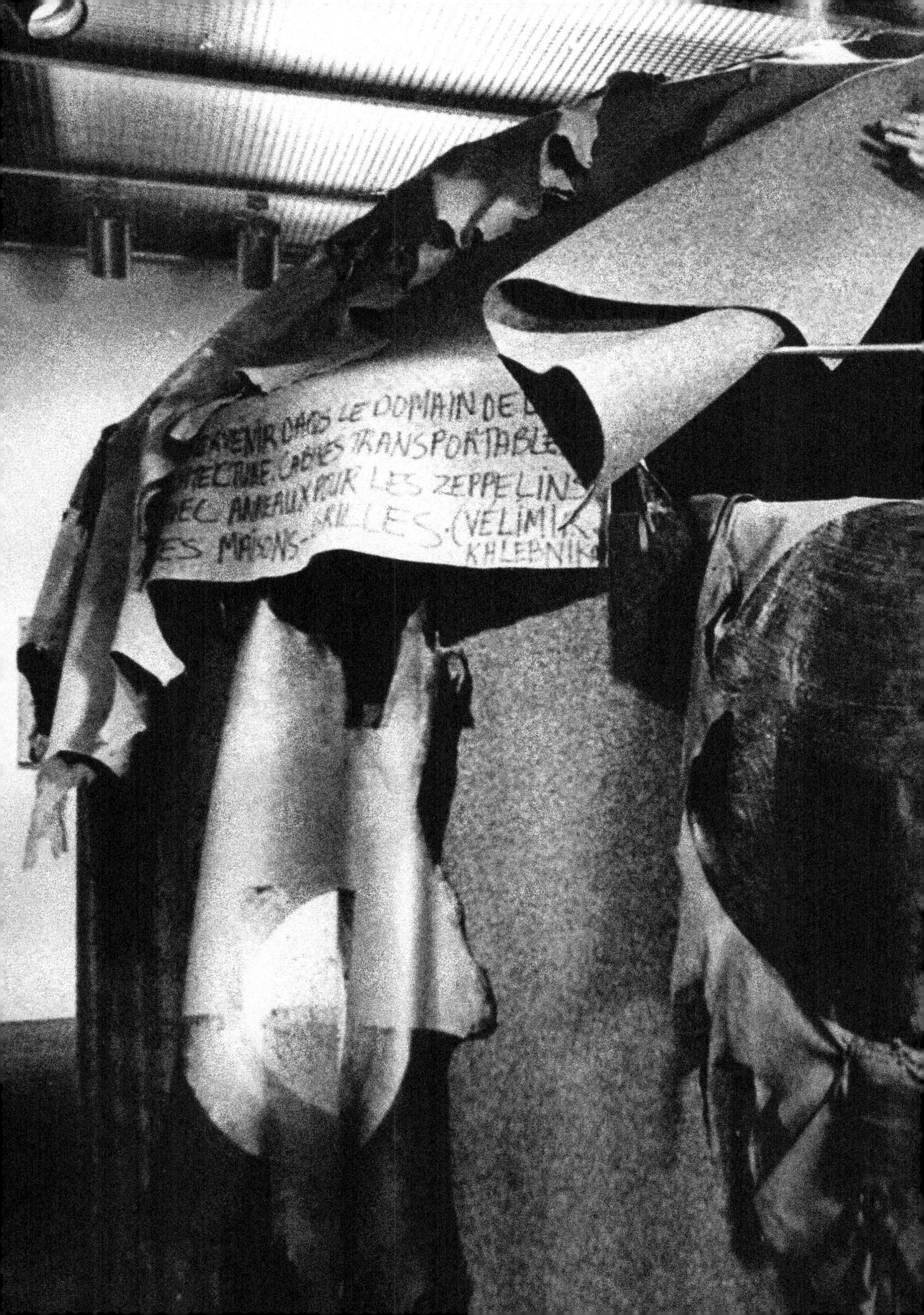

ENIR DANS LE DOMAIN DE L
ECTURE CABLES TRANSPORTABLE
EC ANNEAUX POUR LES ZEPPELINS
ES MAISONS-AILLES. (VELIMIR
KHLEBNIK

MODIFICATION DES
BIDAT DART
CHAMBRE DANS
ET DROIT DE
DE PLACE

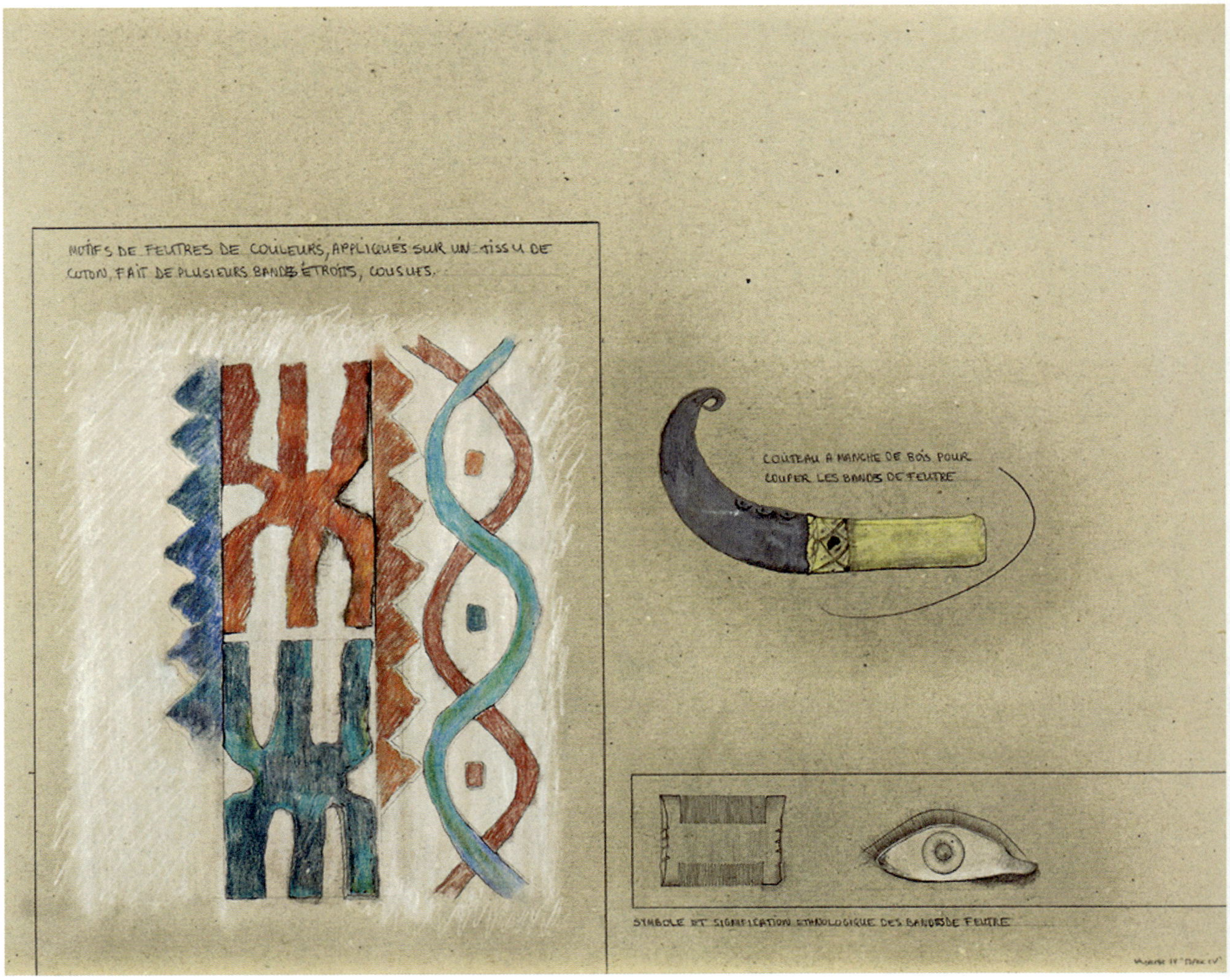

MOTIFS DE FEUTRES DE COULEURS, APPLIQUÉS SUR UN TISSU DE COTON, FAIT DE PLUSIEURS BANDES ÉTROITS, COUSUES.

COUTEAU À MANCHE DE BOIS POUR COUPER LES BANDES DE FEUTRE

SYMBOLE ET SIGNIFICATION ETHNOLOGIQUE DES BANDES DE FEUTRE

DÉTAIL DU NOTIF D'UN TAPIS FEUTRE KURDE.

PHOTOCOPIES DES PHOTOS PRISES PAR BERNARD DUFAIGNE.

MOTIFS DE FEUTRES DE COULEURS, APPLIQUÉS SUR UN TISSU DE COTON, FAIT DE PLUSIEURS BANDES ÉTROITS, COUSUES.

MOTIF DE FEUTRE DE L'ALTAÏ ANCIEN
TÊTE DE BÉLIER PRISE PAR UN LION.
PRÉPARATION DE LA COUCHE DE POILS DE LAINE DE MOUTON POUR FABRIQUER LE FEUTRE

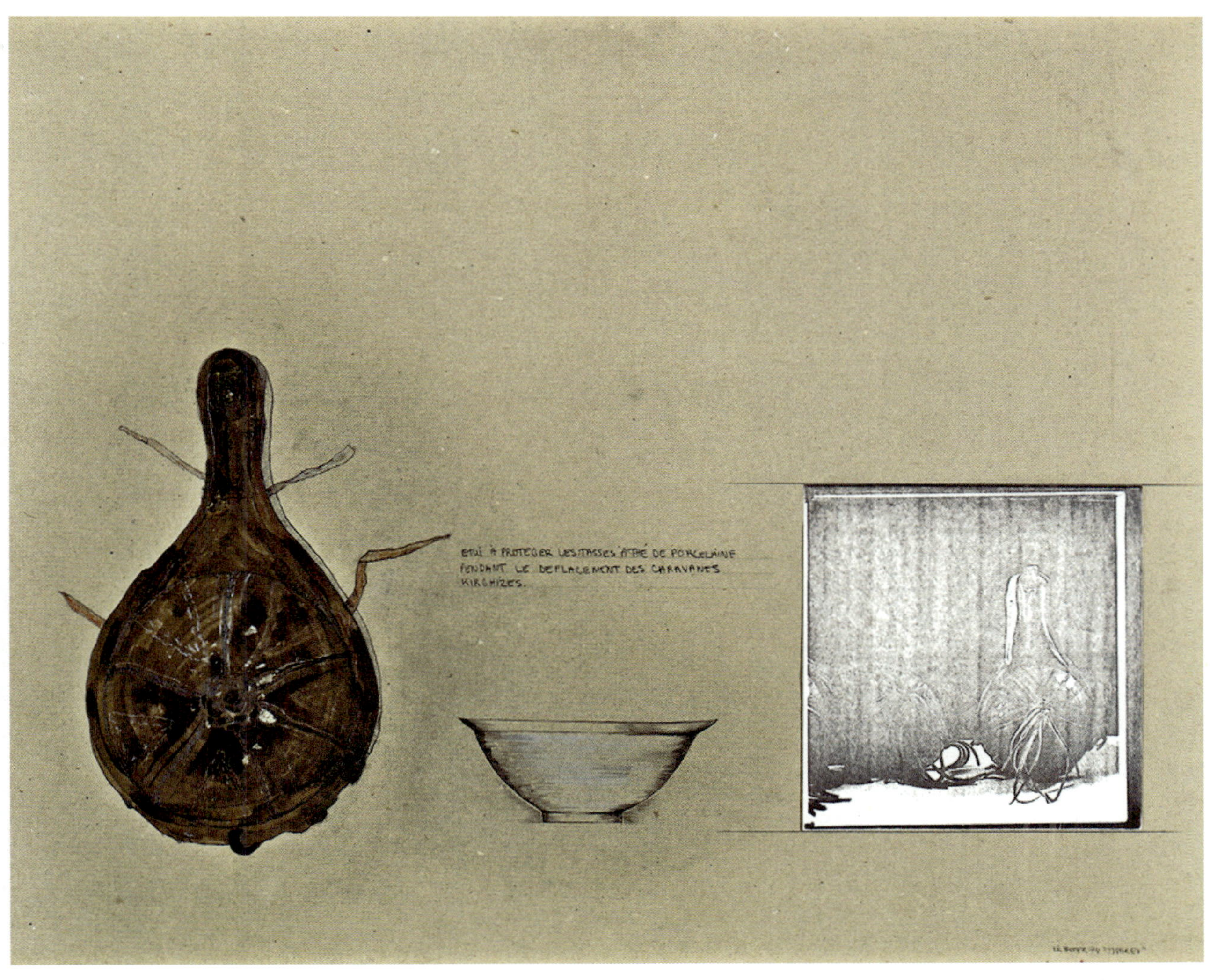

ÉTUI À PROTÉGER LES TASSES À THÉ DE PORCELAINE PENDANT LE DÉPLACEMENT DES CARAVANES KIRGHIZES.

PARIS VILLE LUMIERE
NEUVIEME ARRONDISSEMENT
L'OPERA.. L'OPERA... L'OPERA...
LE DRAPEAU ROUGE DE LA COMMUNE A FLOTTE SUR L'OPERA
L'OPERA EST OCCUPE PAR LES FEDERES EN 1871.
LE FANTOME DE L'OPERA
L'OPERA: REFLET ARCHITECTONIQUE DE L'IDEOLOGIE BOURGEOISE: L'OPERA
FOLIES BERGERE.. CASINO DE PARIS
FOLIES BERGERE
2e ACTE
L'ORIENT
FEERIE DES EAUX
LES FEMMES FEMMES
LA PRISON
LA CATHEDRALE
LOVE STORY
CHARLESTON
LE PONT SOUS LE DIRECTOIRE
IVOIRES DE CHINE
LA FETE DES IVOIRES
LA TOUR EIFFEL AUX FOLIES
LE TOURBILLON DES FOLIES

PARIS VILLE LUMIERE
VINGTIEME ARRONDISSEMENT
MENILMONTANT.. MENILMONTANT..
MENILMONTANT
LA POLICE A MENILMONTANT...
MENIL' MAILLE
GAMBLING ON THE
1 2 3 4
5 6 7 8
9 10 11 12
BLVD. MENILMONTANT
JEU SUR
1 2 3 4
5 6 7 8
9 10 11 12
BLVD MENILMONTANT
CIMETIERE DU PERE L'ACHAISE..
AUX MORTS DE LA COMMUNE
4e DIV. 1re SECT
VISCONTI
ROSSINI
ALFRED DE MUSSET
COLETTE
4e DIV. 2e SECT
ARAGO
LEDRU-ROLLIN
FELIX FAURE
7e DIV
RACHEL
HELOISE ET ABELARD
11e DIV.
CHOPIN
CHERUBINI
12e DIV.
GERICAULT
29e DIV.
MARECHAL NEY
39e DIV
MURAT
PARMENTIER
27e DIV
FAMILLE HUGO
25e DIV.
MOLIERE
LA FONTAINE
44e DIV.
SARAH BERNHARDT
68e DIV.
GEORGES BIZET
89e DIV
OSCAR WILDE
49e DIV
DELACROIX
48e DIV
HONORE DE BALZAC
97e DIV.
EDITH PIAF
MAURICE THOREZ
HENRI BARBUSSE

PARIS VILLE LUMIERE
QUINZIEME ARRONDISSEMENT
TOUR MONTPARNASSE
NOUVEAU QUARTIER DE PARIS...
LA PROMOTION IMMOBILIERE A PARIS
PRIX = 332.000 Frs. PRIX D'ACHAT DU STUDIO
LE STUDIO = 1.950 Fr. PAR MOIS DE LOYER
CUISINE
SEJOUR 42 m²
SALLE DE BAINS
HOTEL

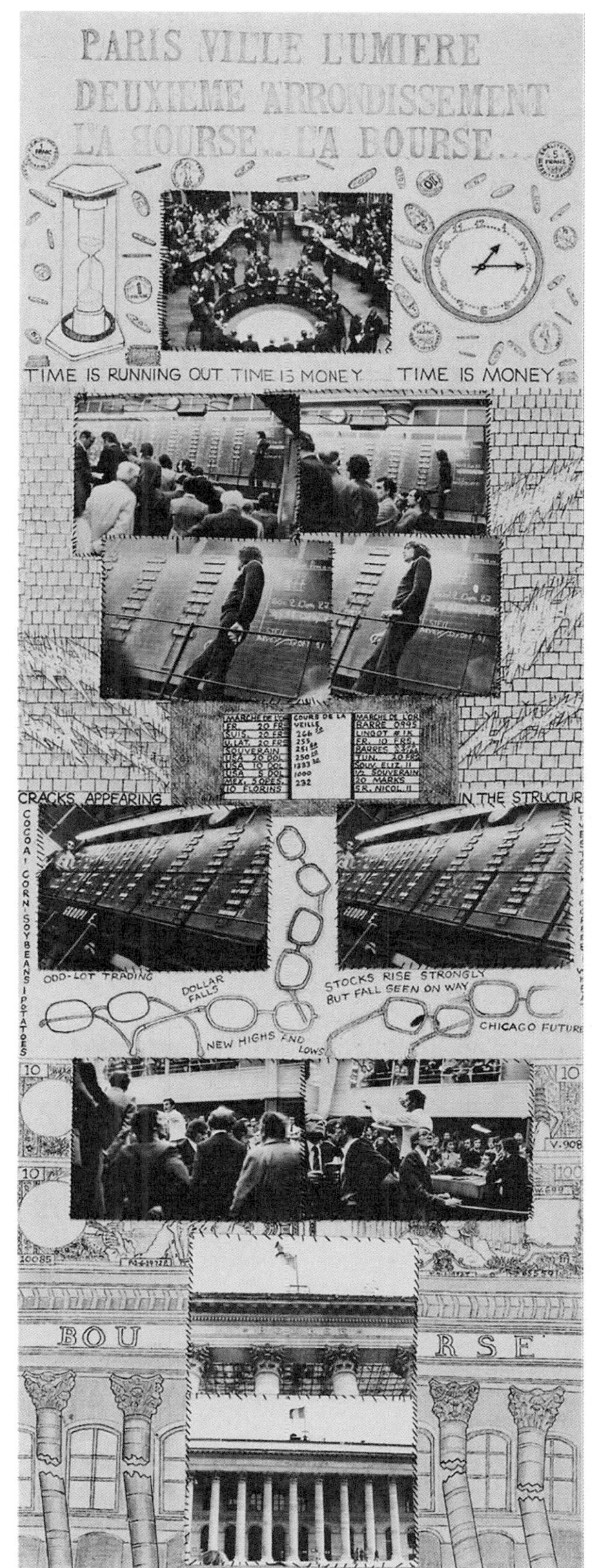

PARIS VILLE LUMIERE
DEUXIEME ARRONDISSEMENT
LA BOURSE... L'A BOURSE...
TIME IS RUNNING OUT TIME IS MONEY   TIME IS MONEY
CRACKS APPEARING
IN THE STRUCTURE
ODD-LOT TRADING
DOLLAR FALLS
STOCKS RISE STRONGLY BUT FALL SEEN ON WAY
NEW HIGHS AND LOWS
CHICAGO FUTURE
BOU RSE

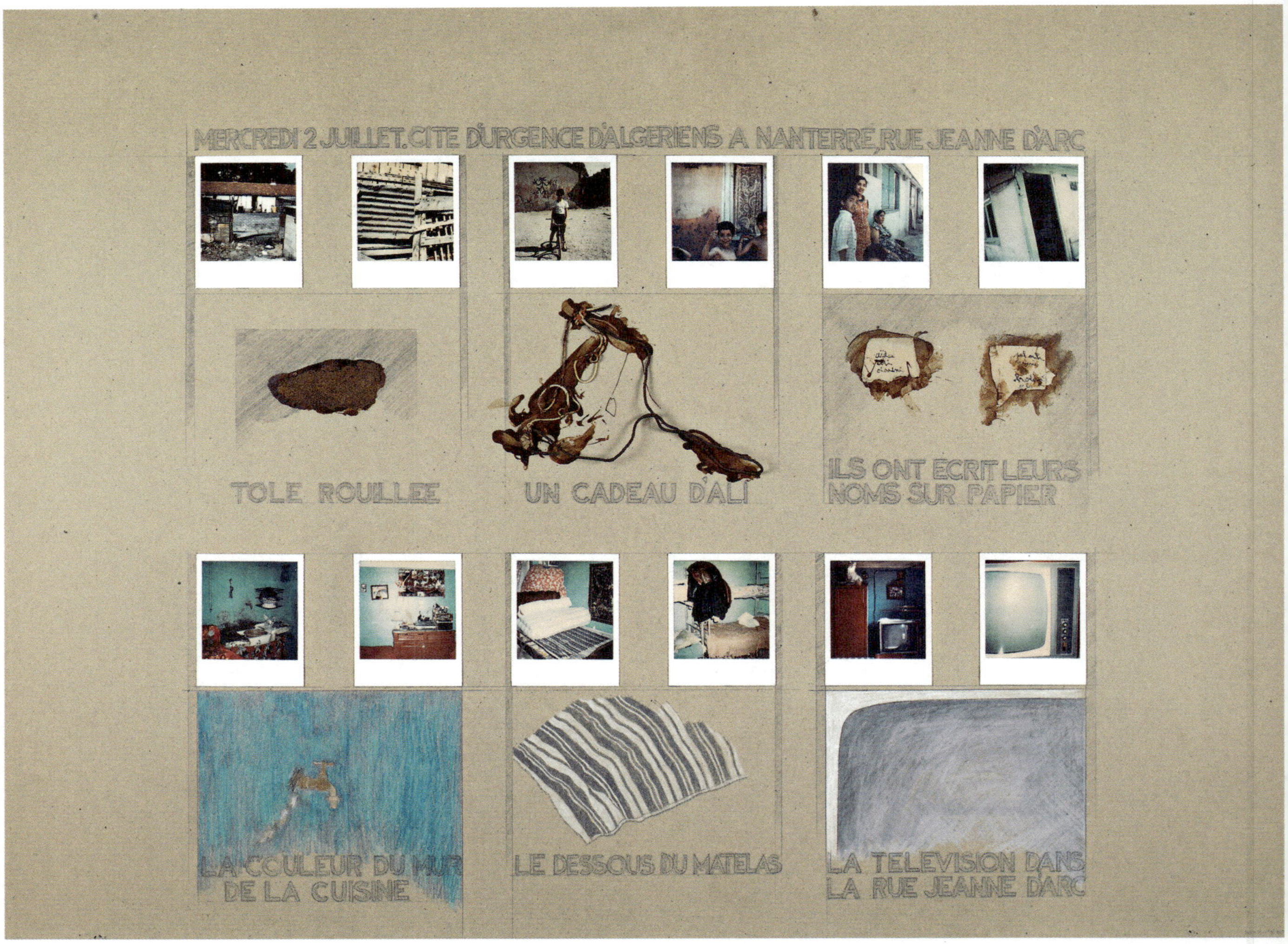
MERCREDI 2 JUILLET. CITE D'URGENCE D'ALGERIENS A NANTERRE RUE JEANNE D'ARC
TOLE ROUILLEE
UN CADEAU D'ALI
ILS ONT ECRIT LEURS NOMS SUR PAPIER
LA COULEUR DU MUR DE LA CUISINE
LE DESSOUS DU MATELAS
LA TELEVISION DANS LA RUE JEANNE D'ARC

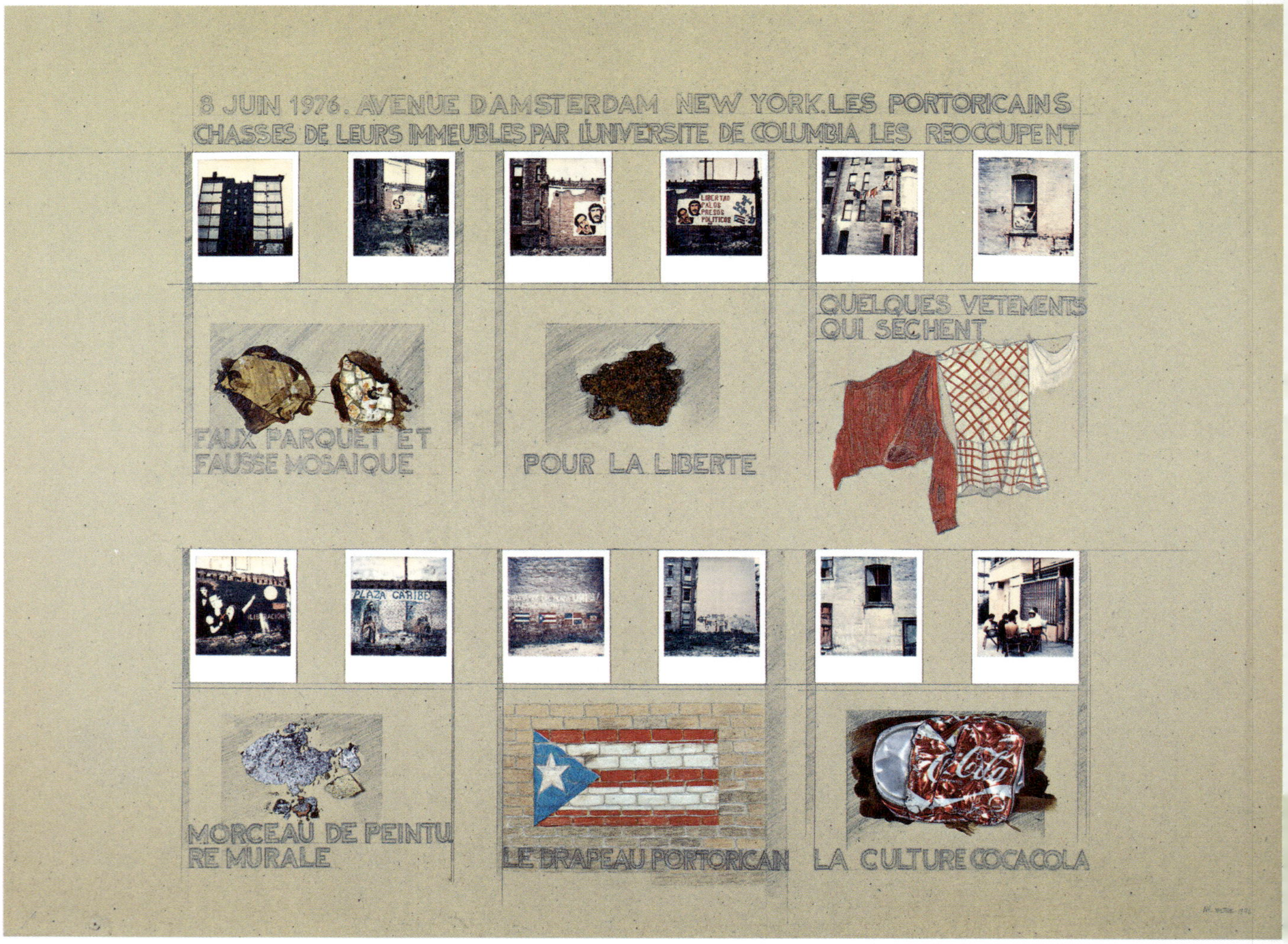
8 JUIN 1976. AVENUE D'AMSTERDAM NEW YORK. LES PORTORICAINS CHASSES DE LEURS IMMEUBLES PAR L'UNIVERSITE DE COLUMBIA LES REOCCUPENT
PLAZA CARIBE
FAUX PARQUET ET FAUSSE MOSAIQUE
POUR LA LIBERTE
QUELQUES VETEMENTS QUI SECHENT
MORCEAU DE PEINTURE MURALE
LE DRAPEAU PORTORICAIN
LA CULTURE COCACOLA
Coca-Cola

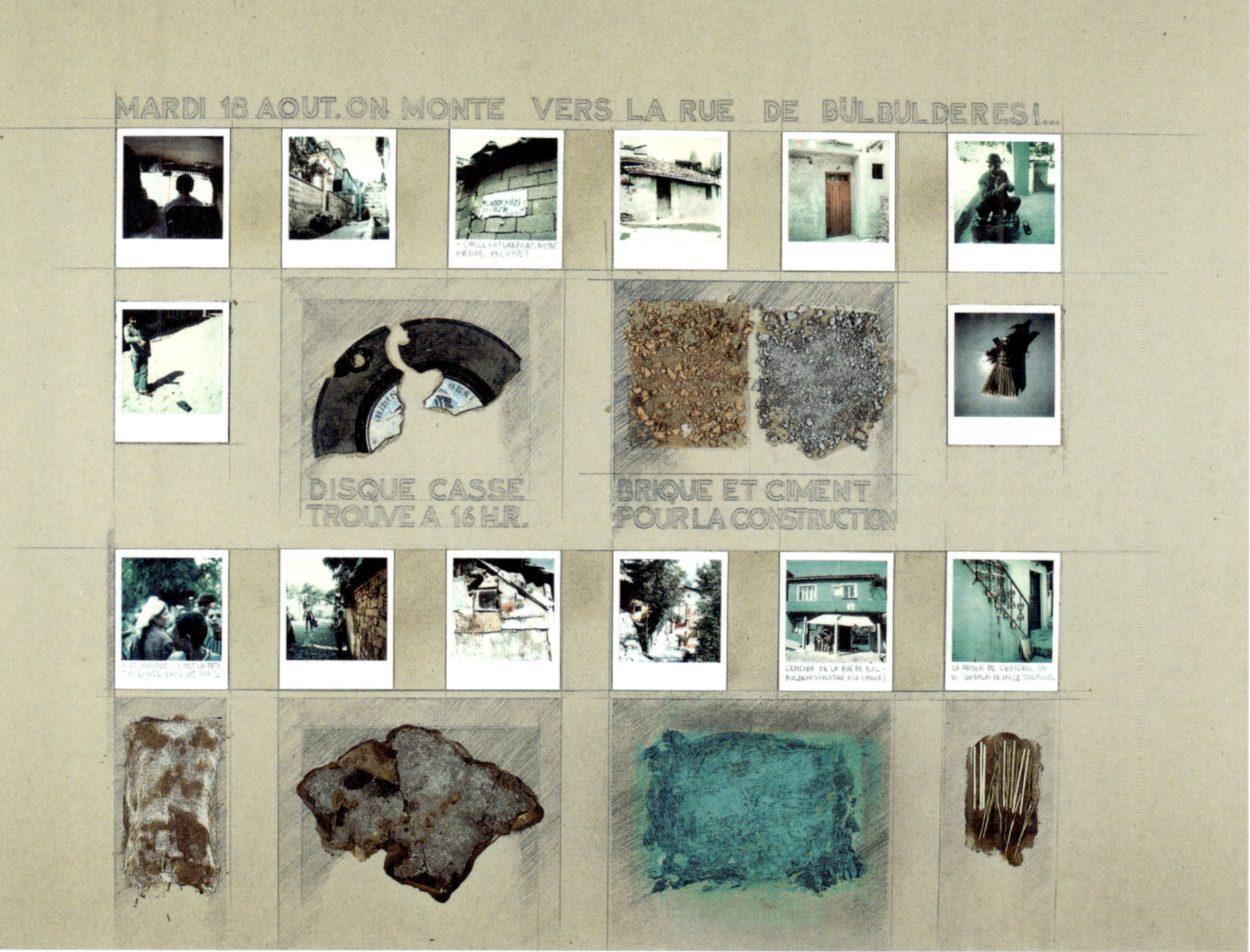
MARDI 18 AOUT. ON MONTE VERS LA RUE DE BULBULDERESI...
DISQUE CASSE TROUVE A 16 H.R.
BRIQUE ET CIMENT POUR LA CONSTRUCTION
6 JUIN 1976. ON COMMENCE NOTRE TRAVAIL SUR LE BOWERY.. 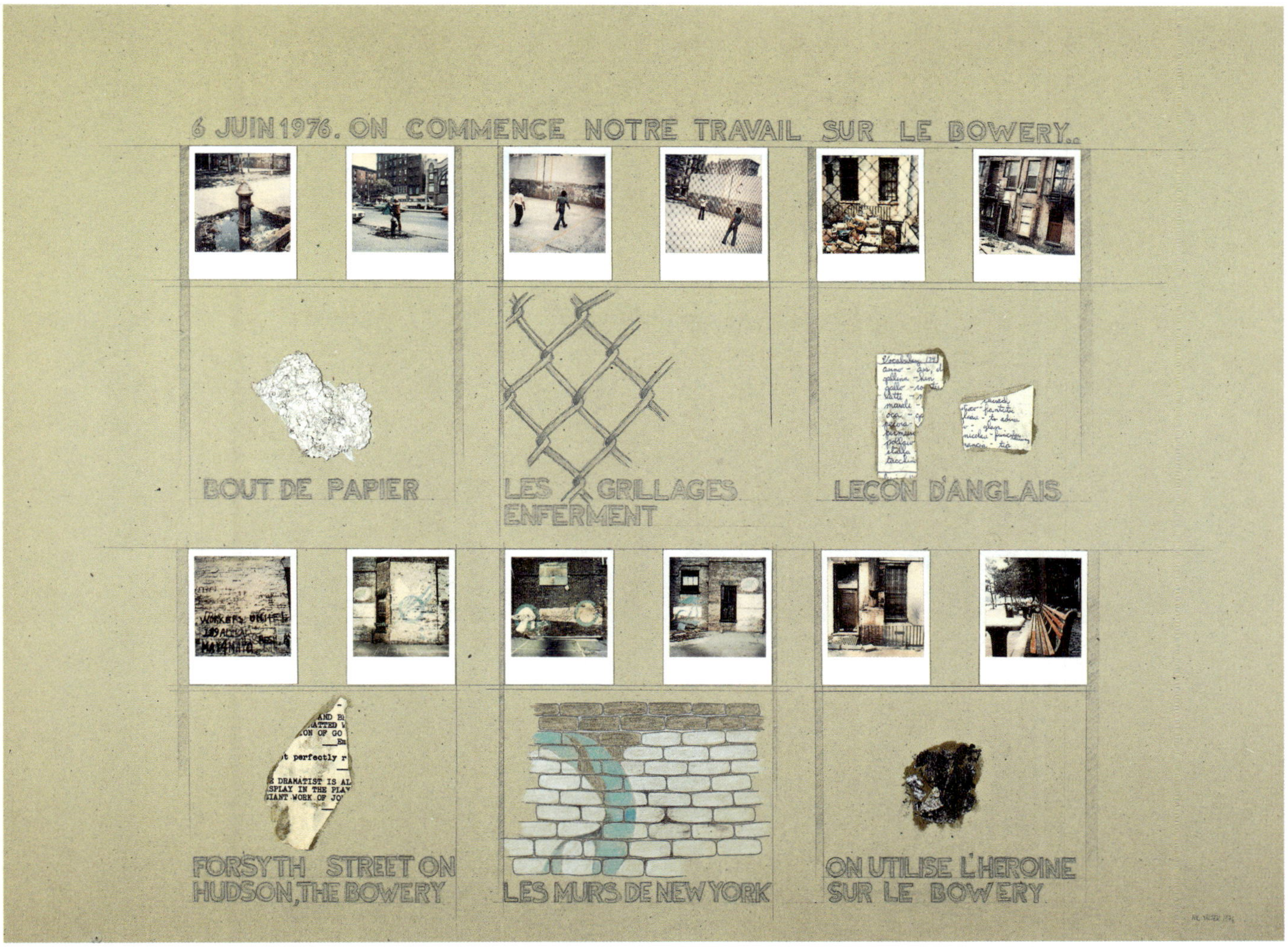
BOUT DE PAPIER
LES GRILLAGES ENFERMENT
LEÇON D'ANGLAIS
FORSYTH STREET ON HUDSON, THE BOWERY
LES MURS DE NEW YORK
ON UTILISE L'HEROINE SUR LE BOWERY

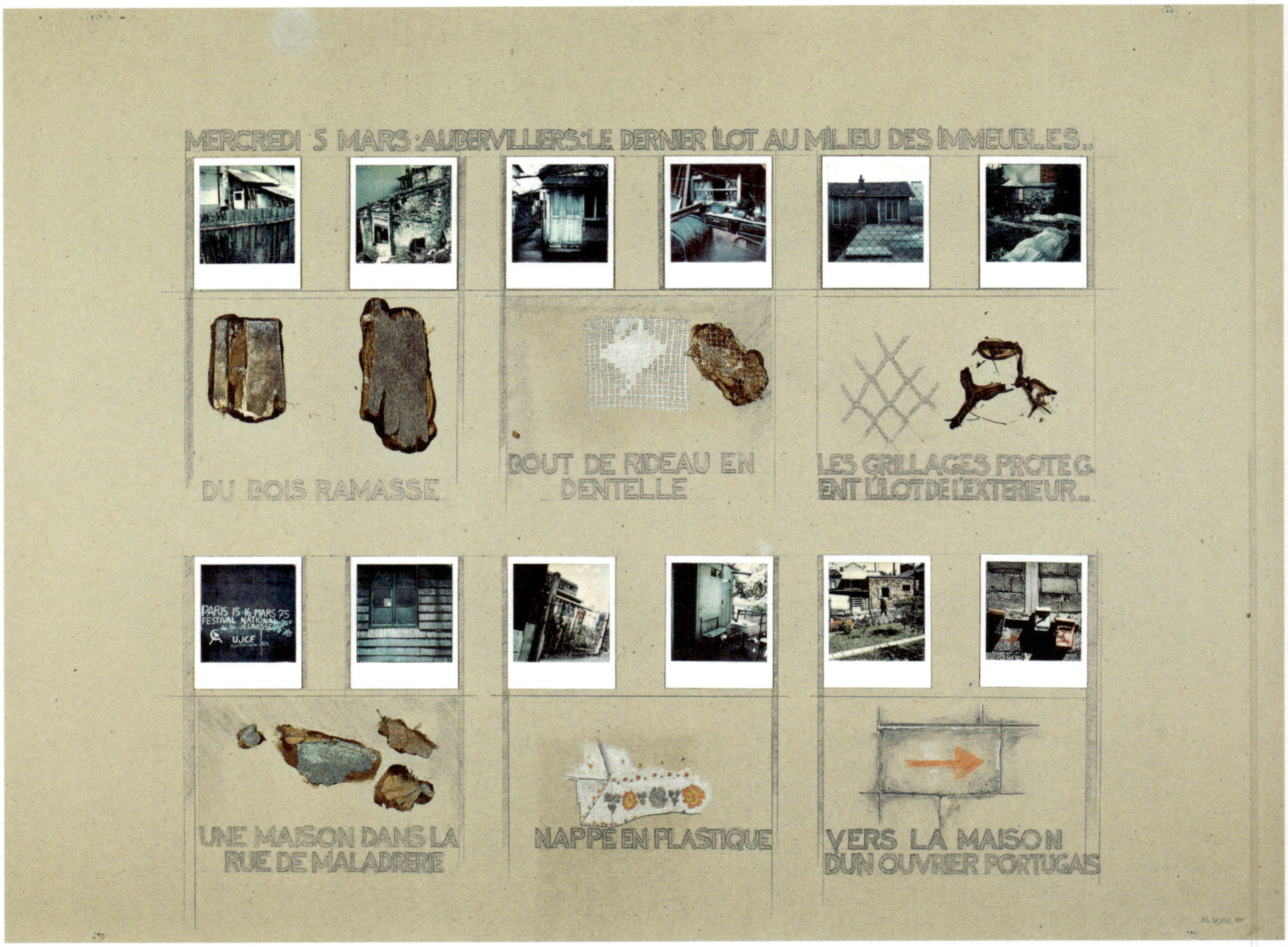
MERCREDI 5 MARS : AUBERVILLIERS: LE DERNIER ILOT AU MILIEU DES IMMEUBLES..
DU BOIS RAMASSE
BOUT DE RIDEAU EN DENTELLE
LES GRILLAGES PROTEG. ENT L'ILOT DE L'EXTERIEUR...
PARIS 15-16 MARS 75 FESTIVAL NATIONAL DE LA JEUNESSE UJCF
UNE MAISON DANS LA RUE DE MALADRERIE
NAPPE EN PLASTIQUE
VERS LA MAISON D'UN OUVRIER PORTUGAIS

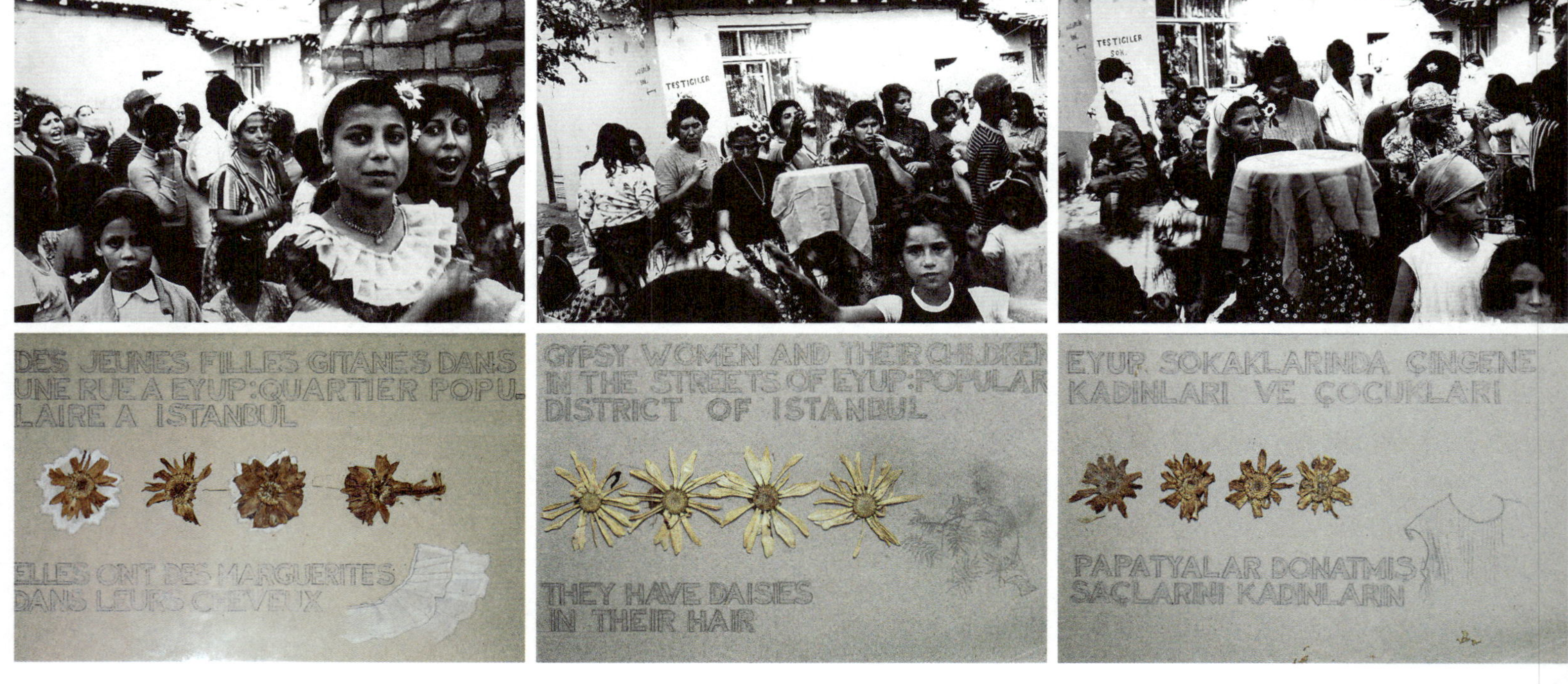

DES JEUNES FILLES GITANES DANS UNE RUE A EYUP: QUARTIER POPULAIRE A ISTANBUL
ELLES ONT DES MARGUERITES DANS LEURS CHEVEUX
GYPSY WOMEN AND THEIR CHILDREN IN THE STREETS OF EYUP: POPULAR DISTRICT OF ISTANBUL
THEY HAVE DAISIES IN THEIR HAIR
EYUP SOKAKLARINDA ÇINGENE KADINLARI VE ÇOCUKLARI
PAPATYALAR DONATMIŞ SAÇLARINI KADINLARIN

7 JUIN 1976.BROOKLYN, OU HABITENT DES NOIRS, DES JUIFS, DES PORTORICAINS
BOUTS DE TISSUS
MORCEAU DE DISQUE TROUVE A BROOKLYN
DU BOIS RAMASSE
DES MORCEAUX DE MUR TROUVES DANS LA RUE
LES MURS EN BRIQUES
EDDIE
GRAFFITIS A BROOKLYN

PIGMENT POUR PEINTURE
A LA CHAUX DES MAISONS
D'EYUP

POSSIBILITÉ MINÉRALE
POUR UN HABITAT A EYUP

PIGMENT POUR PEINTURE
A LA CHAUX DES MAISONS
D'EYUP

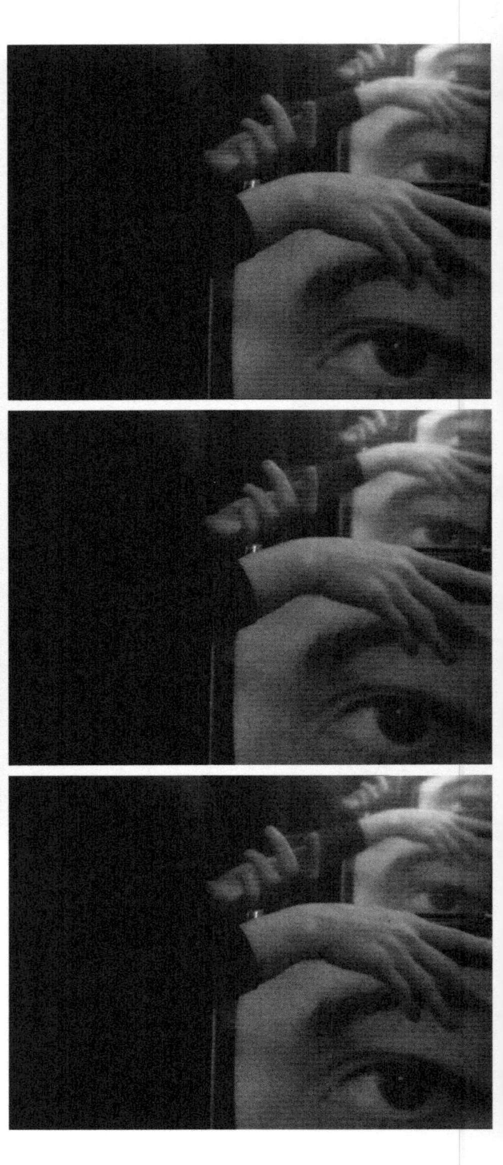
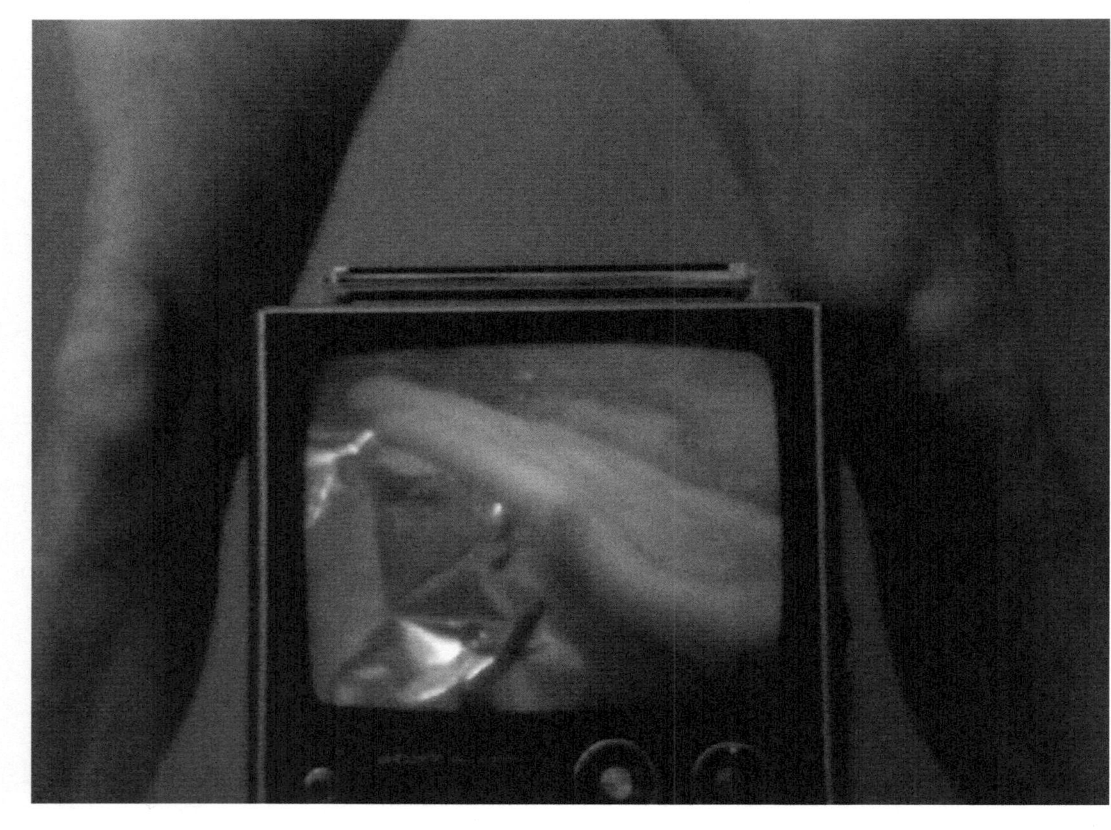

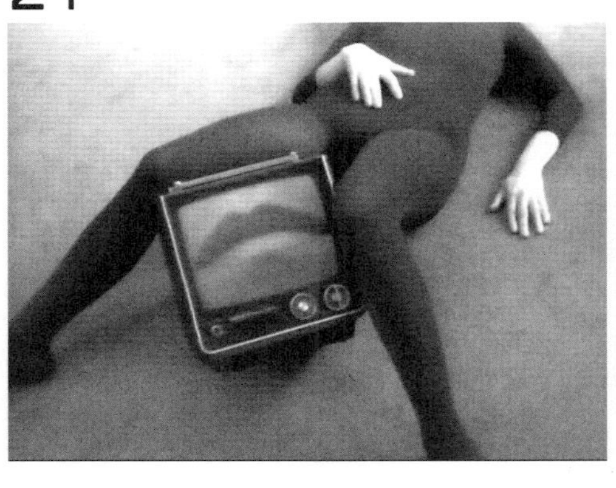

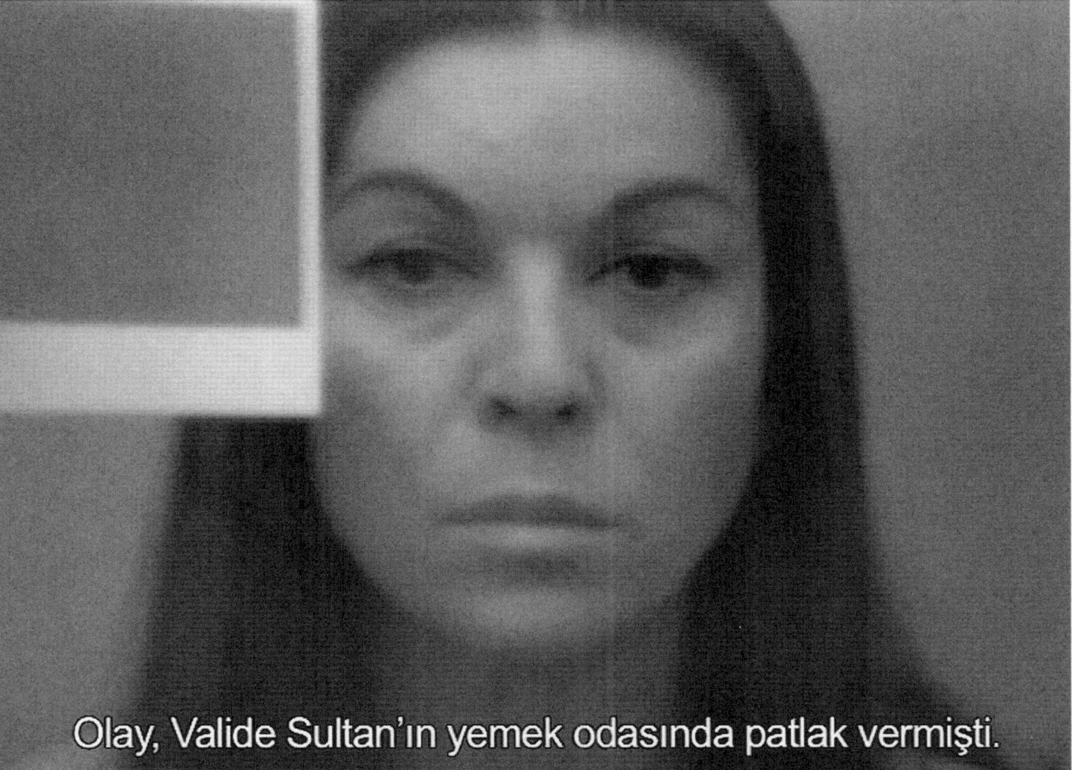
Olay, Valide Sultan'ın yemek odasında patlak vermişti.

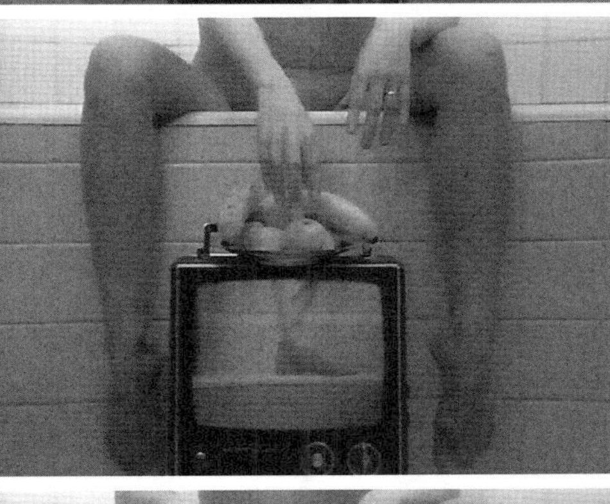

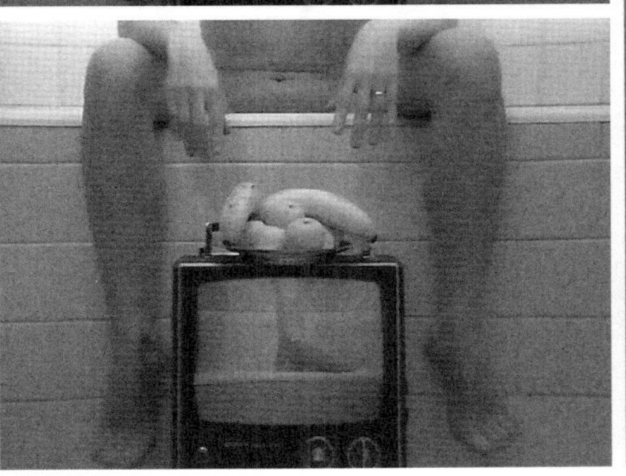

BERABER BÜYÜDÜK, BERABER
KIZIMLA ARKADAŞ OLDUK BÜYÜ
DÜK.. ARTIK BANA BIR ANNE OLDU
BIR FIDAN OLDU, ÜSTÜMDE BIR GÖL
GE BEN ALTINDA OTURAYIM DIYE,
GÖLGESINDE.. MAALESEF IŞTE KA
DER VURDU GITTI...BENDE NASILKI
BIR BALIK KUMUN ÜSTÜNDE KALIP

...DA DENIZ ÇEKILDIMI, BEN
KUMDA KALDIM ÖYLEDIR IŞTE...
JE ME SUIS MARIEE A TREIZE ANS.
A QUATORZE ANS, J'AI EU MA FILLE.
APRES, ELLE A GRANDI. ON EST DE
VENU AMIE, ET J'AI GRANDI AVEC
ELLE. ELLE ETAIT UNE MERE POUR
MOI, ELLE ETAIT COMME UN JEUNE
ARBRE SUR MOI. POUR QUE JE

PUISSE M'ASSEOIR SOUS SON OMB
RE MALHEUREUSEMENT LA DESTI
NEE LA TUEE. MOI JE SUIS COMME
UN POISSON QUI SE DESSECHE
SUR LE SABLE QUAND LA MER SE
RETIRE. JE SUIS RESTEE SUR LE
SABLE.. J'AI JAMAIS SOURI DANS LA
VIE ET JE NE SOURIRAI PLUS JA
MAIS... L'HISTOIRE DE RAHIME.

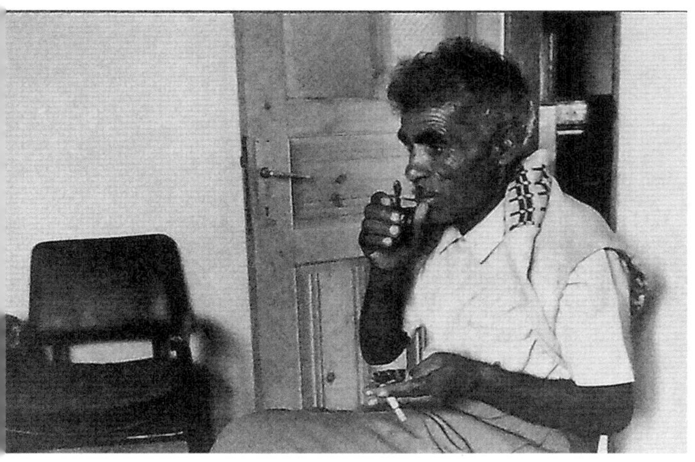

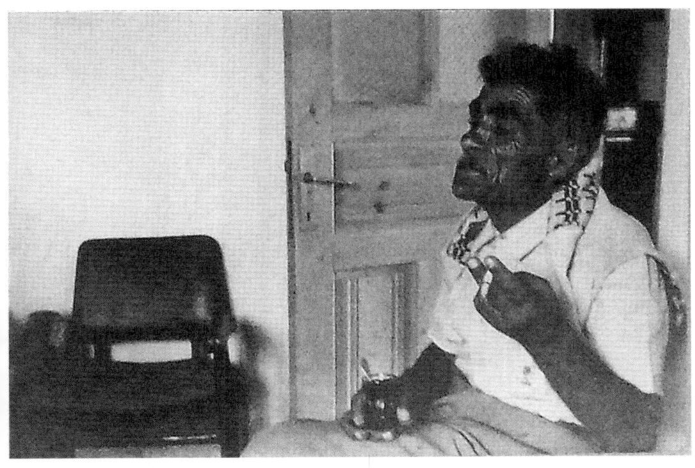

UNE FEMME CHAMANE TOUNGOUZE NERTCHINSK.
PHOTOCOPIE D'UNE PHOTO DE LA COLLECTION DE MUSÉE DE L'HOMME

SACHET DE POUDRE
POUR GUER DES
MALADIES (PRES DE NERTCHINSK)

TAMBOUR DE CHAMAN ALTAÏEN, FAIT D'UNE PEAU DE MOU-
TON PEINTE ET TENDUE SUR UNE LATTE DE BOIS DOUBLE.
DESSIN D'APRES UNE PHOTO (COLLECTION MUSÉE DE L'HOMME)

MORCEAU DE MANDRAGORE
UTILISÉ EN SORCELLERIE

NIL YALTER "TOPAK EV" 1974

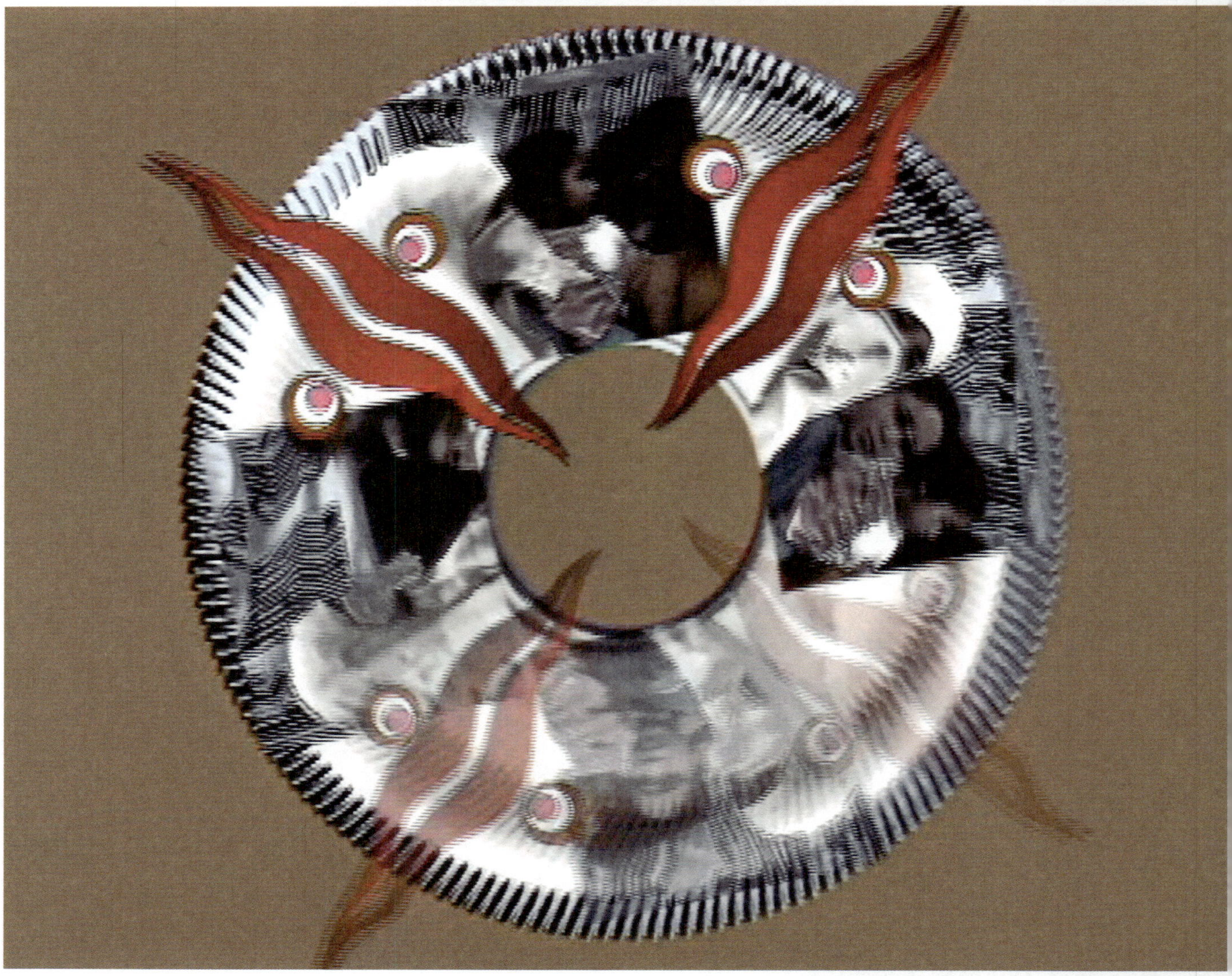

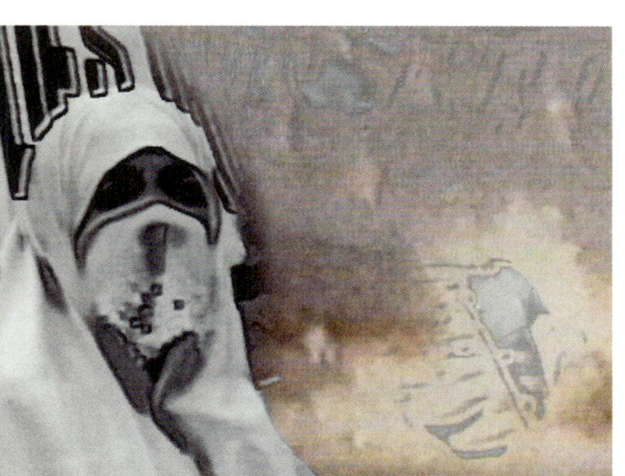

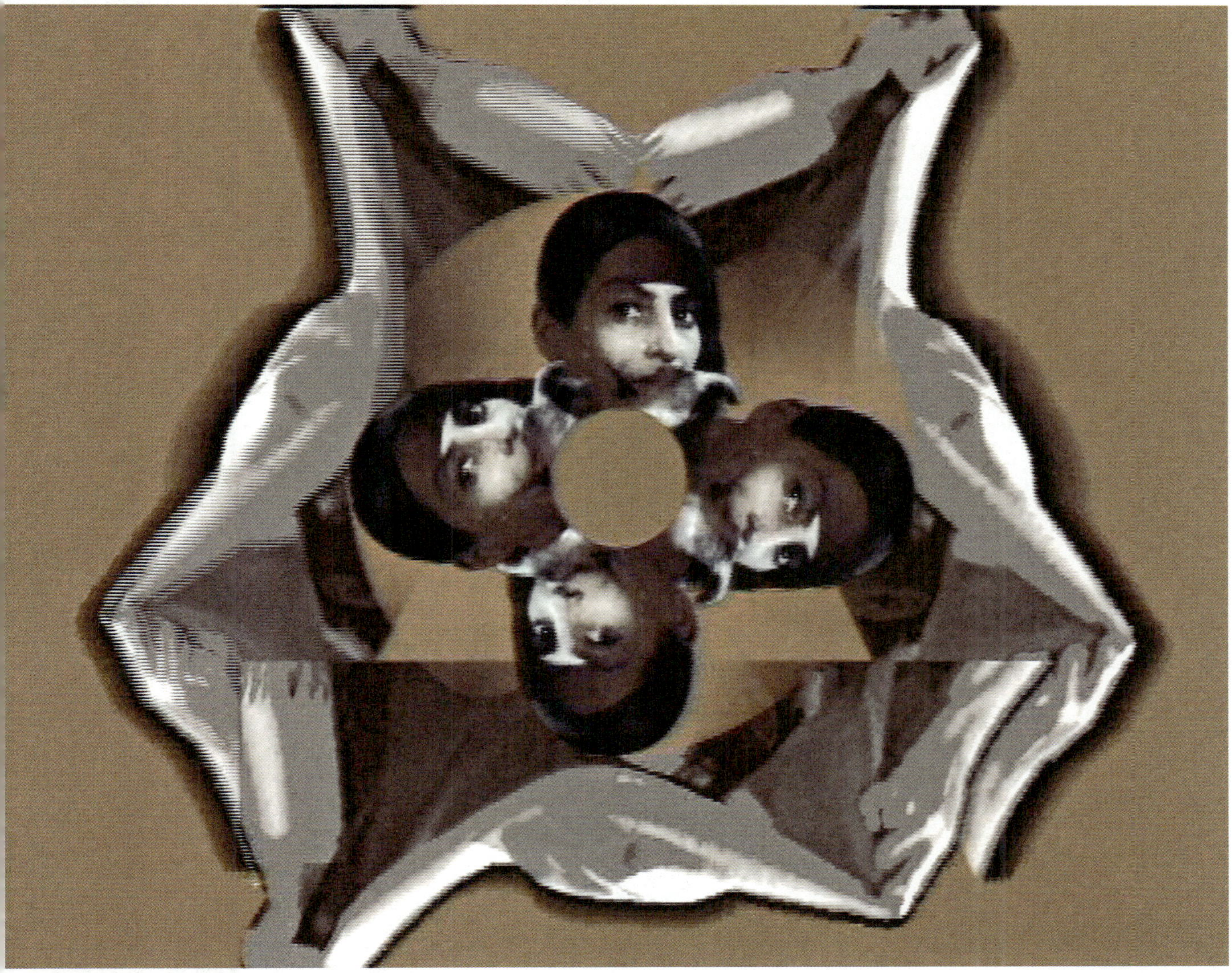

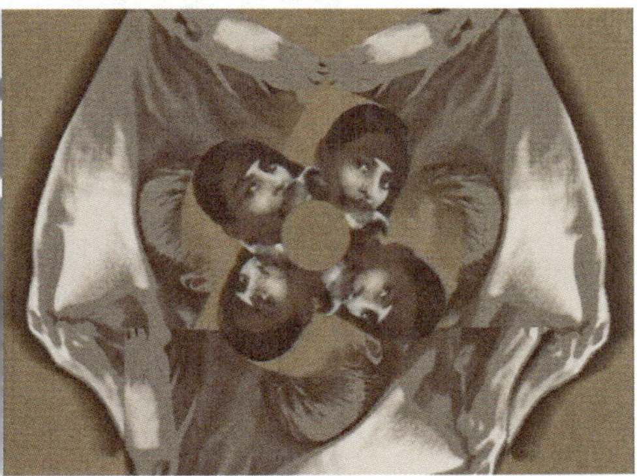

TRAVAIL
MORT

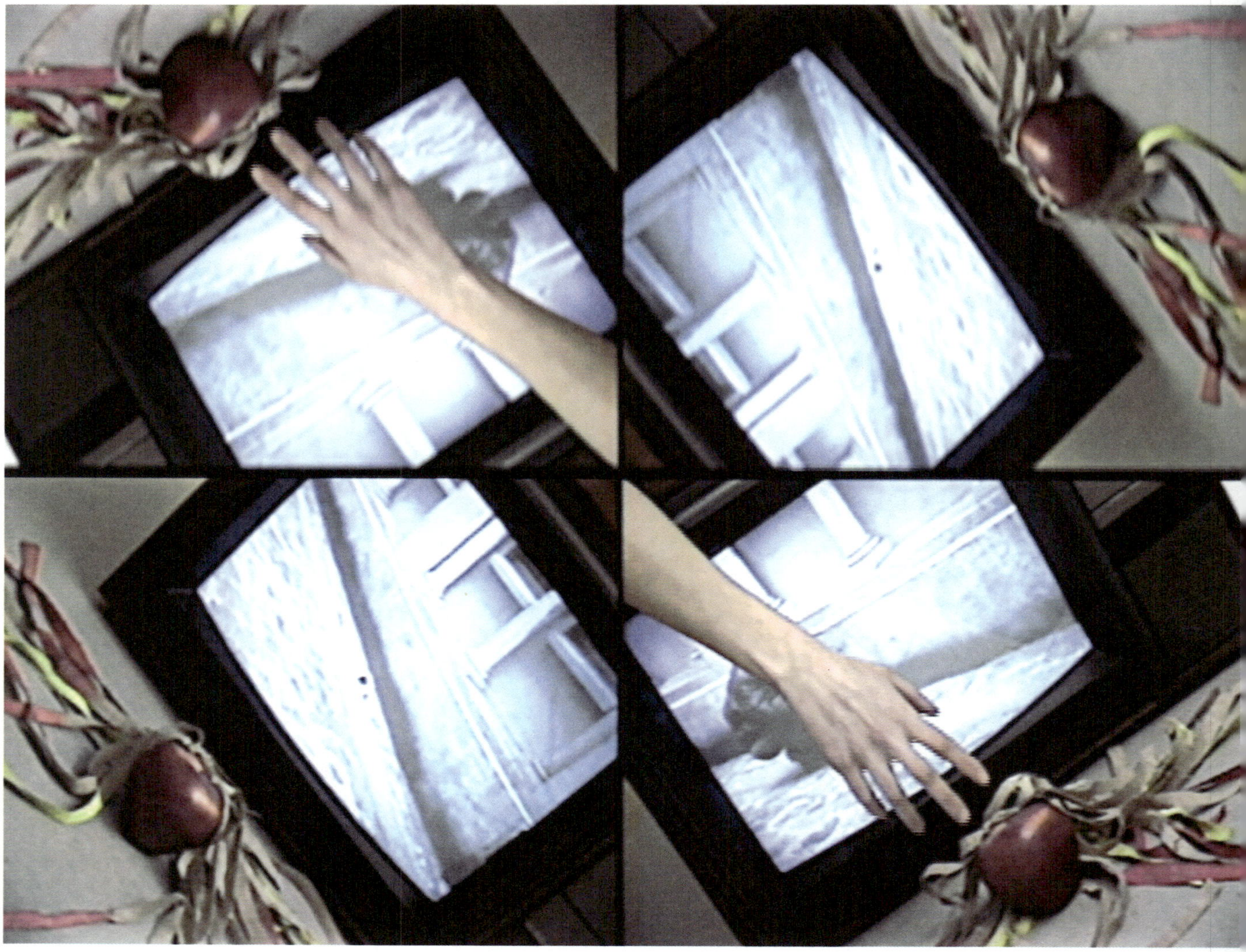

Nil Yalter: Circular Tension
by Omar Kholeif

Mousse Publishing

Nil Yalter as a child in Cairo, Egypt. Date unspecified.
Courtesy of the Nil Yalter archive

The following is a facsimile of a letter authored by the artist's mother, Guzin Berkmen. It was written to the artist in 1987 while she was working on the installation, Pyramis, a major work set in Egypt, where the artist was born.

The year is 1936   . The world is still not quite in the realisation of the
whirlwind and the turmoil that will strike it a few years later. At least life
is still pleasantly sheltered affording a certain stable security for the high
middle class and middle class where its members have the opportunity to receive
both an academic and intellectual as well as a cultural education.

The month is October 1936 . The Basarabia is sailing from Istanbul for Alexandria.
She and her sister ship the Transilvania, the two beautiful Rumanian steamers-
taking turns sail from Constanza every week for Alexandria via Istanbul_ ,Piraeus
and Beirut.

A young girl gets on board from Istanbul. She is accompanied by her mother and
future mother in law. She is going to Egypt to join her fiancé who has had to leave
before they could get married in Turkey. The formalities were not ready and he had
to leave as quickly as possible to be at the head of his job. He has been appointed
general manager of the cigarette factory which the Turkish State Monopolies has
opened in Cairo.

The voyage is smooth, beautifully sunny throughout the crossing except for one
stormy night off the shore of Crete. The passengers are told the sea is almost
always stormy off Crete.

Alexandria is finally reached. A city of wide avenues, sturdy buildings, rocky
beaches with foamy waves. A very cosmopolitan city with western appearance.

And after crossing the desert they finally arrive in Cairo.
Cairo, enchanting, lovely city with its beautiful"meydans" (squares)-Meydan Ismailiye-
-Meydan Soliman Pasha- Cairo with its wide avenues decked with trees, its magnificent
stone buildings reminiscent of old Egyptian architecture, its great luxury hotels
The Shepherd, The Semiramis, its excellent restaurants like the St. James, its
tea rooms like the Groppi and pastry shops L'Americaine.

Cairo the city of shopping with its stores Cicurel, Shemla, Sidnawi, its smaller
boutiques, the Gategno and Rivoli, all profusely rich in what they offer. There is
nothing one can't find in Cairo, the dreamland of the shopper.

Cairo, the city of great cultural activity offering a season of wonderful plays,
ballets, operas, concerts. The best theatre, opera troops, top singers, dancers and
musicians perform in Cairo- classical concerts and ballets to be contrasted by

avant garde plays and stylised oriental dances at the foot of the sphynx
performed by the then very well known Leila Bedirkhani.

First nights are always attended in evening attire with the king and queen or
court members usually present giving a fairy atmosphere to the splendour.

Cairo with its rich, golden museums, its long winding river and queint boats
sailing on it. Its suburbs Heliopolis, Maadi and Helwan where there are
beautiful villas surrounded by luscious gardens which flash with colours of
the bouginvellias, hibiscus, mimozas and greenary everywhere. One has never seen
such an abundance of roses anywhere.

Cairo , the city of Beledi (local people) where the mysterious charm and magic of
the orient reigns. The narrow streets full of the fellahin, noisy street vendors
selling fawa (mashed horse beans)., stands over-burdened with exotic fruit and
vegetables, noisy conjurors -the galli galli-
The beledi women always dressed in black, the men in long white tunics topped by a
red fez or a turban.
Cairo with its covered Bazaar overflowing with everything that the orient can offer.
Cairo with its mosques, its pyramids and sphynx, its vast stretch of desert.

The day is November the 5th 1936
The young couple is getting married at the Turkish consulate. The only people
present are the two mothers and the two witnesses. They celebrate the occasion
with a dinner party at the St. James restaurant. Later in the evening the couple
will go to Mena House for a week's honeymoon. Mena House the beautiful Edwardian
Hotel close to the Pyramids surrounded by a huge lovely park where the deer and
the peecocks gracefully walk about. A hotel where a gracious way of life is
predominant, excellent food, perfect service where people dress up for dinner
and where everything spells luxury and good taste.

The couple has rented a flat in a building in the residential district of Zamalek
on the island of Gheziveh. The flat overlooks the Nile and is joined to the
mainland by the lian recked Kasrel Nil bridge. The famous English Sporting Club
is right behind the building. The grounds of the club is beautifully kept and its
ever-green shade is a joy to watch.

The year is 1937   . It goes by with sight-seeing, visiting Luxor and Carnak,
enjoying everything that this richly endowed city offers- along with the joy
of expectancy of a child. The name has already been chosen for a girl - Nil -

Everyone finds the name enchanting and prayers go for it to be a girl.

The year is 1938 - the month January - the date the 15th. The prayers have
been answered- it is a girl - Nil, the bonny child with slanting eyes is born.
The year is 1939. The world will no longer be quite the same. A certain way of life
is on the verge of decline. The family returns home to Turkey sailing this time
on the Transilvania, which if not on her last trip, will soon be navigating no
longer between Constanza and Alexandria.

Nil Yalter: Circular Tension
by Omar Kholeif

"I was born in Cairo, but I woke up in Istanbul!" professes Nil Yalter. Her arms stretch toward her face, framing it as she holds my gaze, almost as if to task.

This afternoon in Paris and London, mediated across a web platform, Yalter and I are convening for the first time in nearly a decade. Ten years of loss, grief, and subsequent resuscitation have affected us both. As with any biographical tale, origin stories help locate the reader, be they the subject of a particular truth, fiction — or, in the case of many artists, one of constant myth. Yalter was born soon after the end of the Ottoman Empire in what was once perceived to be an artistic and scientific cradle of civilization — Cairo, Egypt. Her father was an economist working under the new regime of Turkish president Mustafa Kemal Atatürk, and was primarily responsible for directing the nation's monopoly of the tobacco trade. As Yalter speaks, mental sketches begin of my late grandfather, a physician and former dean of the Faculty of Medicine at the University of Assiut, exhaling plumes of smoke in the familial living room. He would have been sixteen years old when the artist was born, the same age that I took up smoking.

Growing up amid the tug of celluloid dreams, smoke filled the old Cairene movie theaters that I visited as a teenager. Flooding the proscenium-like venues of El Korba Square near Heliopolis, the haze served to mask the illusion of light and sound heard at the back of the theater, where children were instructed to sit. In most instances, we sat beneath the projection booth. The smog would send me drifting, as my nails dug into the velour seats, dreaming of an elsewhere where I could be the person that I knew I could never be in Egypt itself — an elsewhere where the erotics of poetry and the sublime phenomenology of art could be seen, *felt*, lived. The metaphor of the Freudian dreamhouse that pulls from the subconscious, today, feels like an apt beginning to Yalter's opening salvo.[1] Relocating to Istanbul, aged five, she professes not to have any memories before this time.

Could it be that the Bosporus at sunset, the whirling dervishes, the palaces, mosques, and bazaars of Istanbul,

---

1   Sigmund Freud, The Interpretation of Dreams: The Complete and Definitive Text (1899; repr., New York: Basic Books, 2010).

a place still referred to by some as Constantinople, is more lurid than the picture palace of Cairo, where the Hollywood of the Nile was unfolding its intoxicating Golden Age of cinematic dreams?[2] Which image, I wondered, was more impressionable to the mutable brain of a child who would grow up to become one of the pioneering artists to work in video?

The air conditioning in the room where I sit pushes the plumes of smoke so that it resembles a swarm of insects — the magic of light, that constituent node in photography and video that has fueled much of Yalter's practice, is playing its trick. It is a scene of daring adventure. One city straddles the perceived notion of an "east" and a "west," the other sits at a precipice — its rivers (the "two" Niles) serving as a gateway to the south and north — to the Mediterranean Sea in the case of the latter, and to the south, to the mythologies of African life fueled by literary and cinematic imaginaries conjured in the diaspora.

2   Viola Shafik, *Arab Cinema: History and Cultural Identity* (Cairo and New York: American University in Cairo Press, 1998).

Norman M. Klein and Thom Andersen, two historians of the diagonal cityscape that is Los Angeles, come to mind. In *Los Angeles Plays Itself* (2003–13), Andersen unbuckles the palimpsest of memory and its function. In his collage essay film, a city converted into a stand-in for itself, a postcard of Los Angeles becomes a more credible witness than its perceived artifact. Klein's monumental treatise on the mind and urban erasure, *The History of Forgetting* (1997), expounds through luxuriant prose and critical analysis on how a city can eat itself and wipe away the anchors that distinguish it.[3] Could it be that Yalter's investigations into the inner crevices of the psyche, through the embodied acts of producing performance, film, photography, and painting, serve as a means to cradle history, to unmoor the thresholds of time, to serve as a vestige for the leaking and seeping of reminiscence, to recapitulate these various facets as a means to explore a common diasporic experience?

3   Norman M. Klein, *The History of Forgetting: Los Angeles and the Erasure of Memory* (London: Verso, 1997).

"I was also born in Cairo," I interpolate. "I find it hard to believe that you cannot remember it. Your parents must have kept you locked in a room." I continue, jestingly.

My visits to Paris, where Yalter has primarily lived and worked since 1965, often involve hurried stopovers to museums to lay eyes on the drawings of the late Antonin Artaud. The French author and dramatist's pictures and sketches of interned life in an asylum reveal a cornucopia of entanglements in the mind-body continuum. Often untitled, the orbits of these intricate realms suggest propositions for how one can disentangle, interrogate, and parse one's own sense of constructed ontology. In 1988, the philosopher Jacques Derrida honed in on Artaud's reference of and to "the subjectile" to consider a theorization of madness, "the invisible quotation marks" of the subconscious that bring us to a point where we may query: Is what I say an intentional duplicity?[4] I have always interpreted this analysis as one of reclaimed agency, as a form of subjective realization for the perceived subaltern subject to craft a form of lived experience, to allow them the means to set parameters of their own visuality, one that can exist within the bounds of both narrative and visual hegemony.

Yalter's consciousness surrounding her beginnings locate us within the architecture of her present. Weeks later, when I visit the artist in the nineteenth-century apartment block where she lives, it seems from the outside that her home is housed in an exemplar of Haussmann-style architecture. The mansion-like grandiosity of the building's entrance is contrasted by the loose cable and wire found on the building's staircase.

"The builders have taken to Arte Povera," she says jokingly.

4   Jacques Derrida and Mary Ann Caws, "Maddening the Subjectile," *Yale French Studies*, no. 84 (1988/1994): 154–71.

Her reference to the Italian art movement, translated as "poor art," gives shape to the limitless paper files, the mountain of archival boxes, the seemingly antiquated computer screens and monitors, the stacks of VHS tapes, and the exhibition catalogues, which become architectures themselves, assembled neatly in every corner. They appear akin to leviathan-like bodies, mounted in the nooks of her home — awaiting collection, or confiscation? Limbs that are to be amputated until reunion occurs.

It is only two weeks before Yalter's eighty-sixth birthday, and she has been forced to move. Necessary maintenance. That the artist who popularized the phrase "exile is a hard job" finds herself in such a state is a poignant irony that is not lost on me. Over hours that stretch back and forth across decades, I learn about so many things that require further probing. Interactive artworks produced on DVDs in the early to mid-1990s, such as Pixelismus (1996)[5]

and Terra Nomade **(1997), summon references of nomadic**

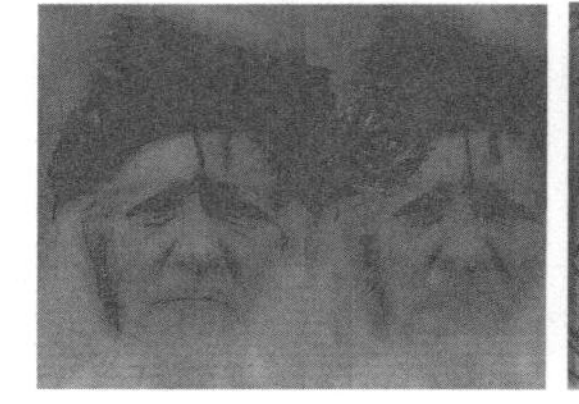

P. 1

modern life. Equally, they foreshadow the testosterone-fueled imagination associated with the violent first-person shooter video games that my siblings would play in strip malls growing up in the deserts of California and Saudi Arabia. Her paintings — contracting and refracting oval forms, the "circular tensions," sit wrapped up in her home, ready to be shipped off for exhibition. Several of these canvasses from the 1960s have been animated by Yalter into videos that present these images as living architectures. Developed in the early 2010s, they usher me back to my first encounters of the Op art of Bridget Riley and Peter Sedgley that I would encounter at what is now known as Tate Britain and the Hayward Gallery, both in London. References in shape and color to Byzantine and Ottoman motifs, such as Oval Form **(1967),** Circular Tension **(of**

5   Pixelismus began as a series of paintings in 1993 created by Nil Yalter, originally titled, La Chora (1993) (illustrated here). In 1996, the work was re-named Pixelismus to reflect the total work including the interactive component.

which there are several variations between 1967–69), and

Optical Cubes (1968), hint to the multidimensional field of

P. 111

Kazimir Malevich and Serge Poliakoff, as well as Japanese

metabolic architecture — womb-like constructions that

protract to offer the human body its space of dwelling.[6]

As I venture further into the different rooms of this

residence, a creaking footstep leads to layer upon layer

of fractious history. Photo collages are hung above picto-

graphic panoramas on the walls, revealing faces redacted

and hand painted that push the folds of memory back to

the migrant crisis in France in the 1970s; its ripples still

linger in the Parisian city air to this moment. Here, the bod-

ies of dispossessed individuals from Syria, Turkey, Sierra

Leone, and Senegal — those whose lives have been lost in

modern-day shipwrecks, gone overboard before life has

6   Yalter has often cited Malevich's concept of a fourth dimension as a driving impetus behind her pictorial composition in art. References to Russian Suprematism, as well as Art Informel, have been cited as well. See Fabienne Dumont, "Nil Yalter: From Constructivism to Socio-Critical Art," in Exile Is a Hard Job, ed. Rita Kersting (Cologne: Koenig Books, 2019), 180–87.

begun—can be heard as an echo, a resonant sound contoured to the shape of an ear.[7]

I want to return to Egypt. To the beginning. And we do, via a library of annotated first-edition volumes of dyssynchronous, fictionalized travelogues by François-René de Chateaubriand, who lent inspiration to Yalter's video installation filmed in Egypt, Pyramis: ou le Voyage d'Eudore (1987). I query the desire to embody her own life, as a Turkish woman through that of a polemical French diplomat and historian. Was it his solitude, his exile, or was it a wish to enter the mind and body of another character that enticed the artist the most?

With Chateaubriand as her avatar, could she venture forth to disentangle her own partiality, to absorb the object of the Egyptian landscape and environment for what it was, is and could be?

P. 7

---

[7] In 2018, a project entitled Fortress Europe documented a list of 34,361 deaths of refugees and migrants attempting to enter Europe who died en route due to the continent's border policies. The report is available at https://uploads.guim.co.uk/2018/06/19/TheList.pdf, accessed January 21, 2024.

The meandering path toward the act of quintessence that produced Pyramis, a work created when Yalter was half a century old, brings us back to the topography of colonial imagination.

"They see all of us from the Middle East in the same way . . ." The words flow from Yalter's mouth into the air before they dissipate.

The fated loop of geography – an imaginary proposition that still presumes a globe as a flat entity to be bisected into borders for military expansionism – is a topic that has preoccupied Yalter's lines of inquiry from the beginning. Her body, an animate vessel, seeks to rupture this notion, and now everyone is looking.

"Then let us not call ourselves Middle Eastern. *They* use this to perpetuate myths, and not least our sense of exile by giving us a name to embody." I agree with her, as I search for another language through which to express our connected experiences.

Yalter, born in 1938 in the continent of Africa, raised in Istanbul, and domiciled in France since the mid 1960s, has always devoured any literature on art available to her. Some of those books – on Expressionism, for instance – still live with her, publications that she has collected since she was fourteen. Having lived in France for much of her adult life, she has circumnavigated in loops, creating a tense counterbalance to notions of cultural hegemony. For decades the French cultural scene looked upon her with skepticism despite her successes. Rather than descend into melancholy, the public's raised eyebrows served to enable an insider-outsider perspective, nurturing a Marxist feminist who sought to convene worlds around constellations of women artists who shared similar interests in exploring perceptions of common difference. Yalter's art fashions worlds, where one's implication in the terrain of capitalism is left looming, prodding at you, interrogating the visitor: What will you do with this life? It is the superimposition of these interleaving lives – the many Yalters of yesteryear and today – that stimulate the skeleton of our collective imagination. For it is here, at the juncture of the communal and the individual, that we begin the labor, the "hard job" of constituting life itself.

In the wake of the events often referred to as May 1968 — a period denoting weeks of civil unrest in France propelled by an economic recession and anti-capitalist protest — came the formation of the women's liberation movement in 1970. This group associated with a branch of second wave feminism focused on the legal, social, and political rights of women. Several artists, authors, and public intellectuals became associated with the movement, but Yalter's intersectional contemporaneous interrogation of gender, ethnicity, and migrant life existed both inside and outside of the scope of public discourses of feminist debates at the time. She found route in communal explorations into the lives of French women, who like herself existed both literally and figuratively on the fringes of the sanctioned societal sphere. In 1973, Suzanne Pagé — who would go on to become one of the most consistently recognizable figures in the nonprofit French art scene — invited Yalter to present Topak Ev (1973) at ARC, the contemporary arena within the Musée d'Art Moderne de la Ville de Paris.[8]

8   Suzanne Pagé interviewed by Saskia Bos (2023) featured in Museums from the Inside: 60 Years of CIMAM. Available at https://cimam.org/news-archive/interview-to-suzanne-pag%C3%A9-by-saskia-bos/, accessed January 26, 2024.

Topak Ev, which translates from the Turkish to "round

P. 8–9

house," is also known as *Nomad's Tent* and *La Yourte*.

At first it may appear as a cover, a shell, a temporary no-

madic dwelling. Linguistically, "yurt" is synonymous with

diasporic exile and the term translates to "homeland" or

"motherland" in Russian and Central Asian languages.[9] The

artwork itself was inspired by shamanic rituals and tradi-

tions that the artist encountered in Anatolia, a peninsula

in western Asia today known as Kyrgyzstan.[10] Invoking the

formal composition of transient lodgings, the installation is

constituted of detailed materials like soft felt placed inside

a collapsible metal shell, annotated and draped in myriad

ways to evoke a simulacrum of multiple possibilities.

Topak Ev is accentuated by painted sheaths of fab-

ric and leather, inscribed with poetry and accompanied

9   A. Zhakupova, B. Suiyerkul, S. Tashimbay, and S. Berkimbayeva, "Kazakh Yurt as a Unique Creation of Nomadic Civilization Architecture," 10, no. 1 (2022): 1–13.

10   Nil Yalter, "The Woman's Tent: Plot for an Imaginary Exhibiton," (Spring 2016), https://artэeast.org/ quarterly/the-womans-tent-plot-for-an-imaginary-exhibition/.

by multi-media drawings of semi-didactic instructions, which forge a metaphor for a conglomeration of interstitial bodies. Animal skins, stitched together via interlocking threads, are suggestive of transversal entities. The perceived instructions reveal a bodily system that delineates the various compartments of the vessel of the hold, the womb — directions authored with a suppressed irony at the anthropological gaze directed at nonwhite women, individuals who were not offered the platform afforded to those of the diaspora in France. The tent's circular form and its detachable roof allow for a lookout. It can simultaneously be conceived as a lighthouse or, from a contemporary reflection, as a panoptical device, a camera that looks outward as much as it does inward.[11]

The roof of the yurt may also be conceived as a retinal detachment, an abstraction of the body from the carpeted meeting space within the lodging. Speaking to Yalter

11   Nil Yalter has often referred to Topak Ev as a "semiotic system."

about this now celebrated artwork, she informs me that at the time of its first presentation, the response to her installation focused on the formal and surface-level aspects of the piece. The consequence was negative. A critical audience was unwilling to make sense of it.

"Several journalists and members of the public were asking why this was in a modern art museum and not in an anthropological museum," she recalled.

Topak Ev exists as a metaphor of diasporic life. The collaborative impulses behind its formation and dissemination laid the groundwork for one of her first collaborative artistic endeavors, Paris Ville Lumière (1973) with artist Judy Blum. Black-and-white photographs form an annotated collage that presents a critical discourse of Paris.

The housing crisis, the subjugation of the female body, the spectacle of the perceived "other" are addressed in texts whose dual authorship allows for possible kinship — a referent that can be found in the apertures afforded to us through exile.

These concerns return to Topak Ev's centrality as a moveable tent, a connective tissue in Yalter's lived practice. It has become synonymous with a living organism. Its continued exhibition through the 1970s to the present as a vestige and arbiter of circular tensions that bind together the multiple forms of agitation that surround the sociopolitical context of exile, and the longing of the diasporic body for a womb to encase and suture it.

In one conversation, the artist articulates how the tent required regular housekeeping, much like any interior organ would. As it traveled, it became a place to receive offerings from visitors, such as dried cheese from nomadic communities. The gifts offered to the yurt led to worms and other organic life to breed in its fabric. It be-came an incubator, a site for gestation.

The thought-picture constructed by Yalter of the yurt as an animate and living vessel links one to Christina Sharpe's metaphor of the womb as a "belly," as a place where the female experience of birth cannot be extracted from the prison industry and its machinations.[12] Writing in the "wake" of "slavery," a framework that Sharpe argues, citing Saidiya Hartman, which continues to define "the cal-culus of Black life" since the perceived ends of slavery, enables a propositional mode to engage with contested histories. Specifically, Sharpe argues for the possibility of "wake work" — a form of aesthetic resistance that allows for a thriving imagining circumnavigating pasts that shall never pass.[13] Through this lens Topak Ev could be considered a

P. 13

womb as well as a hold, a site of capture and surveillance,

12   Christina Sharpe, In the Wake: On Blackness and Being (Durham, NC: Duke University Press, 2016), 42–43.
13   Sharpe, In the Wake, 54–55.

a metaphor for capitalism's mechanism of swallowing citizens into its machinations and subsequently spitting them out. A monument to the living and the undead, it continues a journey set forth into an orbit of its own.

In 1975, Yalter formed a different circular vestige that was dubbed Women in Struggle. The group's creation with artists Dorothée Selz and Isabelle Champion Métadier confronted the unsavory aftermath of gender protests in France, presenting histories that would neither pass nor wield to the contours of what was accepted in the public imagination. The collective sought to construct both a language and a means to negotiate work invented to create and substitute the lexicon that existed due to gender discrimination, not only within France's cultural scenography, but also globally. Constituted to reify, restore, and resuscitate narratives and aesthetic tendencies of women artists, Women in Struggle sought to turn the topographic sphere of art so that it would look inside out.[14] By circumnavigating local and national conditions through the ocular and affective perspectives of women – including artists working in a global, diasporic context – the political strategies deployed by Yalter and others would find new gloves, shelters, and possible outlets for visibility.

In this regard Yalter, whose name does not appear in the history books of creative endeavors related to second wave feminism, should be reconsidered not only a key actor within this movement, but also an errant pioneer – a Marxist feminist whose politics existed to decode the inequities of class, race, and ethnicity. Her gaze parses through the discordant, urban, modular experience of poverty in an installation completed during this time entitled Temporary Dwellings (1974–76). The work follows the

P. 20

lives of migrant communities in three major metropolitan spheres – Istanbul, where she was raised; Paris, where she

14   Derya Yücel, "A World for the Future," in Nil Yalter, ed. Derya Yücel (Berlin: Galerist and Revolver Publishing, 2013), 9–40.

has worked since 1965; and New York. Conceived via close encounters with individuals living in these spaces, the resulting installation is presented across several board panels comprised of Polaroids, drawings, text, and found detritus from each locality. These reliefs are accompanied by single-channel videos, evoking an interdisciplinary province that references the archaeological and sociological. The crumbling air and patches of dirt that annotate her drawings are presented in an indexical grid-like system that resembles the study of people, movement, and planning in wayfinding grid plan drawings and illustrations, a practice that has evolved since the Indus Valley Civilization to the present day in city-planning.[15]

Living among the transitory routes and outskirts of sanctioned society, Yalter often collaborated with ethnographer Bernard Dupaigne, who served as an interlocutor regarding her method and approach within this expanse

15 Mohan Pant and Shjui Fumo, "The Grid and Modular Measures in the Town Planning of Mohenjodaro and Kathmandu Valley: A Study on Modular Measures in Block and Plot Divisions in the Planning of Mohenjodaro and Sirkap (Pakistan), and Thimi (Kathmandu Valley)," Journal of Asian Architecture and Building Engineering 4, no. 1 (2005): 51-59.

of outsized society. In these unofficial neighborhoods, Yalter negotiated relationships with immigrant women who had been subjected to physical, social, and economic desperation at a time when they were expected to "integrate" within the confines of the patriarchal societies that they had come to inhabit. The probing questions that emerged during the mid to late 1970s continue as a red thread throughout the artist's practice to the present day: Turkish Immigrants (1977), Algerian Marriage in France (1977), and Rahime / Kurdish Women from Turkey (1979) through to Terranomade (1997) and Exile Is a Hard Job (an ongoing project that documents the lives of dispossessed immigrants). The latter, staged first in Valencia, seeks to involve migrant workers in the act of staging, looking, and annotating their very life and person. Informal collages drawn from her project Turkish Immigrants serve as

P. 107

a jumping-off point to engage the silenced voices of cities across the globe from Cologne and London to Sharjah and beyond. Constituted as a corporeal and ever-expanding archive, powered by a community-engaged energy, the project of documenting the diasporic body becomes a relentless pursuit.

Aesthetically, these various multidimensional works concerned with women's dispossession do not adopt a singular trope or strategy. Contrary to the anthropological, which focuses on studying the physiological and social aspects of a person over a durational period, these manifestations exist because of a contra-colonial practice of imagination. Multi-vocal and multi-modal, the dimensions of these markers of time, deploy aesthetic tactics that are conjured out of a necessity brought forth by circumstance. The first time that my eyes lingered across the faces of these women was in Turkish Immigrants. My gaze wandered not to the faces of the artist's subjects but to the redactions that she had created. A form of erasure, often painted in gold leaf, served as a gesture not necessarily intended to conceal the identities of subjects, but to speak of the circular tension of their lives — the assumption of invisibility. Who could or could not be known? The gold hues create an efflorescing glow that has a contradictory effect — amplifying the hidden, the marked, the unsanctioned, just as the light of a camera would. I appeal here to Krista Thompson's argument that "the contrapuntal," i.e., the contrasting counterpoint — that the oppositional narrative is both simulated and stimulated by the aesthetics of light.[16] The possibilities of framing sight, as Yalter does, crafts spaces where phantasmic bodies are reified through the camera's ability to both reveal and obfuscate it's given subject.[17]

What at first might appear as an archival impulse to take inventory becomes complicated in the artist's practice through her relentless process of interpreting, revisiting,

6   Krista Thompson, Shine: The Visual Economy of Light (Durham, NC: Duke University Press, 2015), 1–18.
7   Thompson, Shine, 1–18.

and reimagining the frame of immigrant life. The form that these annotations often create contrasts to the images that everyday spectators would traditionally bear witness to. Intimate pictures in a domestic setting are accompanied by large, handwritten messages in capital letters, invoking the aesthetics of the political campaign poster in works such as Paris Ville Lumière. Yalter's materiality complicates each scene. Charcoal, graphite, ink, paint, and chalk are often used to embellish, scaffold, and conceal singular and collective bodies. Cumulatively, her scenes of trans-migrancy can be read as grids, maps, and flags, which evoke multiple forms of consciousness.[18] The motor-sensory, operational, and default positions that humans assume are untethered from formal figurative convention or reproduction. Her art seeks to embody the person by pivoting toward the multidimensional possibilities of political abstraction.[19]

18   William Lycan, Consciousness and Experience (Cambridge, MA: MIT Press, 1996), 1–4.
19   Briony Fer, "Abstraction at war with itself," in Adventures of the Black Square: Abstract Art and Society, ed. Iwona Blazwick (London: Prestel, 2015).

In Algerian Marriage in France, Yalter presents a series

of photographs in the women's quarters of a segregated

Muslim wedding ceremony. In these scenes, photographs

reveal that the walls in these homes are adorned with car-

pets suspended precariously from ramparts, as if to create

the familiar context of being at home while still subjected

to the constant possibility of having to pack up and leave

at any moment. These photographs are mounted within

a colorful geometric pattern reminiscent of Andalusian

tiles, which offer a thin, soft form of protection to her sub-

jects like gauze — women living on thresholds, simulating a

domain of their own making.

In Algerian Marriage in France, Yalter presents a series

Geometric abstraction and its attendant political connotations appear for some – me included – early in life. It first appeared to me in the sanctuary of the church at the age of six: the cruciform sword – the cross, two planks of wood that mark a site of death, a literal and figurative evocation of Jesus Christ's sacrifice. The cross stood at the end of endless empty pews, as the church was occupied only by my mother and me. This womb seemed to offer my mother respite. It sutured the unceasing longing for a home as she negotiated her state as a diasporic citizen. It served as a place where the limits of her spoken language did not matter so much. The building's roof formed a large pyramidal skylight that still stands at the end of the street where we once lived in Glasgow, Scotland. The light was diffuse, seeping through an overcast sky that began to rain, casting shadows on the walls, floors, and our faces. More than a decade later, I returned to Glasgow to visit the Kelvingrove Art Gallery and Museum, where atop a staircase a similarly precarious sighting took place with Salvador Dalí's *Christ of St. John of the Cross* (1951).

On my first encounter, I was less preoccupied by the depiction of Christ himself than the directional perspective through which the cross was painted. As if filmed from a bird's eye view, the two abstract planks of wood resembled a dagger about to shoot into the Earth from above – or was it to ascend? The precarity of the scene filled the air with suspense, fueling a tension around the picture. In Dalí's depiction of the crucifixion, there are no nails or blood that annotate or mark the body. The perceived notion of the Holy Trinity, invoking the Christian doctrine of the Father, the Son, and the Holy Spirit, is evoked commonly, as it was here, via geometric principles that guided Dalí. Specifically, the artist connected circles through triangles – the inscribed incircle within a triangle.[20]

The site of this painting lingers with me as I look to Yalter's *Pixelismus*, a series of twenty-four interlocking paintings of the cross that the artist states were inspired

20   George A. Cevasco, "Dali's Christianized Surrealism," in   4, no. 180
     (Winter 1956): 437–42.

by the architecture of Greek churches. Distinct in this evocation is the fragmentation of the cross into constituent nodes, which come to form a whole when pieced together in a circular fashion. The crucifix appears. The act of deconstructing symbolism and the attendant visual histories associated with them continues in the interactive CD-ROM that Yalter developed alongside artists David Apikian and Nicole Croiset. In this depiction, a tessellating translucent form floats on screen, constituting a three-dimensional cross suspended midair. Above it in French are narrative fragments that the spectator can navigate. "The light of a metallic modernity/the sulfur candle/The temperature of Rubens and Rembrandt." The eyes gaze as this spectacular object rotates. Or rather, it is that the camera inside this game is rotating — never quite at 360 degrees, concealing as much as it chooses to reveal.

The rotations in Yalter's multidimensional works engage the author with the artist's investigative impulse — her desire to consider how the ocular is constructed. For the eyeball in and of itself is in constant rotation across three axes: horizontal, vertical, and torsional. It must find comfort to rest. The hovering abstractions and roving metaphors in Pixelismus seek to destabilize the phallic ver-

P. 4

ticality of the cross, and with it the sanctioned folds of art history evidenced through the painters whom she alludes to in her work, such as the Flemish painter Peter Paul Rubens, known for his compositions of Christian narrative history. Her reference to the Dutch artist Rembrandt — a member of the Dutch Reformed Church whose life and work is one of the most researched in art history — has been subject to a "light" not traditionally afforded to diasporic women artists.[21]

The domain of so-called great white men is one that offers Yalter context to intercede. Early paintings such as Circular Tension, Optical Cubes (1968), and Temple of

21   Amy Crawford, "An Interview with Stephanie Dickey, Author of 'Rembrandt at 400,'" Smithsonian Magazine, December 12, 2006.

Apollon (1969), were produced using acrylic and mixed media on canvas. At first these works might resemble the mechanical components of a technical apparatus. A second glance suggests the formal composition of rooms developed through the interplay of color and line, as in *Circular Tension*. These loops, conjured in *Optical Cubes*, engage contemporaneously. They appear to be in the process of combustion — a metaphor for the political tension that overtook the streets of France at the time as much as it references the circuitous loop of CCTV cameras. *Temple of Apollon* daubs the viewer into Yalter's realm of morphology — one where the Greek god Apollo is summoned — evidencing the artist's interest in literary mythology and the very act of mythmaking itself. She has explained time and again that these forays link back to the Supremacist visual phenomena articulated by the Russian-Ukrainian artist Kazimir Malevich, who argued that objects them-selves are meaningless in the world. Rather, what is of significance is the feeling that the "radical work of art can evoke."[22]

22   Liz Publika, "Black Square: Adventures in the Fourth Dimension With Kazimir Malevich," Artpublika, December 1, 2020, https://www.artpublikamag.com/post/black-square-adventures-in-the-fourth-dimension-with-kazimir-malevich.

A shroud of melancholy subsumes me as I leave Yalter after dinner on this Saturday night in winter. Gusts of wind accelerate. Distracted by the light of my phone, I stumble to the ground after bumping into a bollard, spilling the contents of my half-open backpack onto the pavement. There is a hint of romance afforded by the wet street of Avenue Montaigne where I find myself, named after the French Renaissance author Michel de Montaigne and known for parties at Bal Mabille and its luxury fashion stores. The cinematic imaginary severs me deeper from reality and into a dark sadness. A wandering scribe coughing, anxiously making their way across city grids to dingy hotel rooms, only to teleport into an elsewhere, feigning presence but always unseen. A group of passersby nearly step on my hands as I search for a pill bottle. The unconscious begins to wander back into the corridors of Yalter's apartment and into the fold of a video from 1977–78 that formed part of the installation The AmbassaDRESS. A chiffon wedding dress stands suspended on a round stand, a relic, a body without organs, a specter of an ambassador's wife who the artist professes to have once known.

The mythologies of the missing ambassador's wife are less significant than the stage that is set by her absence. The now-stained gown evokes a memory of a short film entitled Prenses Model (2018), directed by Dilan Engin. Engin's moving film explores the wedding dress as a metaphor for the impermanent state of women in modern-day Turkey. Prenses Model, translated as "The Model Princess," created a bridge to discuss complex acts of femicide occurring in Turkey.[23] Women gather around dresses, unpicking social mores and norms before descending into stories of hopelessness. It was after seeing this film that I was reminded of the tragic life of the Arabesque singer Bergen, who died at the age of thirty-one. I recall thinking that she, like me, had a lazy eye, which she concealed with her meticulously draped hair. In truth, her husband had thrown nitric acid in her face in a fit of rage. He was imprisoned, but

23   Durrie Bouscaren, "For This Turkish Filmmaker, Wedding Dresses Are a Metaphor to Discuss Femicide," World, April 9, 2019, https://theworld.org/stories/2019-04-09/turkish-filmmaker-wedding-dresses-are-metaphor-discuss-femicide.

upon his release the fated loop continued, and he killed Bergen on August 14, 1989. The AmbassaDRESS exists in

P. 116–117

opposition to forms of phallocentric violence, an aspect that is revealed in the accompanying photographs, drawings, and video work.

Monochromatic prints are paired into diptychs. They resemble ultrasound imagery as much as they do a womb that has been bisected. A seven-and-a-half-minute video veers the viewer into the interiority of the dress' fabric and into the private domain of a womb-like structure. This witness feels as though they are gradually entering a portal, a vortex for women only, a route to a prismatic architecture — one that exists justly for invited guests. The picture, like the photographs, is carved down its center, contracting and refracting into a cellular, almost interstellar realm of abstraction. As with her circular tension paintings, the pursuit of a fourth dimension — an alterscape — is choreographed using a multiplicity of media forms, and developed through a visual language that is entirely the artist's own.

The womb in this regard, abstracted and reconstituted, continues to resurface as a central theme in Yalter's art. That evening in Paris, after I'd picked up my belongings, I was haunted by The AmbassaDRESS. The womb's emblematic status, it could be argued, is reflective of the artist's embodied pursuit to disentangle the ocular gaze around the female body in society, as well as functioning as an allegory of gestating and nurturing. This mirrors the artist's pursuit in the 1970s of forging communities of and for women artists who were commonly excised from public narrative debates and discourses. The melancholy that had overtaken me that evening emerged from a sense of perpetual loss that has plagued me throughout life.

P. 114

My own mother believed the womb to be a marker of functional suppression. After the birth of her fourth child, she suffered from severe postpartum depression and accused me, the firstborn, of having scarred her body, and with it the perception of herself as a working woman — a poet, an author, a journalist, and a public figure were no more. At least that is what she believed at the time.

Had she considered the metaphorical potential of the womb, had she lived in a place where modern and contemporary art enabled a different form of fluency, her feelings could have been different. Examining Yalter's work intimately can be seen as an aggressive resistance to cultural hegemony, but all the while infused with love. I look back at The AmbassaDRESS and its connections to the refractive crucifixes. They ignite archives of inspired affection. I return to Simon Critchley's argument that the very crucifixion of Christ is a "quasi suicidal act performed out

P. 114

of love."[24] Before long, I am reading Rainer Maria Rilke's "Requiem for a Friend" (1909), where he wrote, "We need in love, to practice only this: letting each other go. For holding on comes easily; we do not learn from it."[25] The works herein begin to reflect the contradictions of love and loving, a feature in Yalter's *Le Chevalier d'Eon* (1978), one of the artist's most searing artworks, a lyrical treatise on the evolution of the most embodied of forms.

24   Simon Critchley, *Notes on Suicide* (London: Fitzcarraldo Editions, 2015), 6.
25   Rainer Maria Rilke, "Requiem for a Friend," trans. Stephen Mitchell, *Paris Review* (Winter 1981), https://www.theparisreview.org/poetry/3205/requiem-for-a-friend-rainer-maria-rilke.

Like a bruise that changes color from red to purple, love and its infectious pangs evolve, stretching and allowing for tolerance and acceptance. But what do we see of affect? A phantasm that we have chosen to construct, or a person in all their sincerity. Love's manifestation is interminably tethered to the form of the body – its shape, contour – its ultimate form. Our bodies are rarely our own, but the subject of constant contest in the political sphere. Our bodies have become politicized vessels, entities that stand in as allegories: gendered, sexualized, racialized, and, what?

Ta-Nehisi Coates struck a public nerve in 2015 when he wrote that "the black body, its heritage [exists] to be destroyed."[26] The fight or flight response that greets the body in moments of violence leaves it a bulbous and bruised vestige, a remnant that is subject to ceaseless tampering and abuse. The cultural conditions of and for the ethnic Global Majority, for women, for the LGBTQ+ society, are still governed and under the thumb of a particular kind of patriarchy. How foolish have we have become as individuals? In 2022, the US Supreme Court proposed a landmark ruling that overturned Roe v. Wade (1973), a precedent that had protected a woman's right to her own body – to "choose" the context and condition of an abortion. Shock and grief devolved the woman's body in the United States to a two-tier system, subjected to the whims of individual states, private medicine, and fiscal resource.[27] Class and race again emerge more fully into focus. "Who can afford the right to choose?" appears as a recurring point of interrogation. Intersectional feminism's perceived failure due to its "totalizing notion of gender" comes back to the fore of public debate.[28] Are we *truly* here again, nearly forty years later, *totalized* subjects to another form of totalitarian politics? Surely not.

26 Ta-Nehisi Coates, "Letter to My Son," *Atlantic*, July 4, 2015, https://www.theatlantic.com/politics/archive/2015/07/tanehisi-coates-between-the-world-and-me/397619/. See also Ta-Nehisi Coates, *Between the World and Me* (New York: Random House, 2015).

27 Moira Donegan, "A Year Ago Roe v Wade Was Overturned. Grieve for the New America," *Guardian*, June 23, 2023, https://www.theguardian.com/commentisfree/2023/jun/23/roe-v-wade-overturned-abortion.

28 Ashlee Christoffersen and Akwugo Emejulu, "Diversity Within: The Problems with 'Intersectional' White Feminism in Practice," *Social Politics: International Studies in Gender, State & Society* 30, no. 2 (Summer 2023): 630–53.

In his scrutiny of visual Black culture, Greg Tate's writing introduced me to the cultural appropriation of Black aesthetics — an aspect of my person that is not evidently visible to the spectator. In *Everything but the Burden: What White People Are Taking from Black Culture*, an edited anthology published in 2003 after the end of a captive history bookended by the events of 9/11, Tate laid bare a new analysis of embodied extraction and articulated the ongoing risk of fetishization. Here was a culture in captivity, held hostage by the whims of an elite aristocracy.[29] The closing stream to his edited volume, authored by artist and filmmaker Arthur Jafa titled "My Black Death," shook me with its unrelenting argument toward the inevitable loss of self that awaited. "There is no self-determination"; "There is no self-possession." The "double bind" of ontological existence leads us to a "good death."[30] An excising of oneself

29 Greg Tate, "Introduction: Nigs R Us, or How Blackfolk Became Fetish Objects," in *Everything but the Burden: What White People Are Taking from Black Culture*, ed. Greg Tate (New York: Broadway Books, 2003), 1–15.

30 Arthur Jafa, "My Black Death," in *Everything but the Burden*, 246–57.

for another. Two decades pass. Online streamers with their surfeit of programming might suggest a realm of boundless inclusivity. But scratch beneath the surface and the tender skin, and the wound, is ever present.

The right to be human continues to be under threat, from here and there to everywhere. The "hysterical crowds" that ushered the streets in protest — in revolutions from Cairo to Beirut, in what was deemed a worthy celebration of and for freedom, has collapsed societies unto themselves. In the case of Cairo, where both Yalter and myself were born, the nation has once again fallen into the hands of a military dictatorship governed by a demagogue, who refuses to address social inequity, freedom of expression, or LGBTQ+ life. Entire communities have been the subject of severe persecution.[31]

The sweep of violence extends across oceans. In one of the largest surveys of transgender individuals ever

31 Naomi Klein has been a key reference point and an outspoken advocate with regard to government persecution in modern-day Egypt. See Naomi Klein, "Greenwashing a Police State: The Truth Behind Egypt's Cop27 Masquerade," Guardian, October 18, 2022, https://www.theguardian.com/environment/2022/oct/18/greenwashing-police-state-egypt-cop27-masquerade-naomi-klein-climate-crisis.

conducted, revelations of brutal harassment and unrelenting scaremongering persist in both the physical and virtual sphere.[32] In the public realm, women who endure sexual violence continue to be propelled to speak truths that fuel ideological culture wars, only to be forgotten. Applauded for their camaraderie to the cause, they themselves become marked, just like the bodies of Black humans shot and bludgeoned to death without cause.[33] When will they call them by their names? Aura, Breonna, George, Laquan, Janisha . . . ?[34] Can you breathe?[35] There is of course a "supposed" rhyme and reason for everything. Thinking of the current state of complex carceral injustice not only in the United States but also globally, one cannot help but draw parallels to the fated inevitability of the incarceration that one is afforded through the silkscreen, the border

32 Chandelis Duster, "Largest Survey of Transgender People in the US Reveals Key Insights at a Time When Trans Rights Are under Attack," CNN, February 7, 2024, https://edition.cnn.com/2024/02/07/us/us-transgender-2022-survey-reaj/index.html.

33 For considerations of the complex discussions that have emerged in the rise of recent feminist debates—specifically in light of the #MeToo movement—see Katherine Angel, Tomorrow Sex Will Be Good Again: Women and Desire in the Age of Consent (London: Verso, 2021); Mithu Sanyal, Rape: From Lucretia to #MeToo (London: Verso, 2019); Amia Srinivasan, The Right to Sex (London: Bloomsbury Publishing, 2022).

34 Alia Chughtai, "Know Their Names: Black People Killed by the US Police," Al Jazeera, 2020, https://interactive.aljazeera.com/aje/2020/know-their-names/index.html.

35 Direct reference to George Floyd's cry that he could not breathe while being murdered by a US police officer. I discuss the visuality of this scene as well as its aftermath in the following book chapter: Omar Kholeif, "I Can't Breathe: Black Visuality in the Metaverse," in I'll Be Your Mirror: Art and the Digital Screen, ed. Alison Hearst (New York: Delmonico Books, 2023).

that is one's body — the prescriptive, the assigned, the non-aligned. Is this the vestige of a slave's body, or one that re-sisted? Is this a vessel that can *pass*? Was I born, assigned, a gender all my own?

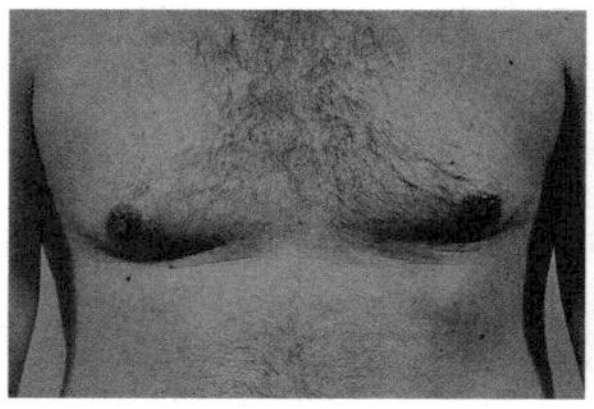

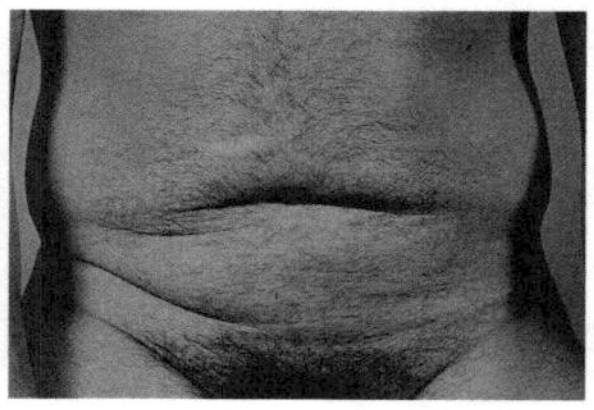

These circular tensions persistently present ques-tions that seem both antiquated and deeply contempora-neous. The sketch of these multiple folds of history finds a perfectly suited outline in the cumulative light of Yalter's landmark installation Le Chavelier d'Eon. A side view, a frontal view, a bird's-eye view of a body in motion. Daily ritu-als accumulate. Are shaving chest hair and stretching a pair of panty hose over legs routine activities, or are they be-ing practiced, learned? These framed black-and-white pic-tures sit in proximity to a pair of images surveying a body in fragments, in close-up — a chest, a torso. Although ambigu-ous, one's ocular sensibilities assume this to be the body of a male figure, a detail revealed as the spectator's gaze wanders to a pair of Polaroids that present a striking figure in a sequined black dress, draped in fur sitting before the camera with their lips plumped and pursed, black mascara, and blush. Their gaze is not tentative but filled with a mixture of innocent excitement and whimsy, their eye summoning as if to say, "Can you see me, I enjoy how *this* situation feels, to be seen like this, even if for the first time."

The form of these narrative gestures is annotated through Yalter's deployment of visual aesthetic metaphors. A blush-colored painting, almost blood red, punctures the orthography of conceived male gender with its metaphor. The painting is could be read as a direct reference to Alexander Rodchenko's Pure Red Color (1921), where he argued that he had "reduced painting to its logical conclusion. . . . It's all over. . . . Every plane is a plane and there is to be no representation."[36] By invoking this disavowal of

36   See The Death of Painting (1998), Museum of Modern Art, New York, https://www.moma.org/interactives/exhibitions /1998/rodchenko/texts/death_of_painting.html, accessed February 12, 2024. Also see Arifa Akbar, "Drawing a Blank: Russian Constructivist Makes Late Tate Debut," Independent, January 2, 2009, https://www.independent. co.uk/arts-entertainment/art/news/drawing-a-blank-russian-constructivist-makes-late-tate-debut-1516801.html.

representation, and in turn the figurative reproduction of bodies, Yalter proposes that gender is not only mutable, but that sexual identity begins from a point of abstraction that is consistently reformulated around *felt* experience, tendencies that evolve through the affecting possibility of encountering one's self throughout life.

In author and philosopher Didier Eribon's blazing sociological memoir Returning to Reims (2009), the author — who is referred to as "openly gay" — makes a poignant and interminable reflection on constructions of the self. He notes that "the process by which you come to belong to yourself or to transform yourself" is reflective of multiple acts of constitution and reconstitution. To assume an identity, for him, is linked to "the refusal of an identity" — a woven "link" whereby the assumed and the refused are to be found "holding each other in check."[37] One interpretation of Eribon's position suggests a double form

37   Didier Eribon, Returning to Reims, trans. Michael Lucy (2009; Los Angeles: Semiotext(e)/Foreign Agents, 2013), 97–98.

of exile – not only from and of one's body, but equally from their environment.

Yalter's subject, a close compatriot, unfurls the interior experience of negotiating their gender duality on screen and in pictures. In a pivotal centerpiece, one can bear witness to her subject as a deconstructed entity. A pair of breasts are draped over a television. Two feet brush across the screen, fracturing the space-time continuum. The picture before us is now split in two. In between our protagonist's legs sit two screens with protruding lips. He is still in jeans, ostensibly man's clothing. We follow him as he traverses streets, dressed in an oversized dress and a large pair of glasses, wandering through thoroughfares like a flaneur. The transitive act occurs when we see him standing in front of what appears to be a shop window filled with multiple television screens. Each of these boxes presents an anterior version of the character that we come to know as the Chevalier — the knight of the round table.

In Yalter's studio in January 2024, she thanks me for drawing attention to this work. When I first approached the artist to present this work a decade prior, she was eager to respect the privacy of the individual whose journey we come to follow. Those of us who have engaged in concerted acts of LGBTQ+ activism understand that the 1970s was notable for the LGBTQ+ community. Same-sex relationships were decriminalized in Great Britain shortly prior in 1967, and homosexuality was removed from the American Psychiatric Association's list of disorders in 1973.[38] In 1972, Sweden became the first country to legally allow its citizens to change their gender, abolishing identification such as "trans-," i.e., "transvestism" and "transgenderism," from being classified as a "disease."[39] Still, cultural discord and alienation persisted — as a policy, arguably, does not alter societal attitudes. One could not mitigate

38   "The APA Ruling on Homosexuality," New York Times, December 23, 1973.
39   Mikael Baaz and Mona Lilja, "I Felt a Little Homosexual Today, So I Called in Sick: The Formation of 'Reverse Discourse' by Swedish Gay Activists in the 1970s," Global Society 36, no. 3 (2022): 330–46.

circumstance – being referred to as "sick" by the legal apparatus of society one moment, only to assume that communal acceptance and coherence was forthcoming, was a troubling notion.[40] The case of transgender identity in France where Yalter worked, was still a cultural taboo. Her compatriot underwent an exploratory gender transition in the mid to late 1970s. It was not until 2010 that the nation stopped violently antagonizing trans people's right to exist, removing the term "identity disorder" from its legal- and state-sanctioned attributions.[41]

I had not spoken with Yalter in nearly a decade when this book began. The first thing that I asked her about was the Chevalier. She shakes her head, as if to suggest that he/she/they may no longer be well or alive. Seeking to offer what comfort I can, I inform her that the work remains one of the most important political works of art ever produced. For the Chevalier is no victim here. The fragments of their

40 Baaz and Lilja, "I Felt a Little Homosexual Today," 330–46.
41 "La transsexualité ne sera plus classée comme affectation psychiatrique," Le Monde, May 16, 2009.

body circumvent categorization, floating across screens and back, reiterating the stringent binaries assumed of gender. The Chevalier's name is naturally of great importance. It is a reference to Charles d'Éon de Beaumont (1728–1810). Based in Versailles, a place steeped in mythology and often referenced by Yalter, Beaumont was a French soldier who also dabbled in writing and spying, as well as serving as a diplomat. He often dressed as a woman stirring confusion, and his corpse was reportedly exhumed to confirm that his biological gender was indeed that of a male.[42] According to certain accounts, Beaumont originated the term "eonism" for "transvestism."[43] The notorious figure also lends Yalter's subject prophetic speech, words to renounce the previous self; we give in to the fight. The male voice professes: "After having been an honest man, a zealous citizen and a brave soldier all my life, I triumph in being a woman."[44]

42   Simon Burrows, The Chevalier D'Eon and His Worlds: Gender, Espionage and Politics in the Eighteenth Century (Edinburgh: A&C Black, 2010), 82-91.
43   Estelle Vallender, Exile Is a Hard Job (Cologne: Koening Books and Museum Ludwig; Annandale-on-Hudson, NY: Hessel Museum of Art, 2019), 137.
44   Omar Kholeif, "Close-Up: Change of Subject," Artforum, April 2017.

The artist's choice to invoke the figure of a historic French aristocrat is emblematic of her penchant for weaving together divergent temporal orders and literary references from the ancient and modern to speak back to her audience — not as an anthropologist, but as a critic of cultural hegemony. Yalter's desire to fracture the ontological construction and visual pleasures associated with the male gaze persists throughout.[45] In *Le Chavelier d'Eon*, she approaches her subject through an empathic lens, affording them a kind of generosity that is inextricably tied to love. Her accumulating portrait of tessellating forms reveals the Chevalier's body not as a subject of sexual objectification, but as an extra-territory subject to its own situated set of aesthetics.

In Yalter's aesthetic experiment, the choice of luminous forms and her visual vocabulary are adventurous.

45   Laura Mulvey and Rachel Rose, (London: Afterall Books, 2016).

The work's mirrored, doubled, and collapsed scenes speak a language that is as audacious as that of Nam June Paik and his circuitous *TV Buddha* (1974) and the cumulative portraits that emerge in works such as *Internet Dream* (1994). Vito Acconci, Vera Molnár, and Lillian F. Schwartz are just a few of the artists who Yalter has exhibited alongside in shows staged across Europe, and Germany in particular. Her profile amid what was at one point referred to as the sphere of "media arts," "interactive art," and "new media" is in no small part due to the encouragement and support of her late partner, Joël Boutteville. Boutteville was a curator of technologically led art who was known, among other things, for cocurating media arts festivals such as Espaces Interactifs Europe.

To construct a life, to fuel a present despite extant pasts that will not go away, proves to be the result of an unrelenting feat of determination. No war zone, no weapon, no vestige of trauma can hold Yalter back from the throes of her creative will to express not only the latent wounds of her own life, but also of those who orbit her. The grieving stranger enters the gallery to find themselves extracted from the buckle of the repressed interiority of mental images that haunt before they short circuit. Here, the fuse of energy is hollowed from its core to the base. There is nothing left here. Feelings linger, the rest of me is on the floor; I wave goodbye.

Walter Benjamin once intimated that seeking his home was one of his greatest "afflictions."[46] We end where we began. In search of a route to a root, to a beginning that the diasporic individual shall rarely find. Expanding on the affliction of longing for home, T. J. Demos has suggested that Benjamin's resilience throughout life was gleaned through a reclamation of exile as a form of narrative resistance, a sensibility also evident in his posthumously published memoir, *Berlin Childhood around 1900*.[47] The punitive delineation of the ostracizing term "exile" as an experience of traumatic displacement is examined by Demos as a possible lens for decoding the ocular in *The Exiles of Marcel Duchamp* (2007). Here, the author expands how we read one of the most recognized artists of expatriation, Marcel Duchamp, using his biographical ostracization and split life across Paris, New York, and Buenos Aires, to narrate — through this — a condition of generative multiplicity assembled and developed through ostracism.[48] Exile in Duchamp's art, from his portable museum suitcases, which assembled miniatures of his works of art, to his constructed sculptures produced for travel, use the peripatetic context of movement as a means to open up possibilities for museology and display as opposed to relying solely on configurations of identarian politics. It enabled the artist a constant mobility — a sustained physical and mental space for becoming.[49]

46   Friedrich Nietzsche, *Thus Spoke Zarathustra*, quoted by Walter Benjamin as an epigraph for "Paris, Capital of the Nineteenth Century," in *The Arcades Project*, trans. Howard Eiland and Kevin McLaughlin (1939; repr., Cambridge, MA: Harvard University Press, 2002), 20.
47   T. J. Demos, *The Exiles of Marcel Duchamp* (Cambridge, MA: MIT Press, 2007); Walter Benjamin, *Berlin Childhood around 1900*, trans. Howard Eiland (Cambridge, MA: Harvard University Press, 2006).
48   Demos, *Exiles of Marcel Duchamp*, 14–17.
49   Demos, *Exiles of Marcel Duchamp*, 3.

When I began this journey, our experiences in constant exile found visuality in the streets of Sharjah, where Yalter had installed a newly imagined construction of her project

P. 120

Exile Is a Hard Job as part of Sharjah Biennial 15: Thinking Historically in the Present. Looking at these pictures affixed to brick and mortar, I was reminded of the imprints of memory and architecture. Chiefly, of the mutable contours of the embodied self, and how it trickles through the palimpsest of time. I continued to search for the shared vessel that linked our lives: Cairo, the city of our birth. Our first discussion began with a reflection of lived pain — of two specific losses in our lives, and the isolation that these engendered, the ruin that a ghost leaves behind. At the time, I was delivering a lecture regarding the concept of pain and its potential to serve as a vessel through which to speak "truth to power" in the sphere of art institutions.[50]

50   Paul A. Bové, "Introduction," Edward Said and the Work of the Critic: Speaking Truth to Power (Durham, NC: Duke University Press, 2002), 2–8.

The example that I was using assumed a particular form of embodiment in the activism against the opioid crisis, principally against the Sackler Family, whose names have historically annotated the wings of several of the world's most influential museums. These efforts were visualized in filmmaker Laura Poitras's award-winning documentary with artist Nan Goldin, All the Beauty and the Bloodshed (2022).

The film moves us through the advocacy efforts of the group PAIN (Prescription Addiction Intervention Now), as narrated through Goldin's sit-ins and protests at museums, to highlight the family's funding of cultural organizations. How could one hold the family that profited from Oxycontin accountable, if the public vessels of people's shared and collective histories were to be forever marked by the imprint of their name? A haunting that does not relent. Three-quarters of the way through the film, in one of the most dramatic scenes, a group of activists are seen taking over the Metropolitan Museum of Art's Egyptian Wing, filling the skylit galleries and expanse of water that sur-rounds the recreated installation of the Temple of Dendur with empty prescription bottles. They fill the artificial scenography and take to the floor, staging a metaphorical "die-in." It appeared to be a very peaceful revolt.

The connotations of this act of naming were suffo-cating, seemingly impossible to reconcile. A connective tissue was formed to the work of artist Lubaina Himid — who I have long studied and worked with — whose land-mark installation Naming the Money (2004) serves as a constant reminder of how language, specifically the act of naming and annotating Black bodies, functioned as a primary ideological weapon in the triangular slave trade.[51] It had not occurred to me that the residues of a shared cul-ture were annotated with the lives of contemporary slave barons — a family who profiteered from the lives of those in despair, and whose financial contributions to museums,

51 See Lubaina Himid, Meticulous Observations and Naming the Money (2017), National Museums Liverpool, https://www.liverpoolmuseums.org.uk/whatson/walker-art-gallery/exhibition/lubaina-himid, accessed February 14, 2024.

or "gifts" [tainted gifts] as they are referred to, served as tax write-offs.[52] Throughout my doctoral study, I was unaware that I was researching fragments of my own culture in museums that bore the Sackler name. The Arthur M. Sackler Museum at Harvard University is recognized for its collections from Egypt and what the museum refers to as the "Near East." The Freer-Sackler Galleries (now the Freer Gallery of Art), part of the US civic museum consortium known as the Smithsonian Institution in Washington, DC, was also a key reference, as it holds one of the most expansive and known constellations of Afro-Asian art from ancient times to the present. At the Louvre, the galleries where I studied Eastern antiquities were also, at the time of my study, named after the Sackler family. If this was a marker of anything, it was of my naivety. Here, I was authoring a thesis on the decolonial possibilities of art, meanwhile, the public imagination was being colonized by a

52   Tom Seymour, "Tainted Gifts: As British Museum and the Met Disavow the Sackler Name, Museums Rethink Donation Deals," Art Newspaper, March 28, 2022, https://www.theartnewspaper.com/2022/03/28/tainted-gifts-museums-rethink-donation-deals.

different vision: one where people's lives were sacrificed by US capitalism and Big Pharma to allow for – and sustain – those ghosts, the ancestors of the dispossessed, to breathe in the vaulted galleries and corridors of the world's most storied museums.

Yalter's work offers relief and a continued space for relearning, as Eribon intimated, the self that we reject and the one we choose or seek to claim as our own. In 1987, Yalter was commissioned to produce a new work. She used her modest budget on a plane ticket to Egypt for her and a friend, borrowing a Portapak camera and attendant resources for her journey. Rather than attempt to reconstruct details of a childhood that she could not remember, she used the writings of the eighteenth-century French author and diplomat François-René De Chateaubriand (1768–1848) to guide her journey up and down the Egyptian landscape. Traveling by foot, car, or boat, the artist used

Chateaubriand's literature, known for collapsing various historical orders, to guide and approach the experience. This led to the creation of Pyramis: ou le Voyage d'Eudore,

P. 6

a twelve-screen, three-channel video installation that was first exhibited in Châtenay-Malabry, a suburb of Paris where Chateaubriand lived during his lifetime.

Yalter explained that the title, Pyramis, was inspired by a pyramid-shaped pastry sold in ancient Greece. The reference connects the viewer to narrator, Eudore, a nineteenth-century Roman officer of Greek origin who features in Chateaubriand's dyssynchronous novel The Martyrs, first published in 1859. In Greek mythology, it is said that Eudore or Eudora was a name given to three nymphs who presented "good gifts."[53] Yalter does not parse Chateaubriand's folklore but rather enters the city-scape and the country, constructing a phantasmagoric

53   Carl Kerényi, The Gods of the Greeks (London: Thames and Hudson, 1951), 41, 64.

dreamscape where differently located CRT monitors serve as mirrors, creating a disorienting view of space and time. Heard overhead is Yalter's voice, embodied through Chateaubriand's prose and his infamous annotations.

The abstraction of Yalter's person within the work — who is neither seen nor embodied on film but rather through an avatar — affords the artist visual possibilities that elide the conventions of the colonial tourist gaze.[54] Scenes of beaches, Pharaonic figures, and Bedouin faces are contorted in the closed-circuit loop of the video installation, withholding the ethnographic tendencies often associated with such acts of looking. Here, local gossip and the monastic life of the Egyptian countryside are suggested without being articulated. Yalter obscures. Her work is akin to a travel book that has fallen out of print — we can reference it but are unable to cite it ourselves. What Yalter can

54   Peter Lyth, "Carry On up the Nile: The Tourist Gaze and the British Experience of Egypt, 1818–1932," in *The British Abroad Since the Eighteenth Century, Volume 1. Britain and the World*, eds. M. Farr, and X Guégan (London: Palgrave Macmillan), 176–78.

offer the spectator is a profound negotiation of narrative reclamation, one that embodies the chaotic madness of a place through an act of visual refraction; it cannot be objectively or numerically assessed. It demands its own coherence. Nil Yalter's work extends into multiple spheres, ones that mandate a distinct form of labor from her audience. Convening worlds, hers is a practice that proffers a manifesto for living — one that expands the silhouette of exile. For her, as for us, the diaspore is the seed that, through dispersal, finds its own logic and function in the world — rightfully or not, on its own.[55]

55   The term "diaspore," as referred to in botany, is one that I began to use thanks to my engagement with the art and life of Otobong Nkanga. See Louisa Elderton, "Interview with Otobong Nkanga," White Review, October 2014, https://www.thewhitereview.org/feature/interview-with-otobong-nkanga/.

ARTIST BIOGRAPHY

Nil Yalter (b. 1938, Cairo, Egypt), lives and works in Paris. A pioneer in the French feminist art movement of the 1970s, Yalter was educated at Robert College, the prestigious American secondary school in Istanbul. While she was engaged in dance, theater, and painting during this time, she also practiced pantomime and traveled by foot to India as a pantomime artist. Yalter has lived in Paris since 1965. She participated in the French counterculture and revolutionary political movement of the late 1960s, immersing herself in the debates around gender, migrant workers from Turkey, and other issues of the era. These social movements and ethnographic science have influenced the artist's videos, performances, and installations from the 1970s onward in the form of idiosyncratic, pluralistic aesthetics. The influence of abstract traditions – particularly Russian Constructivism – can be observed in her paintings and digital works since her early years. Her work reflects a style that blends all these influences, along with autobiographical elements where the personal and the political intertwine.

Yalter is the recipient of the Golden Lion for Lifetime Achievement at the 60th Venice Biennale in 2024. Her works are part of institutional collections such as Tate Modern, the Centre Pompidou, Museum Ludwig, and the Long Beach Museum of Art, among others, as well as private collections such as the Art Collection Telekom, colección olorVISUAL, the Reydan Weiss Collection, and Fundación Foto Colectania. She has participated in international art fairs such as Art Basel, ARCO Madrid, Art Cologne, FIAC, Frieze Masters, the Armory Show, and Frieze New York, to cite a few, as well as the 10th Gwangju Biennial in 2014, the 15th Sharjah Biennial in 2023, the 13th Istanbul Biennial in 2013, and she has been selected for the 60th Venice Biennale. Some of her most recent solo exhibitions have been held at Museum Ludwig, MAC VAL, and the Hessel Museum of Art in 2019; FRAC Lorraine and ARTER Space for Art in 2016; and the Centre Pompidou in 2012 and 2010. Her work has been part of group exhibitions at the Museum of Modern Art, New York, in 2023, Palais de Beaux-Arts in Paris in 2018, WIELS – The Absent Museum in 2017, Tate Modern in 2016, the Centre Pompidou in 2013 and 2009, the Long Beach Museum of Art in 2011, MoMA PS1 in 2008, and CGAC in 2007.

AUTHOR BIOGRAPHY

Prof. Dr Omar Kholeif CF FRSA was born in Cairo to Egyptian and Sudanese parents. They are a British artist, author, curator, historian, and broadcaster who has curated more than seventy exhibitions and produced over 100 commissions of film and visual art on five continents. Kholeif's work explores vanquished archives, with a view to reanimating that which was lost.

Dr. Kholeif has worked at Sharjah Art Foundation (SAF), UAE since 2017 and from 2018 has served as director of collections and senior curator. They are the founder and codirector of artPost21, a cultural agency with a focus on art, technology, and social justice. The author or co-author of dozens of books, their writing has been translated into seventeen languages to date.

Their hybrid social history of art and technology, *Internet_Art: From the Birth of the Web to the Rise of NFTs*, was published by Phaidon in 2023.

Dr. Kholeif is a visiting professor in critical race, visual art history and the creative industries at the Middlesbrough Institute of Modern Art Research Unit at Teesside University, Tees Valley, UK. They are the series editor of *imagine/otherwise* for Sternberg Press – the only peer-reviewed publishing book series focused on critically examining the lives of overlooked female artists.

Forthcoming volumes are focused on artists, Magda Stawarska, Lalitha Lajmi, and Simone Fattal. In summer 2024, Lisson Gallery and D.A.P. will publish Kholeif's catalogue raisonné of Otobong Nkanga's drawings.

Photo: Ashkan Noroozkhani

Image: Caroline von Grone, *Omar, The Prophet*, 2022.
Courtesy of the artist and the Blake Gallacher Trust, Los Angeles

AUTHOR ACKNOWLEDGMENTS

The story of this volume may have come about too quickly—much more so than I would have liked, but its routes were seeded a decade prior to its writing when I was privileged to present Nil Yalter's work in an exhibition that I curated at the Armory Show in New York, which was realized in 2015. I am grateful to Galerist, Istanbul, for supporting this initial endeavor.

The complex task of constellating such a rich life in art in a condensed volume required the support of a small group of individuals. Thanks are due to Salman Matinfar, Azadeh Zaferini, and Mayssa Fattouh of Ab-Anbar Gallery, London, as was the support of Mia Rigo through her venture Archivorum. Special gratitude is afforded to Ilaria Bombelli, head of publishing who was unwavering in her pursuit to complete this book, and Francesco Valtolina, art director at Mousse Publishing who gave this project a home.

My prose would not have made its way to the page without the support of my husband, Frank Gallacher, who despite several personal hurdles oversaw and negotiated the process of our constant exile. He lovingly packed my books and artworks when we were forced to move (again) during the writing of this manuscript. I don't say it enough: Thank you. Artists—well they don't make them like Nil Yalter anymore. Her belief in me, her invitation to write this book is demonstrative of her faith in my voice as an author, an artist, and as a historian. It is this trust that salvages me in moments of darkness. Her invitation into her world and her caring and inquisitive nature made this project deeply rewarding.

Nil: We have yet to complete our journey together, but I hope that with these kernels I have been able to offer further scope and perspective to your life's work. I look forward to our continued conversations.

Special gratitude is afforded to Blake Karim Mitchell of artPost21 who stepped in at the 11th hour to assist with credits and captions.

This book is dedicated to the women of my life, who never cease to inspire: Lubaina Himid, Hannah Feldman, Otobong Nkanga, Sarah Perks, Magda Stawarska, and Sofia Victorino.

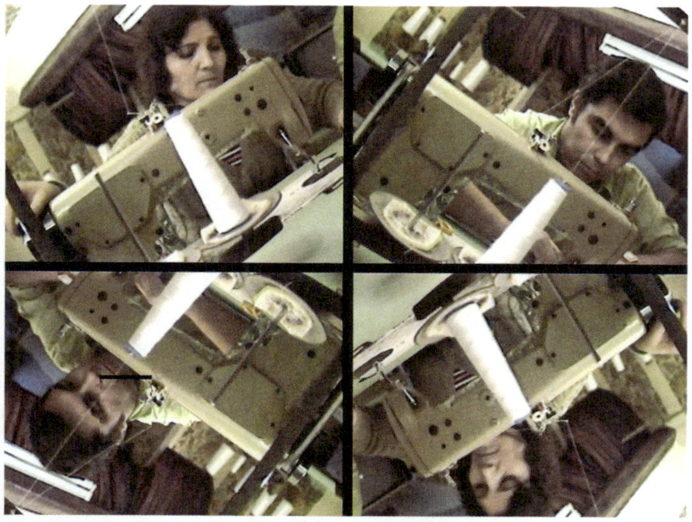

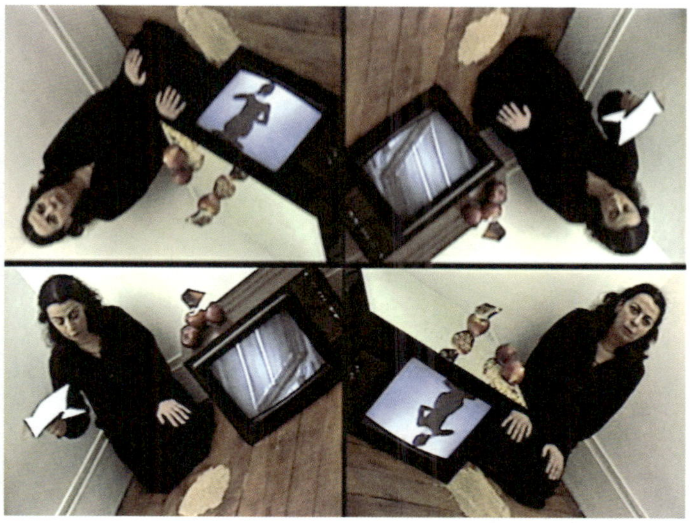

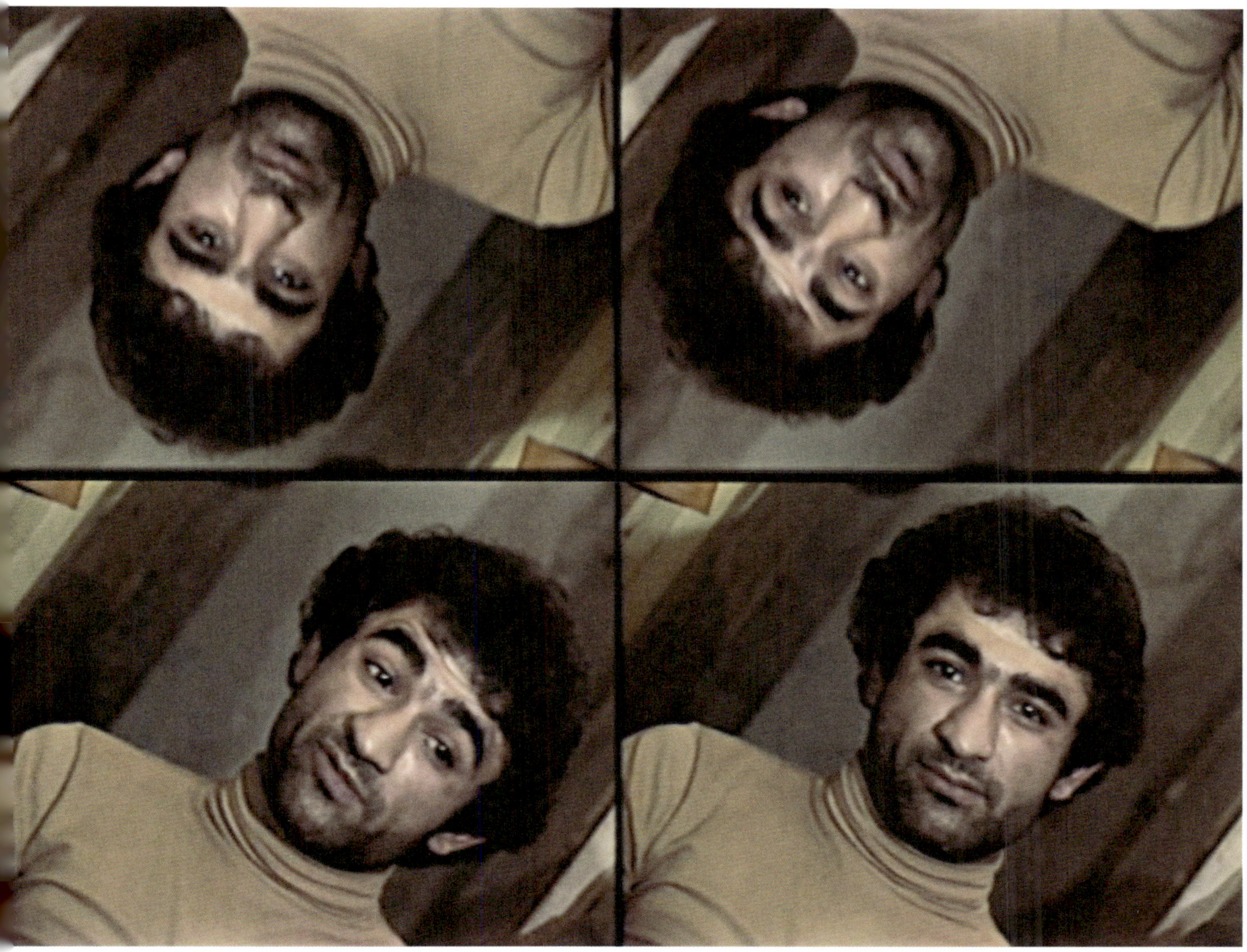

INTO PIECES. DOCTORS CAME, THEY SAID, THIS MAN CANNOT LIVE HE WILL DIE.

WE HAD TO WALK THROUGH A FOREST TO CROSS THE BORDER.
WE GOT LOST IN THE IN THE WILD BUSHES, I WAS SICK.

WHEN I TRIED TO MOVE, I HAD PAIN IN MY LEG AND SHOULDER. MY LEG WAS BROKEN

IT WAS VERY DIFFICULT, WE STILL DON'T HAVE OUR PAPERS.

BETWEEN HUSBAND AND WIFE THERE IS NO INTIMACY

I HAD A WORK ACCIDENT. I WAS FALLING DOWN, I SAID 'FRIENDS, I AM FALLING SAVE ME'

I COULD NOT FIND WORK HERE IN MY PROFESSION, BECAUSE I CAN NOT SPEAK FRENCH

I COULD NOT FIND WORK HERE IN MY PROFESSION, BECAUSE I CAN NOT SPEAK FRENCH

BECAUSE OF LANGUAGE PROBLEMS, I HAVE TO WORK IN A GALVANIZATION FACTORY

işçinin dostu
FRANSA'DAKİ TÜRKİYE İŞÇİLERİNİN GAZETÈSİ
DE L'OUVRIER / MENSUEL POUR LES TRÀVAILLEURS TURCS EN F
AFFICHES
P
la GR VE
GEN RALE

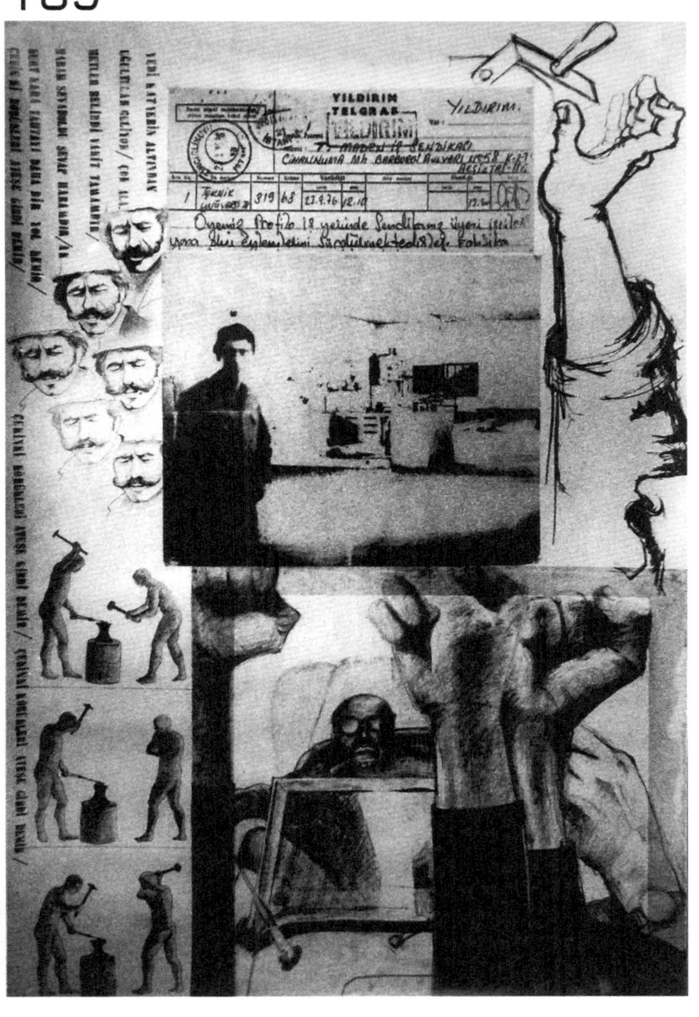

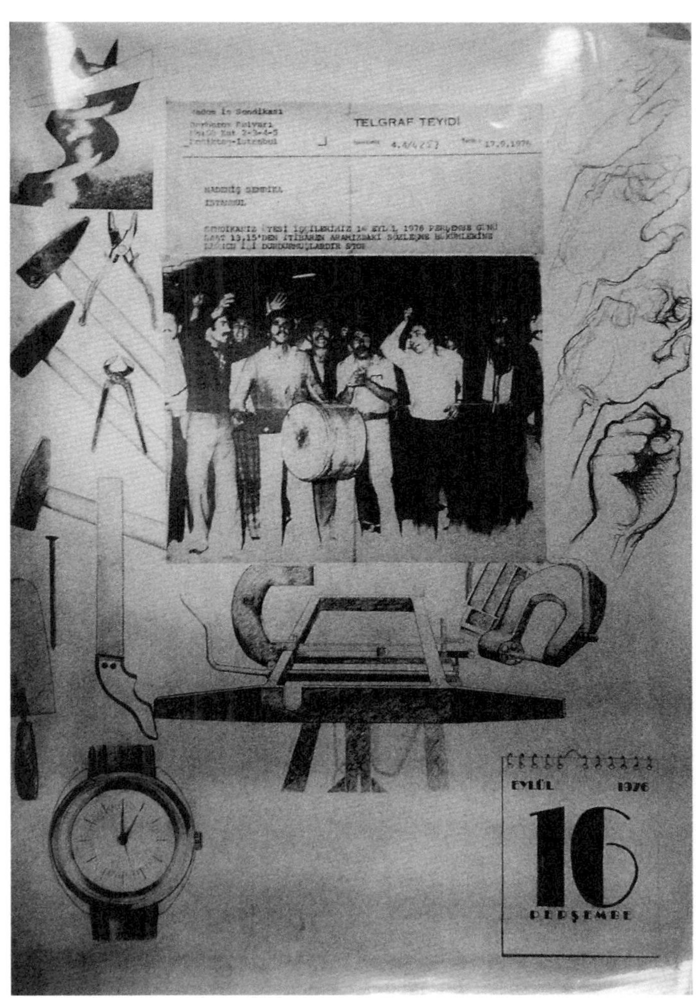

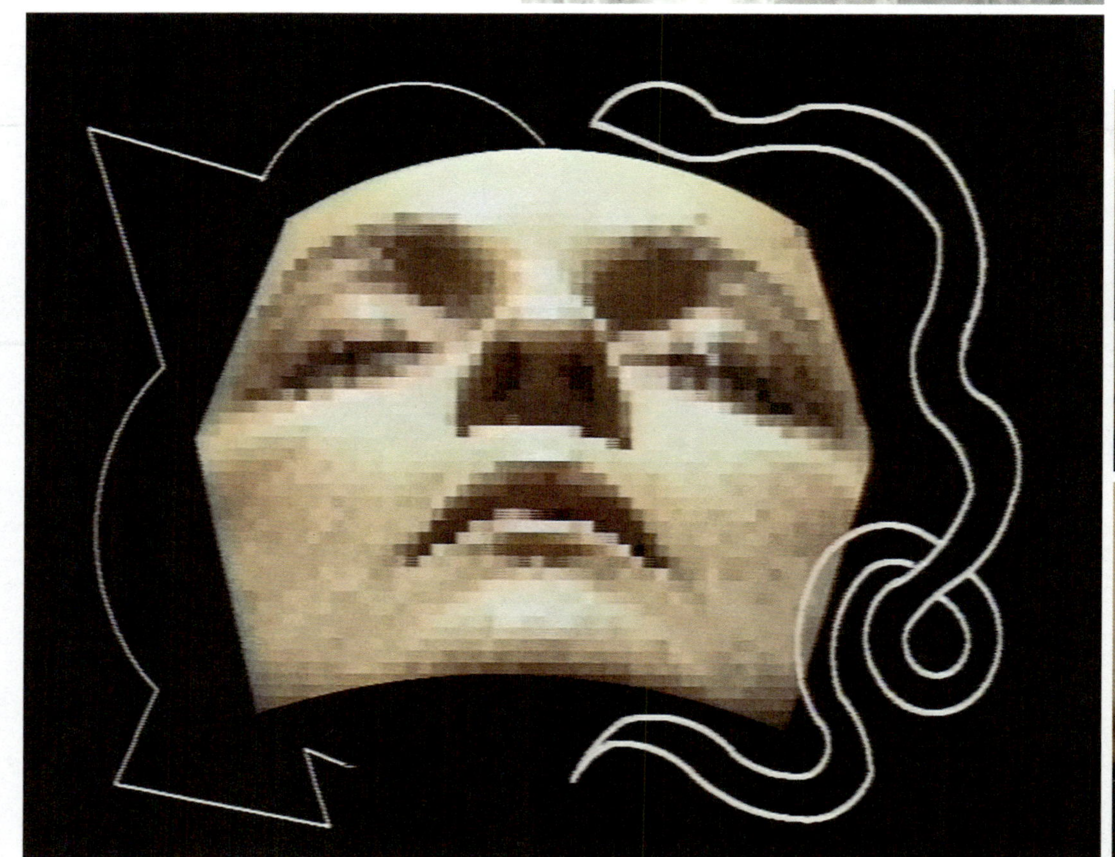

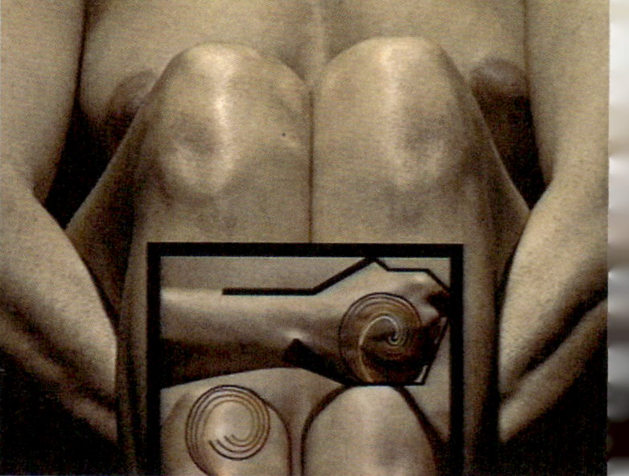

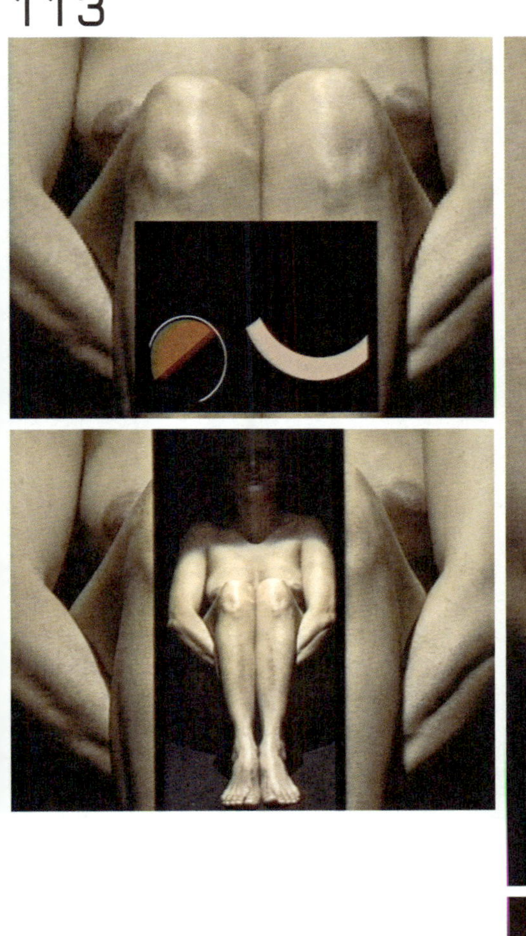

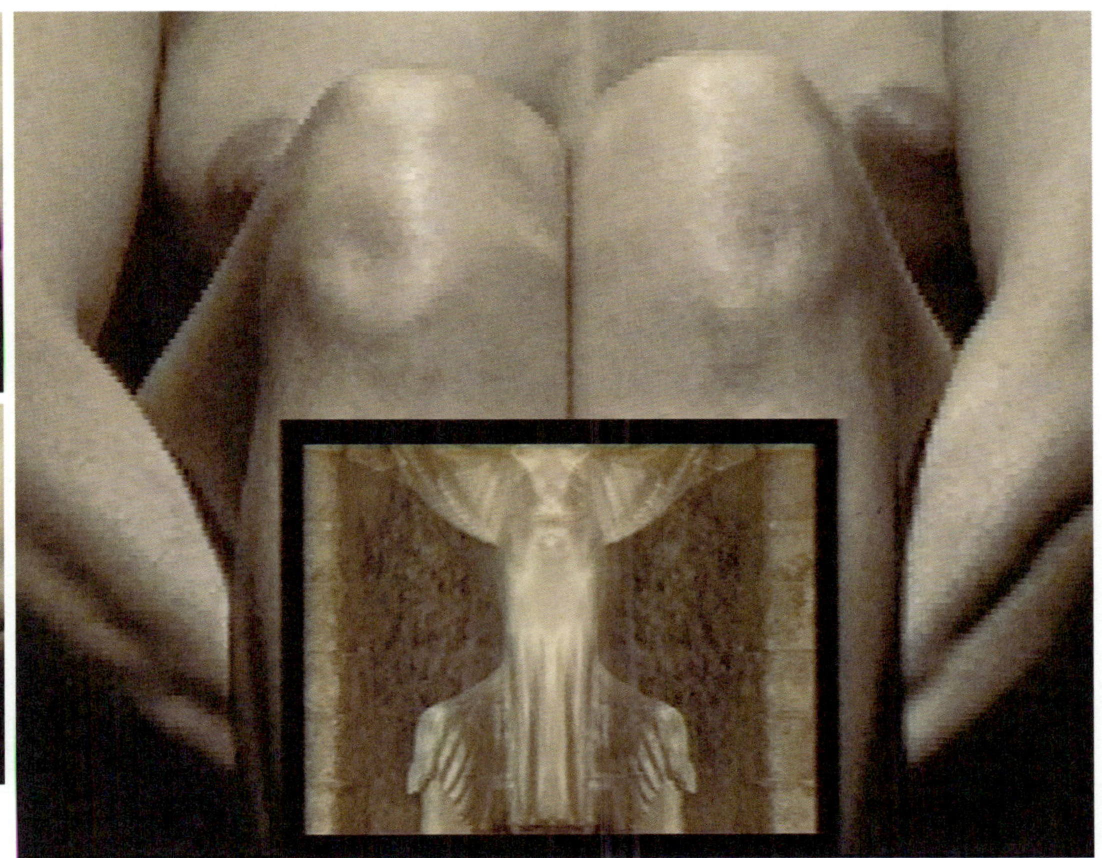

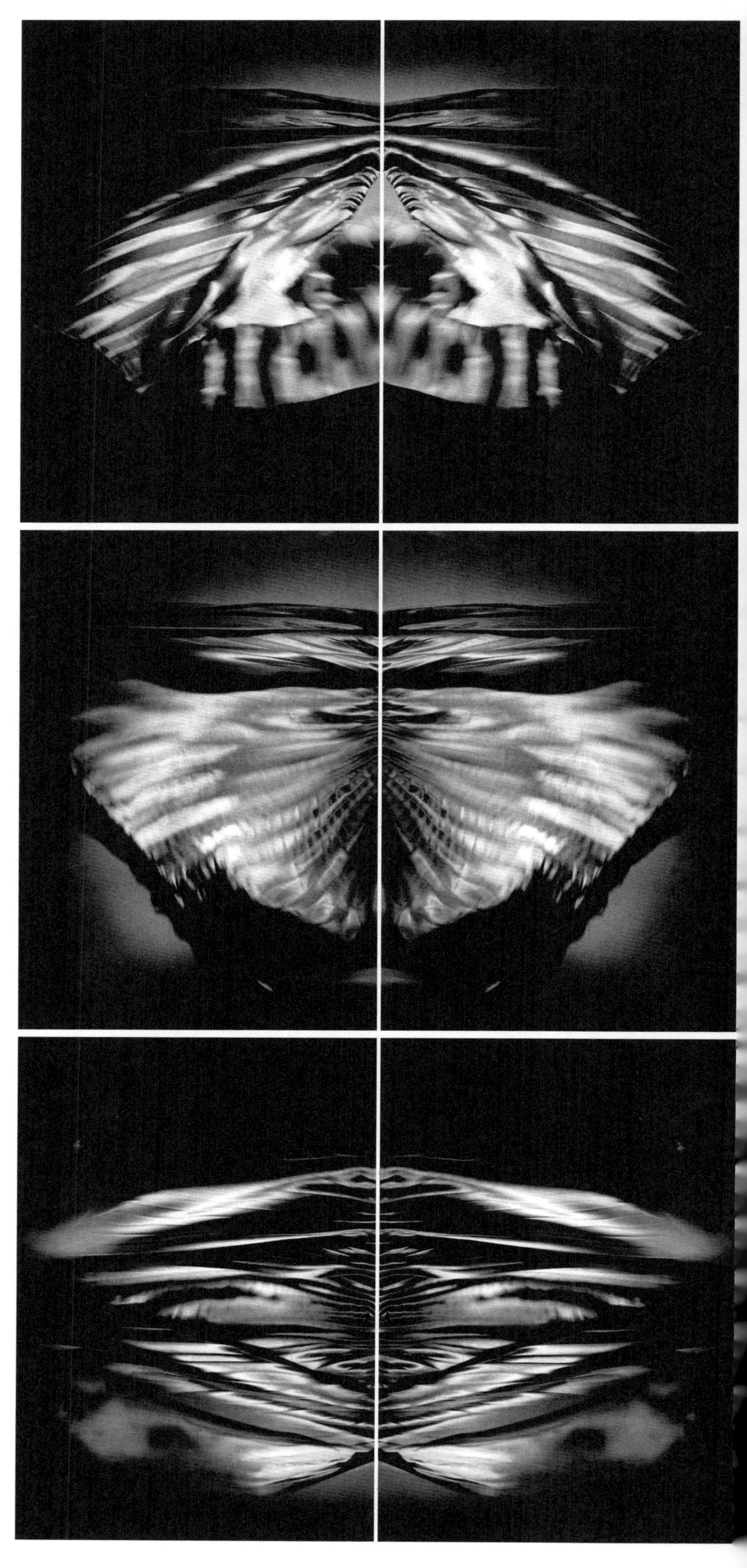

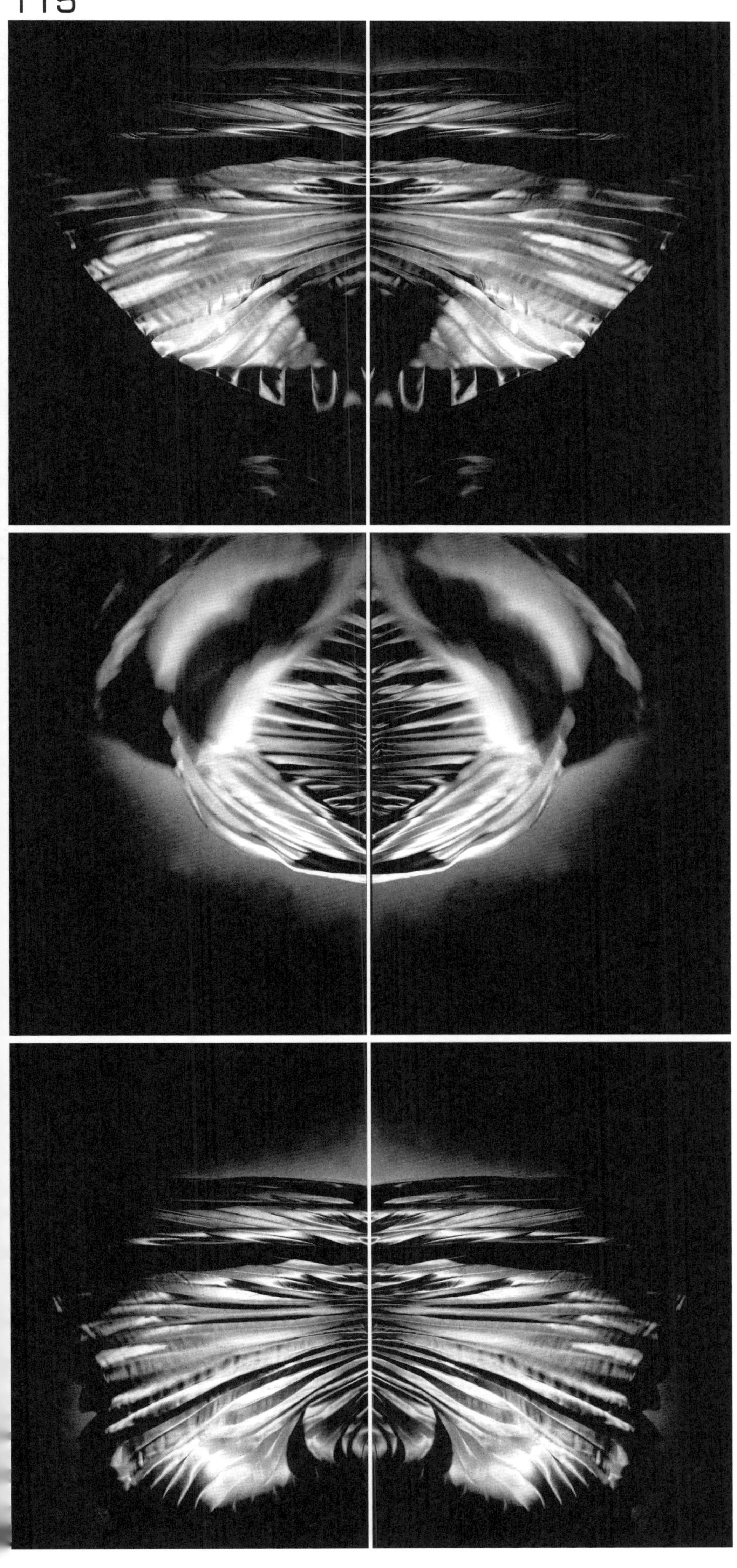

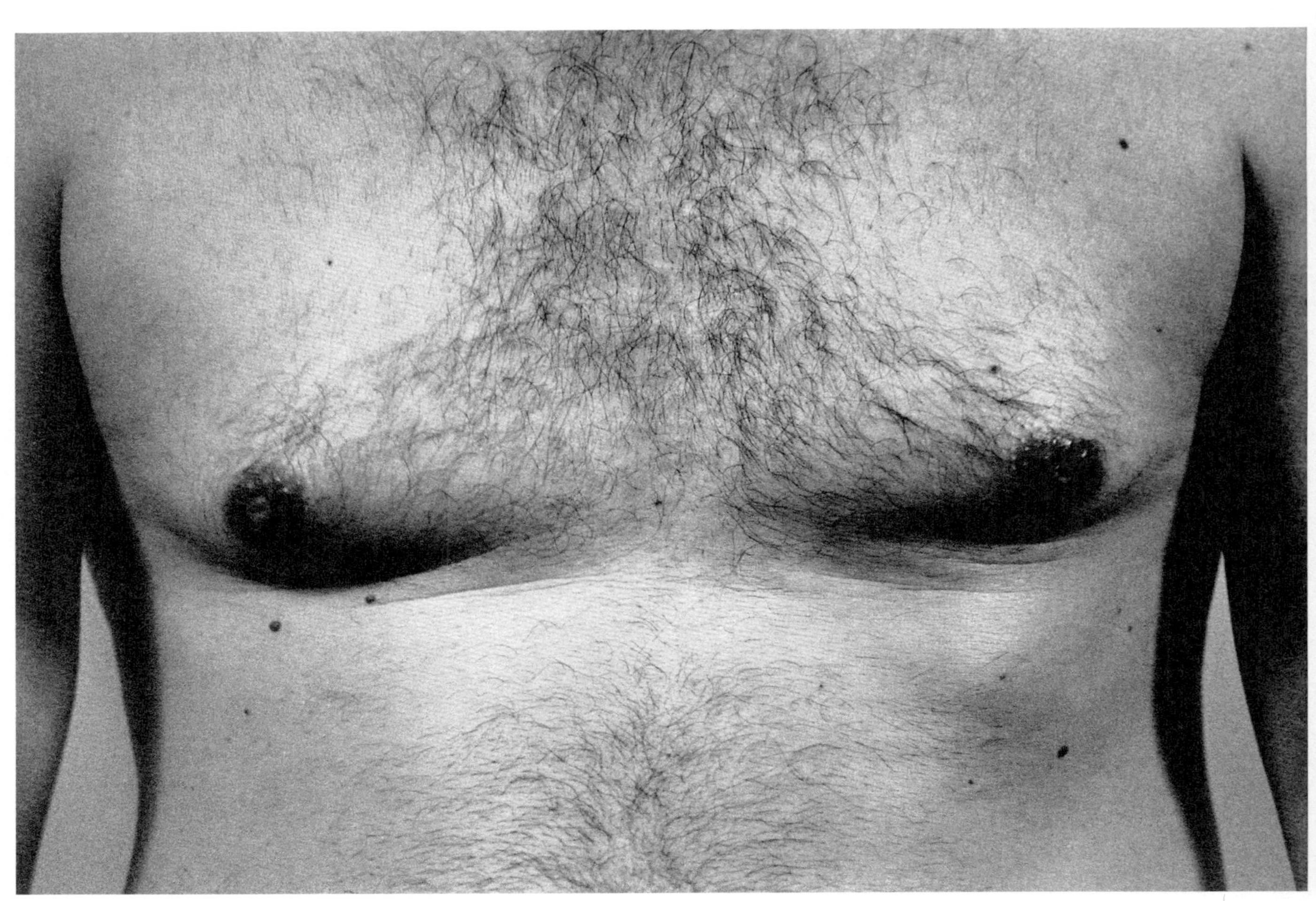

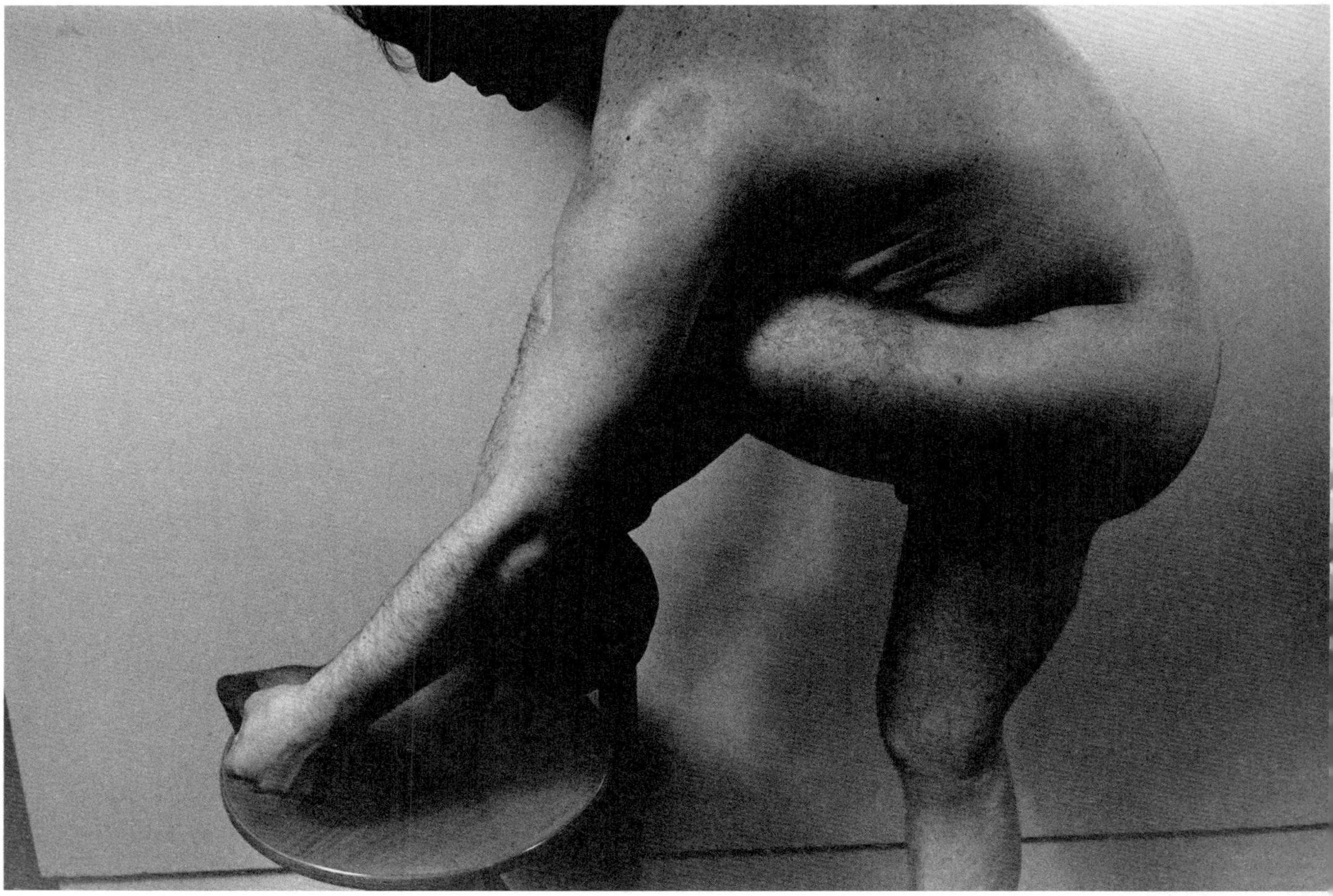

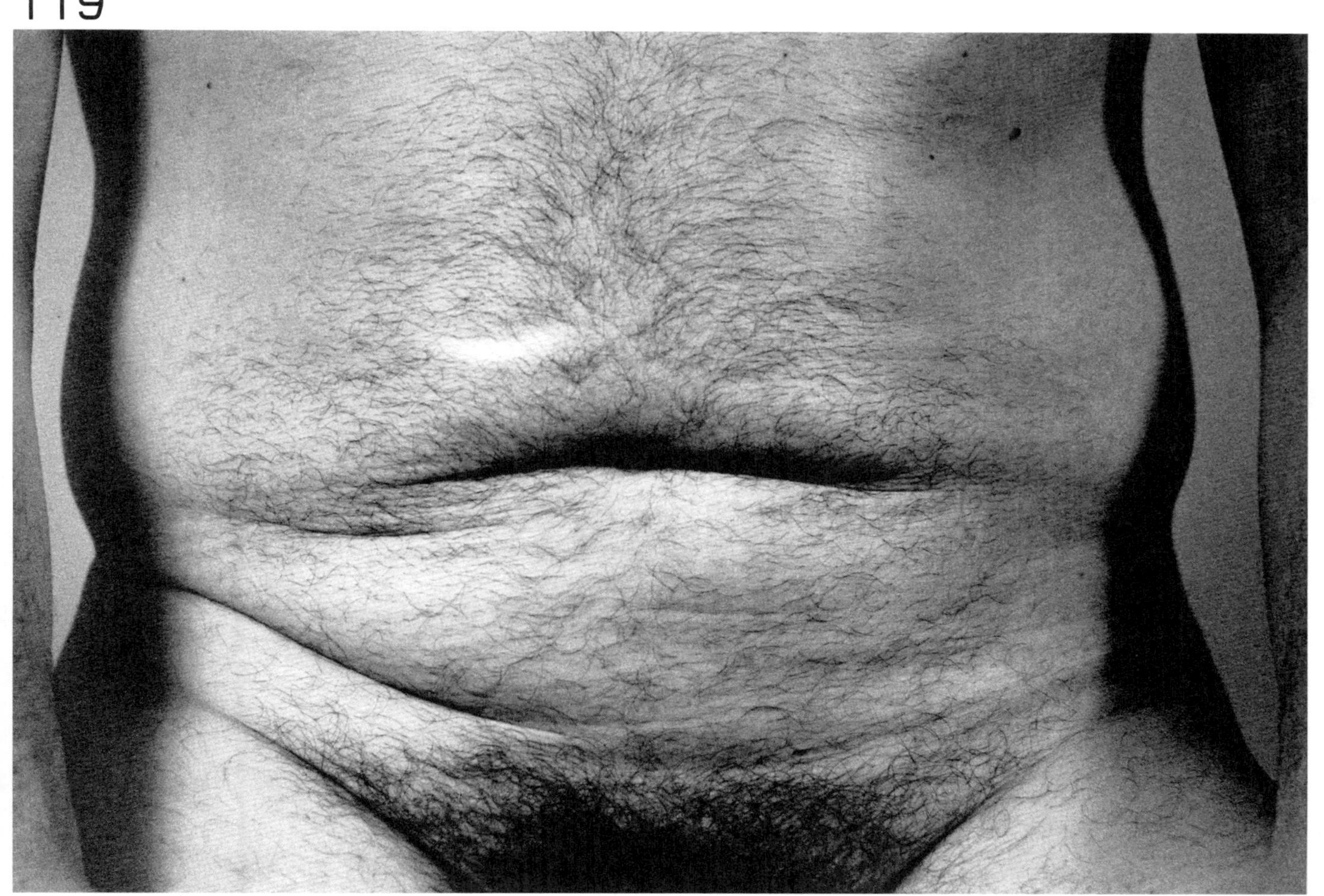

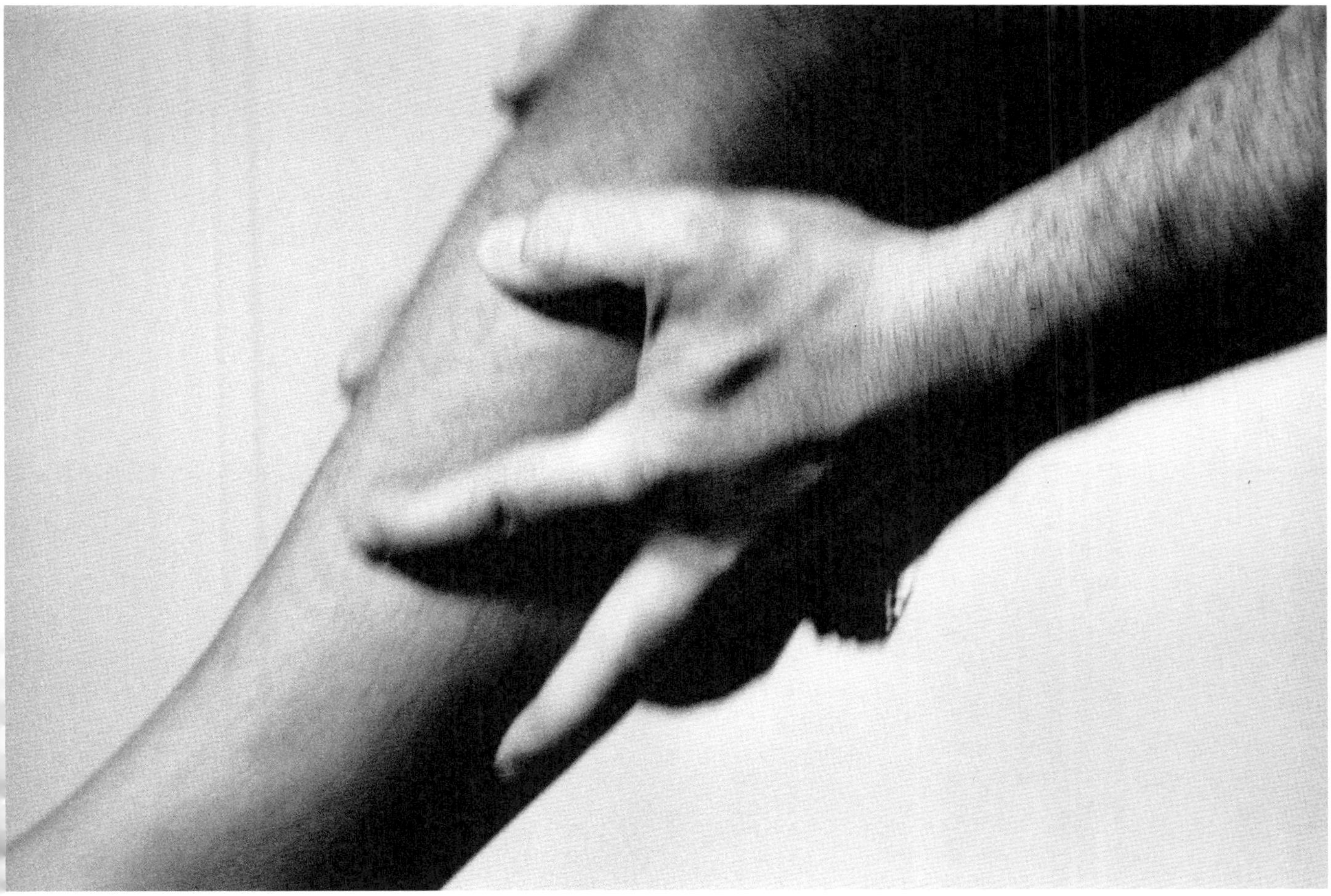

ZOR

പ്രവാസം ഒരു കഠിനമായ ജോലിയാണ്
جلاوطنی ایک مشکل کام ہے
المنفى مهمة شاقة

ŞU GURBETLIK
ZOR ZANAAT ZOR

ŞU GURBETLIK ZOR ZANAAT ZOR

ŞU GURBE

N.Y. 1977

N.Y. 1977

LE CAR ENTRE DANS UNE GRANDE COUR, UN GARDIEN NOUS AMENE DANS UNE SALLE OU IL Y A DEUX FEMMES EN BLANC. JAC KY FUT APPELEE ET RESORTIT UNE DEMI-HEURE APRES AVEC UNE ROBE.

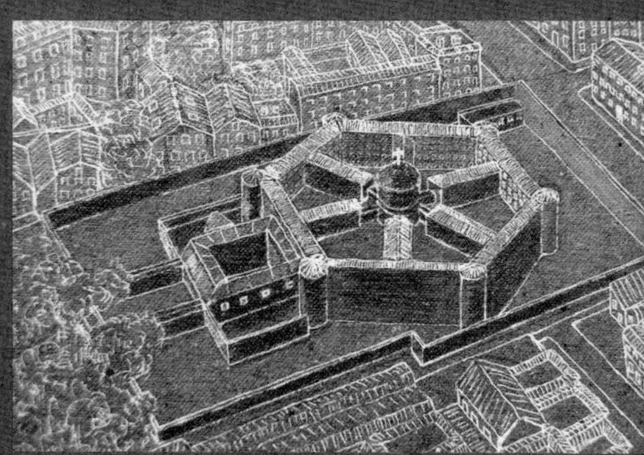

LA PRISON: UNE CASERNE UN PEU STRICTE, UNE ECOLE SANS INDULGENCE, UN SOMBRE ATELIER, MAIS, A LA LIMITE, RIEN DE QUALITATIVEMENT DIFFERENT." (MICHEL FOUCAULT SURVEILLER ET PUNIR)

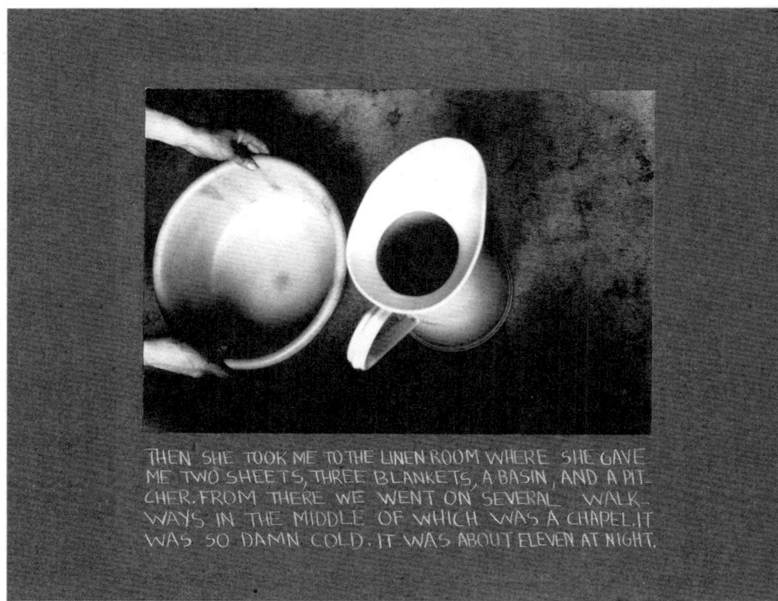

THEN SHE TOOK ME TO THE LINEN ROOM WHERE SHE GAVE ME TWO SHEETS, THREE BLANKETS, A BASIN, AND A PIT CHER. FROM THERE WE WENT ON SEVERAL WALK WAYS IN THE MIDDLE OF WHICH WAS A CHAPEL. IT WAS SO DAMN COLD. IT WAS ABOUT ELEVEN AT NIGHT.

LES DIMANCHES ETAIENT DIFFERENTS DES AUTRES JOURS. LE MATIN CERTAINES ALLAIENT A LA MESSE, ET LES AUTRES RESTAIENT ENFER MEES DANS LES CELLULES, MAIS ON POUVAIT ETRE DANS LES CEL LULES DE NOS AMIES. CE JOUR LA ON NE TRAVAILLAIT PAS.

A NEUF HEURES LES SOEURS FERMAIENT LES CELLULES. ELLES REGAR DAIENT PAR LE JUDAS ET COGNAIENT SUR LA PORTE JUSQU'A CE QU'ON MONTRE SA TETE OU QU'ON BOUGE, POUR VOIR SI ON NE S'ETAIT PAS EVADE.

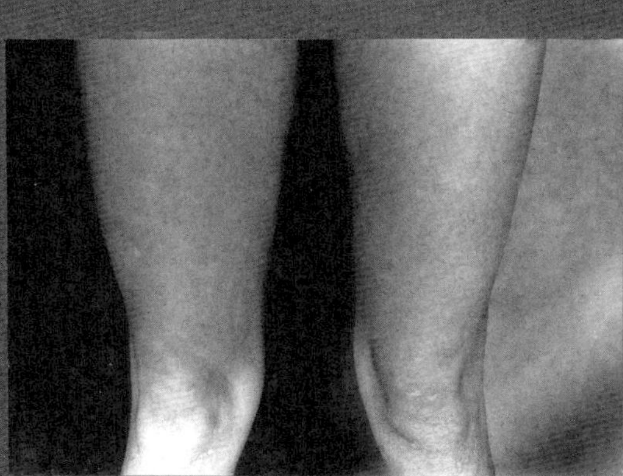

IT WAS MY TURN TO JOIN. NAME, DATE OF BIRTH, ETC... "TAKE YOUR CLOTHES OFF!" I START UNDRESSING IN FRONT OF THE TWO OLD BAGS. I WAS ALL IN BLACK, BOOTS, LEVI'S, AND MY COSSACK SHIRT. AND, ON TOP OF IT ALL, I HAD THE CURSE. THEY MADE ME TAKE EVERYTHING OFF ANYWAY. "SPREAD YOUR LEGS! COUGH!" AND SO ON.... THEY SEARCHED MY CLOTHES,

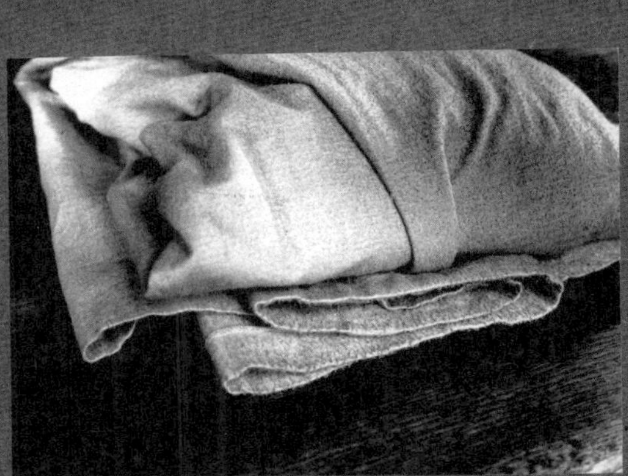

AND THEY KEPT MY JEANS. PANTS WEREN'T ALLOWED IN THE PRISON. THEY MADE ME PUT ON A THING THAT LOOKED MORE LIKE A POTATO SACK THAN A DRESS. IT WAS BROWN, WITH BUTTONS IN FRONT, AND LONG SLEEVES. THEY CALLED THIS RAG A BURLAP DRESS, AND UNTIL I GOT

A WOMAN (CONVICT) CAME TO GET ME, AND I WALKED DOWN A LONG HALL. SHE LED ME INTO A ROOM WHERE THERE WAS A TABLE AND A BENCH. SHE ASKED ME IF I WANTED SOME SOUP. I ACCEPTED WITH PLEASURE, SEEING IT WAS COLD AS HELL; I TOOK TWO HELPINGS. THEN SHE

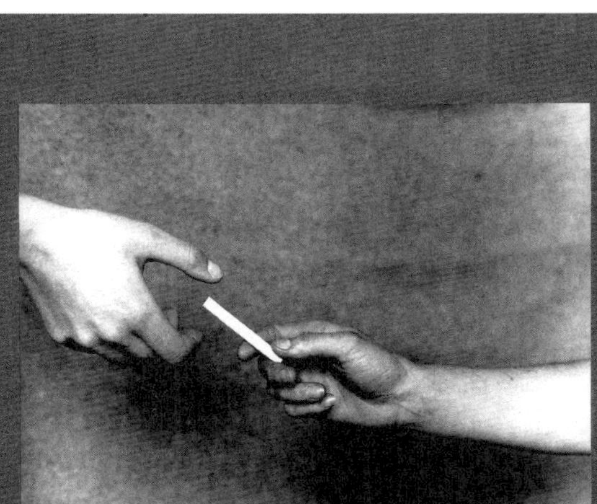

BUT I DIDN'T HAVE TIME TO THINK ABOUT IT LONG BECAUSE THE CHICK NEXT TO ME STARTED TALKING. FIRST SHE ASKED ME FOR A SMOKE, WHICH I GAVE HER, AND SHE ADVISED ME TO KEEP THE BUTT. THEN SHE TOLD ME TO HIDE ALL MY STUFF BECAUSE MARIE-THERESE WAS A THIEF!

...BECAUSE MARIE-THERESE WAS A THIEF!.. THE OTHER ONE HEARD THIS AND THEY STARTED QUARELLING IN WHISPERS. FOR MY FIRST NIGHT THERE WAS ATMOSPHERE. FINALLY THEY SHUT UP PROMISING TO SETTLE THE FIGHT THE NEXT DAY. I FELL ASLEEP BECAUSE I WAS SO TIRED. 10 DECEMBER 66.

The military regime has carried out a harsh commando operation in the south - eastern part of Anatolia where the Kurdish people live. The photograph below shows a young girl who was attacked and wounded by rightist fascist bands. The military regime has carried out in the south - eastern part of Anatolia where the Kurdish people live. The photograph below shows a young girl who was attacked and wounded by rightist fascist bands. The military regime has carried in the south - eastern part of Anatolia where the Kurdish people live. The photograph below shows a young girl who was attacked and wounded by rightist fascist bands. The military regime has carried in the south - eastern part of Anatolia where the Kurdish people live. The photograph below shows a young girl who was attacked and wounded by rightist fascist bands. The military regime has carried in the south - eastern part of Anatolia where the Kurdish people live. The photograph below shows a young girl who was attacked and wounded by rightist fascist bands. The military regime has carried in the south - eastern part of Anatolia where the Kurdish people live. The photograph below shows a young girl who was attacked

DESIRE

ABSTRACTION

R-A· № 484969
К.П.П-К-Н-Я-ВО
FRANCE
15.МАЙ 1965
ИЗЛЕЗЕ - 3

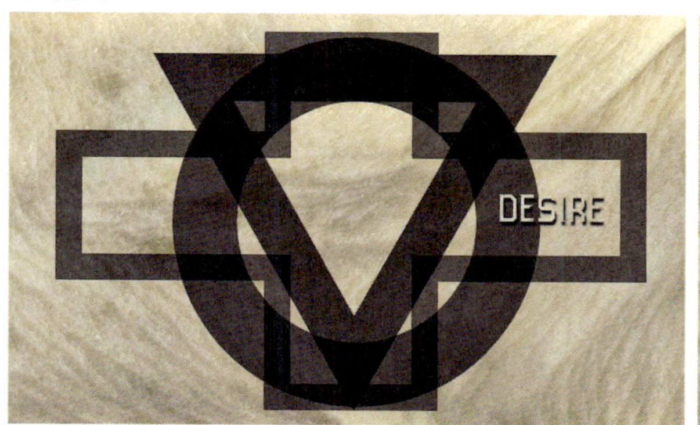

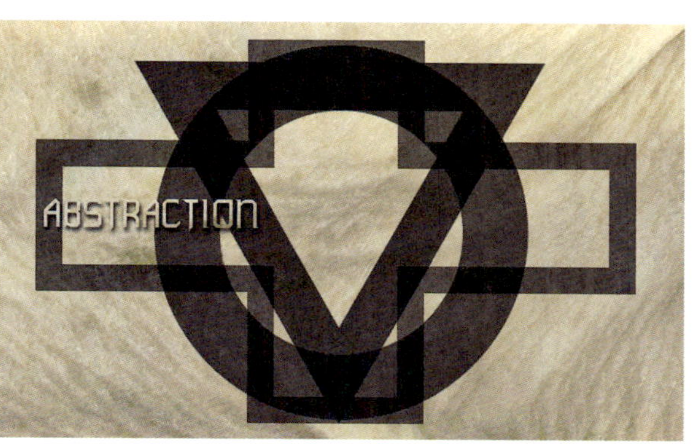

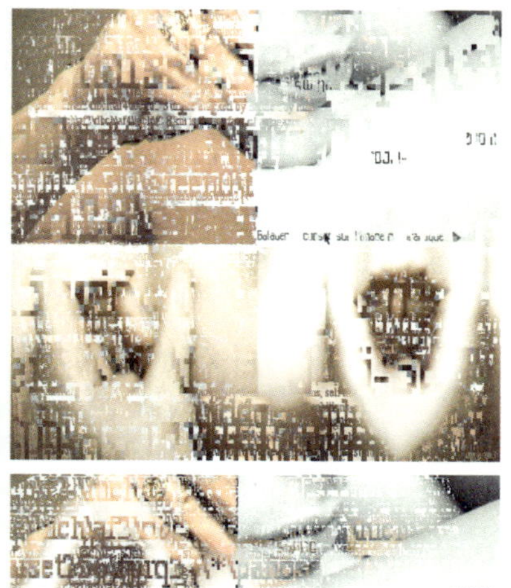

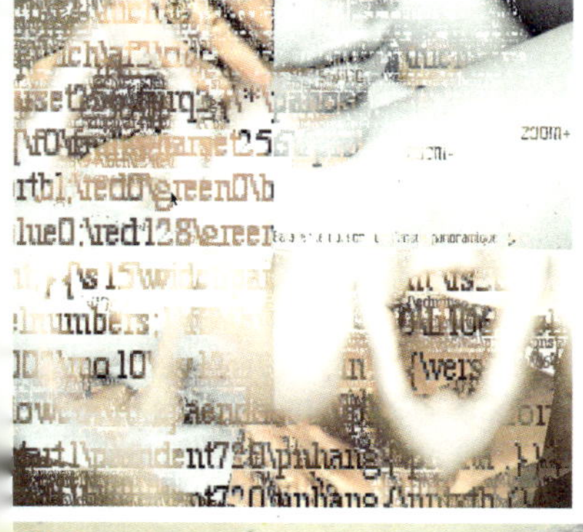

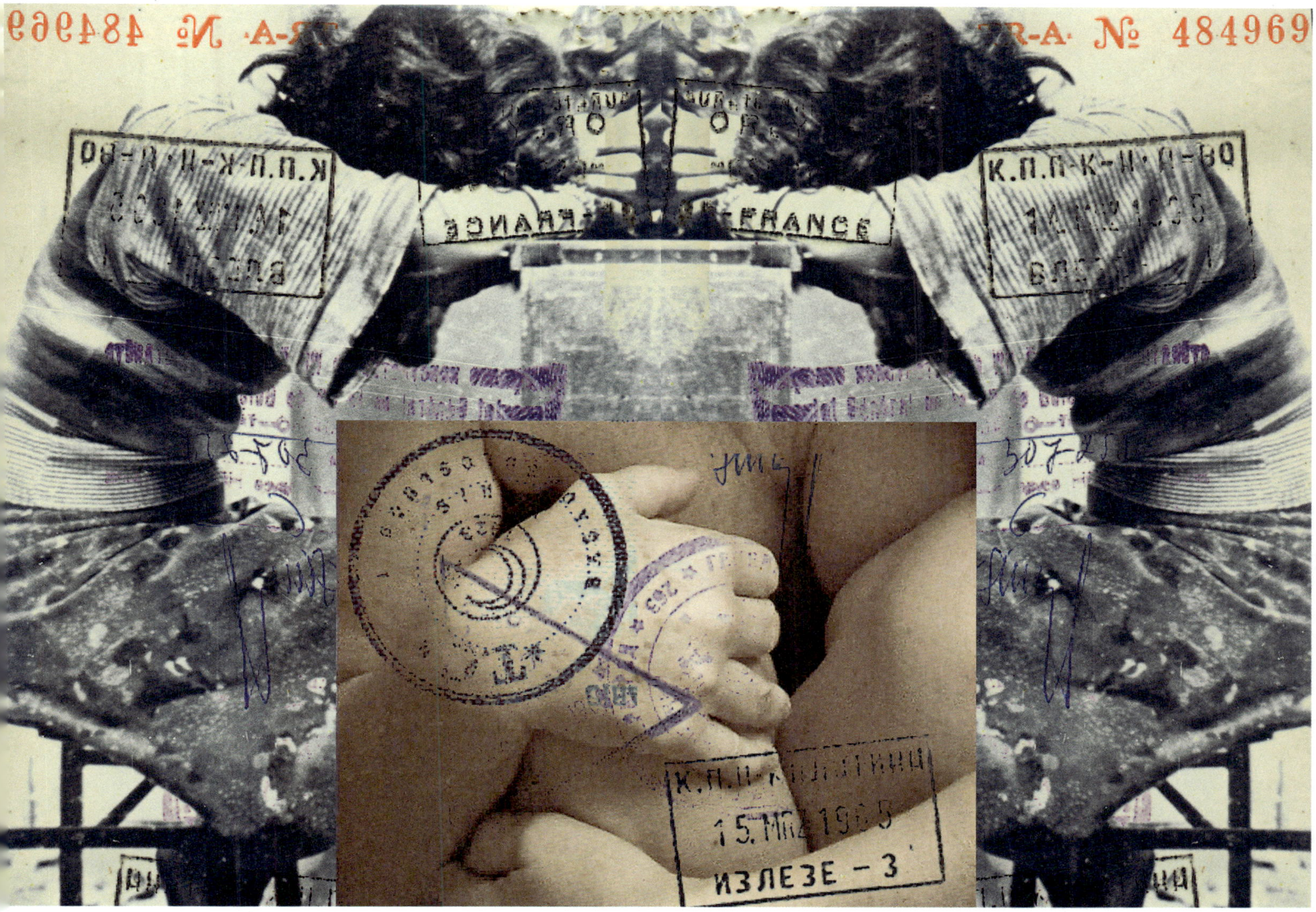

Nil Yalter: Circular Tension
by Omar Kholeif

Published and distributed by

**Mousse Publishing
Contrappunto s.r.l.
Via Pier Candido Decembrio 28,
20137, Milan–Italy**

Available through

**Mousse Publishing, Milan
moussemagazine.it**

**Archivorum
Archivorum.org**

**Design: Francesco Valtolina (Mousse)
Design Assistant: Nicola Narbone
Publishing editor: Ilaria Bombelli (Mousse)
Editorial Coordinator: Emma Passarella (Mousse)
Copyeditor: Rachel Walther**

**Printed in Italy
First edition: 2024
ISBN 978-88-6749-629-7**

€ 30 / $ 35

**The publisher would like to thank all those
who have kindly given their permission
for the reproduction of material for this
book. Every effort has been made to obtain
permission to reproduce the images
and texts in this catalogue. However,
as is standard editorial policy, the publisher
is at the disposal of copyright holders
and undertakes to correct any omissions
or errors in future editions.**

**This publication was sponsored
by Archivorum. Their generous support
has been invaluable in bringing this
project to fruition.**

Nil Yalter: Circular Tension **was produced
in collaboration with: artPost21
www.artpost21.com**

artPost21

P. 1
, 1997. Interactive CD-ROM.
Courtesy of the artist

P. 2–3
From , 1992. Video tape accompanied by 3D images.
The work has been re-imagined, and is illustrated here as
, 1997. Interactive CD-ROM. The work revolves around
a journey through 17th century popular Turkish poetry. Courtesy
of the artist and Ab-Anbar, London

P. 4–5
Paintings of , 1994. Mixed-media paintings
on canvas. Twenty-four variations on the Byzantine Church of the
Chora in Istanbul. Courtesy of the artist

, 1967/1970. Painting on canvas, animated here
througha computer generated single-channel video, 2011.
Courtesy of the artist and Ab-Anbar, London

P. 6–7
Details from , 1987. Three-channel video
installation broadcast on twelve monitors. Courtesy of the artist

P. 8–9
Installation view , 1973. A nomad's tent used to examine
private and public feminine spaces. Installation: tent, drawings,
and photographic works. Dimensions variable. Installation view:
Musée d'Art Moderne de Paris, 1973. Courtesy of the artist

P. 10–11
Details from , 1973. Drawings realized on butcher paper.
Courtesy of the artist and Ab-Anbar, London

P. 12–13
Nil Yalter in-situ with , 1973. Installation: nomad's tent,
drawings, and photographic works. Dimensions variable. Courtesy
of the artist

P. 14–15
Drawings from , 1974. An installation by Nil Yalter
and Judy Blum. Installation: 12 meters high. The work is constituted
of photographs, drawings, and texts focusing on twenty districts
of Paris. Courtesy of the artists

P. 16–19
From , 1974. Twelve collages with Polaroid
pictures, objects, and drawings of temporary dwellings of immigrants
in Paris, New York, and Istanbul. Courtesy of the artist

P. 20–21
Stills from , 1980. Installation: B&W video tape, 45 minutes.
Video accompanied by drawings and photographs, dimensions
variable. The work is a transposition of the daily life of two women
in the spaces of an Ottoman Harem. Courtesy of the artist

P. 22–23
, 1977. Six hand-painted photographs
of an Algerian marriage in Dreux, France. Courtesy of the artist

P. 24–27
From the expanded body of work: , 1974–79;
featuring works: , 1977, and
, 1979. Installation consists of photos and drawings by
Nil Yalter. Dimensions variable. Video by Nil Yalter and Nicole Croiset,
1979. Courtesy of the artists

P. 28
Stills from , 1979. Installation: B&W video, 7 minutes.
Courtesy of the artist

P. 29
, 1974. Silver gelatin print on collage, 51 cm x 65 cm.
Courtesy of the artist and Ab-Anbar, London

P. 30–31
, 1992. Installation: video tape with 3D images, 3 minutes.
The video draws attention to the relationship between Ottoman
calligraphy and to the computer environment, emphasizing
the concept of "technology as a medium." Courtesy of the artist